# THE HERSCHEL CHRONICLE

*The more a man hath unity and simplicity in himself the more things and the deeper things he understandeth. The spirit which is pure, sincere and steadfast is not distracted though it hath many works to do, because it doth all to the honour of God.*

THOMAS À KEMPIS, *Imitation of Christ*, Book I, 3. 3

SIR WILLIAM HERSCHEL
at the age of 46
*From a crayon copy of the oil painting by L. T. Abbott
in the National Portrait Gallery*

# THE HERSCHEL CHRONICLE

THE LIFE-STORY OF
WILLIAM HERSCHEL AND HIS SISTER
CAROLINE HERSCHEL

Edited by his Granddaughter
CONSTANCE A. LUBBOCK

CAMBRIDGE
AT THE UNIVERSITY PRESS
1933

CAMBRIDGE UNIVERSITY PRESS
Cambridge, New York, Melbourne, Madrid, Cape Town,
Singapore, São Paulo, Delhi, Mexico City

Cambridge University Press
The Edinburgh Building, Cambridge CB2 8RU, UK

Published in the United States of America by Cambridge University Press, New York

www.cambridge.org
Information on this title: www.cambridge.org/9781107650015

© Cambridge University Press 1933

This publication is in copyright. Subject to statutory exception
and to the provisions of relevant collective licensing agreements,
no reproduction of any part may take place without the written
permission of Cambridge University Press.

First published 1933
First paperback edition 2013

*A catalogue record for this publication is available from the British Library*

ISBN 978-1-107-65001-5 Paperback

Cambridge University Press has no responsibility for the persistence or
accuracy of URLs for external or third-party internet websites referred to in
this publication, and does not guarantee that any content on such websites is,
or will remain, accurate or appropriate.

# PREFACE

THE aim of the present work is to present the life-story of William and Caroline Herschel in the setting of their family circle and of contemporary history, giving it as far as possible in their own words, by extracts from William Herschel's brief autobiographical notes and from Caroline's very full Journals, as well as from letters.

It may be thought that too many details concerning William Herschel's brothers have been included, but the bonds of family affection were so strong in the Herschel family that no true picture of the two principal characters can be formed if the other members are excluded.

William Herschel's discoveries and theories have also been presented in his own words, quoted either from the Papers which he communicated to the Royal Society or from his letters. Letters also are given from contemporary men of science to show how these discoveries and theories were regarded in his own time; their ultimate value must be decided by the wider knowledge of later generations. It was as an intrepid pioneer, breaking a way to new fields of discovery and to new trains of thought that William Herschel appeared in his lifetime and as such it is sought to represent him in this *Chronicle*.

A few words are needed to explain the peculiarities of spelling to be met with in Caroline Herschel's narrative. She wrote, and no doubt spoke, English with sufficient fluency, but there are certain spellings, as for example: "infitation", "oftens", "wimsical", which she invariably used, and as her version gives an indication of her pronunciation and seems to bring the reader into closer touch with her personality, these words have been left as she wrote them. Occasional slips of the pen, which are more frequent in the autobiography which she wrote in her old age, have, however, been corrected.

The word, now spelt "gauge", was invariably written "gage" by William Herschel, and this form has been retained in the quotations made from his Papers.

The editor of this *Chronicle* would wish to take this opportunity of thanking all those who have so kindly helped her in her task; especially Professor Newall for much assistance in the portions relating to

Herschel's Papers: *On the Construction of the Heavens*, and *On Nebulous Stars*; and Dr Crommelin for some useful suggestions on the same. To Dr Steavenson she is greatly indebted for help in the description of Herschel's telescopes, and to Professor Wright, of the Lick Observatory, California, for the interesting photograph of the Nebula, *N.G.C.* 1514, taken expressly for this work.

Lastly she wishes to express her thanks to the Syndics of the Cambridge University Press for undertaking the publication of this record of the life and work of William Herschel, and to acknowledge her indebtedness to the expert assistance of the Staff and the kind encouragement given her by the Secretary, Mr S. C. Roberts.

C. A. L.

*May* 1933

# CONTENTS

*Chapter* I: 1651–1757     *page* 1

Origin of the Herschel Family. Isaac Herschel's autobiography. Early life of William and Caroline Herschel. William joins the band of the Hanoverian Guards. Historical sketch. William's account of his visit to England with his father and brother. Battle of Hastenbeck. Retreat and surrender of the Hanoverian army. William and Jacob come to England.

*Chapter* II: 1757–1761     *page* 12

William and Jacob in London. Jacob returns to Hanover. William goes to Yorkshire. Lord Darlington's Militia band. Extract from Southey's book *The Doctor*. Letters to Jacob begin. William goes to Edinburgh—meets Hume—studies mathematics.

*Chapter* III: 1762–1766     *page* 29

William engaged as director of concerts at Leeds. Gets his discharge from the Guards Band. Visits Hanover. The Herschel family at Hanover. Isaac Herschel's death. William's Memorandums. Chosen organist at Halifax. Miller's story. William accepts the post of organist at Bath. The Octagon Chapel. Organ built by Snetzler.

*Chapter* IV: 1767–1772     *page* 42

William's life at Bath—Jacob joins him. Dietrich sent to Bath. Alexander settles at Bath. Caroline's life at home—William goes to Hanover—brings her to Bath—her life at Bath.

*Chapter* V: 1773–1780     *page* 59

William Herschel turns to astronomy. Letter to Dr Hutton. Condition of astronomical knowledge. Ferguson. William begins to make telescopes. Caroline's autobiography. William goes to Hanover in search of Dietrich. Houses occupied by the Herschels at Bath. Meeting with Dr Watson—Maskelyne—Aubert. Papers communicated to the Royal Society.

*Chapter* VI: 1781     *page* 78

Discovery of Uranus. Letters from Hornsby, Maskelyne. Paper "Account of a Comet" read at the Royal Society. Letter from Messier. William's attempt to cast a mirror for a 30-foot telescope. Experiments on metals.

*Chapter* VII: 1781–1782     *page* 93

Reappearance of the "comet". Letters from Hornsby, Gerstner. William Herschel receives the Copley medal. Paper "On the Parallax of the Fixed Stars"—its reception by the Royal Society. Dr Watson's letter. Correspondence with Watson, Hornsby, Maskelyne, Sir Joseph Banks.

CONTENTS

*Chapter* VIII: 1782                                         *page* 112

William Herschel's prospect of becoming private astronomer to the King. Letters. William goes to London. His letters to Caroline and Alexander. The King's offer accepted. Congratulatory letters. Name for the new planet. Letter from Bode. Sign for Uranus. Herschel dedicates it to the King as "Georgium Sidus". Letters from Lalande, C. Mayer, Magellan, Lichtenberg, Schroeter.

*Chapter* IX: 1782–1786                              *page* 133

William and Caroline settle at Datchet. Caroline's Autobiography. Damp at Datchet. William has ague. Making the large 20-foot telescope. Move to Clay Hall. The King's gift of £2000 for the 40-foot telescope. Move to Slough. Mrs Papendiek's Journal.

*Chapter* X: 1786–1789                                *page* 147

William Herschel goes to Göttingen—his letters to Caroline. Caroline discovers her first comet. Letters to Blagden and Aubert. William returns to Slough. Making the 40-foot telescope. The King gives another £2000. Discovery of the satellites of the Georgium Sidus, and of two new ones of Saturn. History of the 40-foot telescope.

*Chapter* XI: 1786–1788                               *page* 169

Fanny Burney's *Diary*. Caroline's Autobiography. She receives a salary from the King as assistant to her brother. Mrs Papendiek's Journal. Her account of William Herschel's marriage. Character of Mrs Herschel. Letters—Aubert, Sir W. Watson. Portraits of William Herschel.

*Chapter* XII: 1783–1785                            *page* 183

William Herschel's astronomical work. State of science. Sir Joseph Banks, Priestley, Cavendish. Variability of Algol. Letters—William Herschel to Prof. A. Wilson with summary of "First Paper on the Motion of the Sun", Aubert. Former theories on Construction of the Universe, Thomas Wright. Extracts from Herschel's first and second papers "On the Construction of the Heavens"—reception by the Royal Society. John Herschel's remarks on Prof. Pictet's disapproval. William Herschel as a mathematician.

*Chapter* XIII: 1784–1794                          *page* 200

Foreign correspondence—Lalande, Schroeter, Bode, etc. Abstract of Herschel's first paper "On the Construction of the Heavens" in letter to Bode. Lichtenberg's appreciation. Abstract of second paper "On the Construction etc." in letter to Lalande. Correspondence with Schroeter. Observations on Venus. Volcanoes in the moon.

*Chapter* XIV: 1786–1802                        *page* 221

William Herschel's discovery of nebulous matter. "First and Second Catalogues of New Nebulae and Clusters." Laplace's remarks on Herschel's nebular theory. Present view.

CONTENTS ix

*Chapter* XV: 1792–1802 *page* 233
Birth of John F. W. Herschel. William goes on a tour to Glasgow with Count Komarzewski. Death of Paul Adee Pitt. Letters—Lalande, Méchain. William Herschel made a member of the French Institute. Letters—B. Greatheed, De Luc.

*Chapter* XVI: 1788–1798 *page* 245
Caroline Herschel's astronomical work. Extracts from her *Book of Observations*. She discovers seven comets. Letters—to and from Maskelyne, William Herschel to the Earl of Salisbury, Aubert, Sir J. Banks, Lalande. Caroline's Index to Flamsteed's Catalogue. Letter to Dr Maskelyne.

*Chapter* XVII: 1793–1804 *page* 259
William Herschel's scientific work. Paper "On the Nature of the Sun and fixed Stars". Letter to Prof. Wilson. Discovery of heat rays. Letters—Sir J. Banks. Asteroids. Letters—Piazzi, Bode, Olbers, Laplace. William Herschel's experiments for determining the diameters of small spheres.

*Chapter* XVIII: 1804–1817 *page* 278
Second paper "On the Nature of the Sun". Periodicity of sunspots—effect on climate. Letters—William Herschel to Bode, W. Watson. Brougham's article in the *Edinburgh Review*. Paper "On Concentric Rings"—adverse opinion of the Royal Society. Letter from Sir J. Banks. William Herschel's later papers "On the Construction of the Heavens". Sir J. F. W. Herschel's list of his father's most important discoveries.

*Chapter* XIX: 1797–1802 *page* 291
Family life at Slough. Dr Burney's visit. Caroline's daily life—she visits Dr Maskelyne at Greenwich. John Herschel at Eton. Caroline goes to Bath. Letters—Mrs Herschel and William to Caroline. Mme Beckedorff.

*Chapter* XX: 1802–1806 *page* 304
William Herschel's Diary of his visit to Paris—interview with Bonaparte, Laplace, etc. Outbreak of war. State of Germany. Dietrich's letters from Hanover. Caroline's Journal—the Spanish telescope. Dietrich visits England—his letters.

*Chapter* XXI: 1807–1813 *page* 322
Dietrich comes to England. Caroline's Journal. William Herschel and his family make a tour to the north of England—his letters. Trip to Scotland, 1810. John Herschel's diary. James Watt. Trip to Scotland, 1811. Visit to Brighton. Thomas Campbell.

*Chapter* XXII: 1812–1816 *page* 337
Caroline's Journal. John Herschel Senior Wrangler. Letter from William Herschel to Sir Joseph Banks proposing his son as F.R.S. Last observations with the 40-foot. Alexander's accident. Letters. Alexander returns to Hanover. William Herschel receives the Guelphic Order of Knighthood.

# CONTENTS

*Chapter* XXIII: 1813–1822      *page* 348

Letters from William Herschel to his son on the choice of a profession. John Herschel leaves Cambridge to take up astronomy. Caroline's Journal. Sophia Baldwin's marriage and her death. Death of Queen Charlotte. William Herschel's declining health. Death of King George III. Foundation of the Astronomical Society. William Herschel first President. Caroline's narrative of Sir William Herschel's last days. Letters of condolence. Epitaph in Upton Church.

*Chapter* XXIV: 1822–1848      *page* 364

Caroline retires to Hanover. Letters to Lady Herschel and to John Herschel. Her Zone Catalogue. Becomes an Honorary Member of the Astronomical Society and of the Royal Irish Academy. Her last years and death.

*Appendix* I: Fourier's Eloge of William Herschel      *page* 377

"    II: Epitaph on Sir William Herschel      *page* 380

"    III: Epitaph of Miss Caroline Herschel      *page* 382

"    IV: Papers communicated to the Royal Society by William Herschel, which are referred to in the foregoing chapters      *page* 383

*Index*      *page* 385

# ILLUSTRATIONS

| | |
|---|---:|
| Sir William Herschel | *Frontispiece* |
| [From a crayon copy of the oil painting by L. T. Abbott] | |
| Sir William Herschel | *facing page* 162 |
| [From the pastel portrait by J. Russell, R.A.] | |
| The Ten-foot Telescope | 79 |
| [From a drawing by Sir W. Watson] | |
| The Twenty-foot Telescope, probably at Datchet | 137 |
| [From a water-colour sketch] | |
| The Forty-foot Telescope, at Slough | 167 |
| [From an engraving dedicated to King George III] | |
| The Nebulous Star, described by William Herschel | 231 |
| A Globular Cluster (M. 3 Can. Ven.) | 231 |
| Sir William Herschel | 355 |
| [From the oil painting by Artaud] | |
| Caroline Lucretia Herschel | 375 |
| [From an engraving made at Hanover] | |

## CHAPTER I

## 1651   1757

Origin of the Herschel Family. Isaac Herschel's autobiography. Early life of William and Caroline Herschel. William joins the band of the Hanoverian Guards. Historical sketch. William's account of his visit to England with his father and brother. Battle of Hastenbeck. Retreat and surrender of the Hanoverian army. William and Jacob come to England.

LITTLE positive information is available concerning the progenitors of the Herschel family. The name suggests a Jewish origin, but as far back as they can be traced they were Christian Protestants and had even, according to family tradition, suffered persecution and been driven out of Moravia on account of their adherence to the Reformed Faith. The same tradition attributes the recurrence of Jewish names in the family to Scriptural rather than to racial associations. A certain Carl Herschel, who wrote to Sir William Herschel in 1790, claiming relationship, mentioned that there were in Germany many Jewish families bearing a similar name, but that it was usually, in their case, spelt with an "i", Hirschel. Whatever the truth of this matter may be, if, as seems probable, the Herschels were originally Jews, certain it is that they had inherited the best qualities of their race—industry, intelligence and loyalty to their religion; but, having no talent for money-making, they remained poor.

The authentic chronicles of the Herschel family reach back to the middle of the seventeenth century, when a certain Hans Herschel was living at Pirna, near Dresden in Saxony; he was by trade a brewer. Hans Herschel had two sons; what became of the eldest is not known; the second son, Abraham, was the grandfather of William Herschel, the astronomer.

From this time onwards we are on firm ground, as William's father, Isaac Herschel, left a short account of his own life and parentage which is here given in full, translated from the German. It was written in 1764.

## ISAAC HERSCHEL'S AUTOBIOGRAPHY

"Abraham Herschel was born in 1651; he learnt gardening at Dresden in the Elector's gardens, and was besides very fond of arithmetic, writing and drawing, as well as of music; especially he was very far advanced in arithmetic and writing.

"He became what was called a pleasure gardener, on the estate of Hohensatz, which is part of Zerbst, and three or four miles from Magdeburg. He married Eva Meves, daughter of a dyer in the little town of Loburg; they had four children. The eldest son was Eusebius; he learnt gardening in Gotha, but afterwards took to farming. He married and had five sons who, I hope, are still living.

"The second child was a daughter, Apollonia; she married, after a somewhat romantic manner, a country gentleman named Herr von Thumen; they lived together very happily and had one son, who died young, and one daughter, who married a Prussian officer of good family.

"The third child of Abraham Herschel was a son, Benjamin, who, in the third year of his age, fell into a well and was drowned.

"The fourth and last am I, Isaac Herschel; I was born at Hohensatz on 4th January, 1707. My parents wished me to devote myself to gardening, but my father died in 1718, before I could take it up. My brother Eusebius, who had finished his apprenticeship in gardening, took up my father's work for some years. After he had married he devoted himself to farming in Altenburg, in Chursachsen, and was very successful.

"As I had by that time learnt something of gardening, my brother got me a place at Zerbst, with an old widow lady, as gardener.

"While I was at Hohensatz I had procured a violin and learnt to play it by ear, and at Zerbst I took proper lessons in music from an oboist in the Court Band. I then procured an oboe for myself and was never so happy as when I was occupied with music. I worked day and night as much as I could get time, as my wish was to become an oboist; for I had lost all taste for gardening.

"When I had reached the age of 21, and considered that I had learnt enough to become an oboist, I gave up my situation and went to Berlin to seek service in a regimental band, but, as I found the Prussian service as a bandsman very bad and slavish, I went to Potsdam for a whole year and took lessons in music from the

Prussian conductor, Pabush. My dear brother and sister supplied me with the necessary funds; my brother, however, urged me in his letters to come and study agriculture with him. I obeyed his call, but I could not long resist my love of music and travelling, and in July, 1731, I went to Brunswick, and there I could have taken a position as oboist, but the situation appeared to me too Prussian; I therefore went to Hanover. There it was my destiny to stay, and I was engaged as oboist in the Foot-Guards. In the following year I married Anna Ilse Moritzen, daughter of a citizen of Neustadt on the Rubenberg, three miles from Hanover. On the 12th April, 1733, God sent us our first daughter, Sophia Elisabeth."

Isaac Herschel had ten children, four of whom died young, one of whooping-cough and one of small-pox. His daughter Caroline had small-pox at the same time, being then four years old. She says long afterwards: "Although (I) recovered, I did not escape being totally disfigured and suffering some injury to my left eye".

The surviving children were:

| | | |
|---|---|---|
| Sophia Elisabeth | born | 12th April, 1733, |
| Heinrich Anton Jacob | ,, | 20th November, 1734, |
| Friedrich Wilhelm | ,, | 15th November, 1738, |
| Johann Alexander | ,, | 13th November, 1745, |
| Caroline Lucretia | ,, | 16th March, 1750, |
| Johann Dietrich | ,, | 13th September, 1755. |

All these, except perhaps the eldest daughter, inherited their father's musical talent. Jacob became an accomplished musician at an early age, and William told his son that he could remember himself, as a child of four years of age, being set on a table to play a solo on the small violin which his father had had made for his children.

There is an interesting note in the Commonplace Book of Sir J. F. W. Herschel, William's son, concerning the Garrison school where William Herschel and his brothers and sisters received their first teaching. Under the date 26th June, 1817, he wrote:

"Baron Strandman and four young Russians visited Slough. Baron S. brought them to England to learn the Lancaster and Bell system[1] and

---

[1] The schools for the education of the children of the poorer classes started independently, in the beginning of the nineteenth century, by Joseph Lancaster in Southwark and by Dr Andrew Bell in India, had this feature in common, that the instruction of the younger children was entrusted to the older pupils.

introduce it into Russia, by order of the Emperor. My father says that in Hanover a system of the kind had long been established; he himself was taught to read and write and sum at a school of the kind where were 500 boys, and remembers at 8 years of age being made to teach the boys of 6 what he had learnt himself".

Caroline wrote concerning her brothers Jacob and William:

"There was a great difference in their dispositions; the eldest esteemed music as the only science worth cultivating, whereas William gave early proofs of an eager desire for knowledge in general.... As our Preceptor (at the Garrison school) was a well-informed man, capable of teaching as much Latin as boys in general carry away from preparatory schools, and was besides reckoned an excellent arithmetician, my brother attended his private hours. And with joy Mr Antonius (the name of our Preceptor) informed my father that his son knew not *all* but *more* than he could teach him.

"And when the two brothers began at the same time to receive lessons in French, my brother William was master of it in less than half the time the elder wanted; but he continued to attend the lessons till his brother was also perfect in order to benefit by the learning of their language-master (Mr Hofschlager), whom, many years after, I have heard named as a man well informed in Mathematics".

Isaac Herschel was able to engage the services of this teacher for his two boys, by applying for this purpose the salary which William earned after he was fifteen as a hautboy player in the Hanoverian Guards, and that which Jacob received at the same time as organist at the Garrison chapel. Hofschlager was a well-read man, and he imparted a taste for philosophy and metaphysics to his young pupils. Nothing could be further from the truth than to speak of William Herschel's education as meagre. It was in fact far more liberal and purposeful than that given at the time to English boys in public schools. Having been supplied with the necessary instruments for acquiring knowledge, reading, writing and arithmetic, as well as the rudiments of Latin and logic, and having himself an insatiable thirst for knowledge, he was stimulated by his father and tutor to proceed along the only sure path of learning—that of reading and discussion with those better informed than himself.

Caroline has preserved for us her recollection of the evening talks, when she lay awake listening to her father and brothers debating on metaphysical or mathematical subjects. William seems to have been the most fluent in these discussions.

"My brother William", she says, "and his Father were often arguing with such warmth that my Mother's interference became necessary, when the names of Leibnitz, Newton and Euler sounded rather too loud for the repose of her little ones, who ought to be in school by seven in the morning. But it seems that on the brothers retiring to their room, where they shared the same bed, my brother William had still a great deal to say; and frequently it happened that when he stopped for an assent or reply he found that his hearer had gone to sleep.

"My Father was a great admirer of astronomy and had some knowledge of that science; for I remember his taking me on a clear frosty night into the street, to make me acquainted with several of the beautiful constellations, after we had been gazing at a comet which was then visible. And I well remember with what delight he used to assist my brother William in his philosophical studies, among which was a neatly turned globe, upon which the equator and the ecliptic were engraved by my brother."

The simultaneous training of hand and mind is coming to be recognized as part of a sound education, so this encouragement given by the father to the mechanical tastes of his sons must also be regarded as an educational advantage. It was of course of great practical use to William in after life, when he was constructing his marvellous instruments. The practice of the violin may also have developed his extreme delicacy of touch. Years afterwards, at Datchet, when he was superintending and helping the workmen engaged on the erection of his telescope, one of them, amazed at his dexterity with the turning-lathe and also at forging iron, asked him what he had been brought up to. "To fiddling", was the reply.

It may therefore truly be said that, if William Herschel had not received much definite instruction in his youth, he had yet been well educated; for he had been taught freely to use all the powers of mind and body with which nature had endowed him. Nor must we forget the moral influence of his father's sincere and practical piety.

It may be useful here to take a cursory glance at the condition of Europe at this time, in order to understand how it affected the lives of the poor bandsman, Isaac Herschel, and that of his sons. When Isaac

Herschel went to Berlin and found the Prussian service so little to his taste, he came into touch with that growing military power which was soon to throw the whole of Europe into confusion. The year 1740 was a momentous one, for in May of that year Frederick I of Prussia died and was succeeded by his son Frederick II, called in history Frederick the Great. Three months later Charles, the Emperor, also died, and then the seething cauldron of European troubles boiled over; Frederick seized Silesia and so began the series of wars which devastated Germany for twelve years.

England was drawn into the struggle and sided with Prussia. In June, 1743, a considerable army of mixed English and Hanoverian troops was dispatched to the Rhine; King George II, accompanied by his son, the Duke of Cumberland, went over to take command and was present at the battle of Dettingen, when the French were severely defeated. The English troops were, however, withdrawn to stem the Jacobite rebellion, but the war was carried on for over two more years, though Hanover itself remained unmolested. Isaac Herschel accompanied the band of the Foot-Guards throughout this campaign. He carried in his knapsack a devotional book, *Vom Wahren Christenthum*, by Johann Arndts; and on the inside of the covers of this book he jotted down the dates of the principal events of his life. From these notes it appears that he had been with the army throughout the campaign which led to the battle of Dettingen. Caroline in her Memoirs relates that he contracted rheumatism and asthma from lying in a wet field all night after the battle, and was invalided for some months, but then returned to the regiment. He does not appear to have been near the scene of fighting when the troops were defeated by the French at Fontenoy. His notes conclude with the words:

"1746, *Feb.* 13. Wieder mit dem ganzen Regiment in Hanover einmarschiert, Gott sÿ Dank".

This was the end of the war; Isaac Herschel was now able to settle down to a quiet family life, superintending the musical education of his sons, with good hopes that they might, after serving an apprenticeship in the Guards band, be engaged in the Court Orchestra. The only events which broke the monotony of the next ten years were the births of two more children, Caroline and Dietrich, and the marriage of the eldest daughter, Sophia, to G. Griesbach, another member of the regimental band. Caroline mentions that her father was by no means satisfied about the match, and his fears were justified in the event, as Griesbach turned out to be both lazy and ill-tempered. But the wedding was very gay. Caroline wrote in her Memoirs:

> "I remember how delighted I was when they showed me the pretty framed pictures with which my Brother William had decorated his sister's room and heard my Mother relate afterwards that the Brothers had taken two months' pay in advance for the wedding entertainment, without which in those days it would have been scandalous to get married".

William Herschel joined the regimental band as an oboist when he was fourteen years of age and his brother Jacob when even younger. In the year 1756, when William was in his eighteenth year, the peaceful current of the family life was rudely and, for him, permanently broken up.

England had been nominally at war with France since the previous year and her people lived in constant fear of invasion. Hanoverian troops were brought over to assist in the defence, amongst them the regiment of Foot-Guards in which Isaac Herschel and his two sons were serving. Caroline gives a graphic account in her Memoirs of the confusion and distress in her home when her father and both her elder brothers, as well as her sister's husband, were all thus unexpectedly snatched away. William Herschel has left, in his Biographical Memorandums, some account of his experiences during this, his first, visit to England. These Memorandums begin with the following entry:

> "1753, *May* 1. I was engaged as Musician in the Hanoverian Guards, being then 14 years and some months of age, and I remember playing for this purpose on the Hautboy and on the violin before General Sommerfeld, who approved my performance.
>
> "This engagement furnished the means for my improvement not only in music, which was my profession, but also in acquiring a knowledge of the French language; with the advantage of studying above two years under a very well informed teacher [1755], who, taking a great liking to me, did not confine his instructions to language only, but encouraged the taste he found in his pupil for the study of philosophy, especially Logic, Ethics and Metaphysics, which were his own favourite pursuits.

"Soon after the great Earthquake which the 1st of November destroyed Lisbon, the Guards marched out of Hanover, and about the end of March, 1756, we were at Ritzbüttel, and embarked at Cuxhaven for England, where after a passage of 16 days we arrived in April. We disembarked at Chatham and marched to Maidstone in Kent and were quartered in that town. Here I applied myself to learn the English language and soon was enabled to read Locke on the Human Understanding".

(Caroline remarks that to obtain this book "had put him to great shifts, as he could not bear the thought of asking his Father for the smallest trifle".)

"From Maidstone we marched to Coxheath where the Hanoverian troops were encamped. Here as well as at Maidstone my Father, my eldest brother and myself made several valuable acquaintances with families that were fond of music, and which on mine and my brother's return to England proved of great service to us. During our stay in camp, we took leave of absence for a short visit to view London.

"While we were encamped on Coxheath my brother obtained his wished-for dismission from the regiment and returned to Hanover.

"In autumn when the camp broke up the Guards marched to Rochester, where again I made some valuable acquaintances, but our stay there was not of long duration, as we soon after received orders to embark again for Germany."

A little incident recorded by Caroline in her Memoirs shows how early in her life the kindness of her big brother William had inspired her with grateful attachment.

"My Mother", she says, "being very busy in preparing dinner had suffered me to go to the parade to meet my Father, but I missed him, and continued my search till I was spent with cold and fatigue; and on coming home I found them all at table; nobody greeting me but my brother William, who came running and crouched down to me, which made me forget all my grievances. I mention this instance of my brother's attention as it was the last I should receive from him (though it was not the first) for years to come."

The Hanoverian Guards, after some quiet months at home, were now again to experience the hardships of actual service. The campaign in which they now took part had a disastrous ending. In July, 1757, the

Duke of Cumberland's army, consisting mainly of Hanoverian and Hessian troops, was utterly routed by the French at Hastenbeck, not far from Hanover, and retreated in confusion to the mouth of the Elbe, where in September the Duke surrendered to the French commander. Under the terms of the capitulation, known in history as the Convention of Klosterzeven, the Hessian troops were allowed to return home, but the Hanoverian army was detained, and remained encamped before Stade, on the Elbe, while their country was left to the mercy of the French, who occupied Hanover for nearly two years.

As the charge of being a deserter from the army, which is so repeatedly brought against William Herschel in popular biographies, rests on his conduct after the defeat at Hastenbeck, it is only fair to his memory to examine closely the records which have come down to us and to bear in mind the condition of the country at the time.

After recording his return from England with his father in the autumn of 1756, William Herschel continued his Biographical Memorandums as follows:

"1757. Early in the spring my Father and I went with the regiment into a campaign which proved very harassing by many forced marches and bad accommodations. We were obliged, after a fatiguing day, to erect our tents in a ploughed field, the furrows of which were full of water.

"*July* 26. About the time of the battle of Astenbeck we were so near the field of action as to be within reach of gun-shot; when this happened my Father advised me to look to my own safety; accordingly I left the regiment and took the road to Hanover; but when arrived there I found that, having no passport, I was in danger of being pressed for a soldier; it was therefore thought proper for me to return to the army".

Caroline, then a child of seven years, describes in her recollections how the poor distracted mother hurried her son out of the town, feeling he would be safer with his father and the army as a bandsman than liable to be enrolled in the hasty defence force which the panic-stricken burghers were trying to raise. Jacob, who was no longer in the band, went into hiding and, a couple of months later, managed to join William at Hamburg and to escape with him to England. To return to William's own narrative.

"When I had rejoined the regiment I found that nobody had time to look after the musicians; they did not seem to be wanted. The weather was uncommonly hot and the continual marches were very harassing. At last my Father's opinion was that as, on account of my youth, I had not been sworn in when I was admitted in the Guards I might leave the military service; indeed he had no doubt but that he could obtain my dismission; and this he afterwards procured from General Sporken, who succeeded General Sommerfeld.

"I found no difficulty to leave the army as nobody seemed to mind whether the musicians were present or absent. My intention being to go to England, I took the road to Hamburg; and the French having taken Hanover, my brother Jacob left the place and came to me at Hamburg about the end of October and soon after went with me to England."

It will be seen clearly enough from these extracts, indeed William himself makes no secret of it, that his father aided and abetted him in leaving the regiment without waiting for a formal dismissal, but the circumstances must be taken into account and also the fact that he had never been enrolled as a regular soldier; the word "soldat" is scratched out in the formal discharge which he later received and the word "hautboist" is substituted for it, showing that he was engaged as a musician and non-combatant.

The insinuation which has crept into some biographies that he sneaked away from the army in peace time is probably derived from a careless perusal of Caroline's account of his leaving Hanover to *rejoin* the regiment.

After William and his brother had gone to England, their father remained with the Hanoverian army and shared the banishment enforced upon it by the terms of the Capitulation of Klosterzeven. The men did not return to their homes for two years; such would have been William's fate had he remained with the army.

The poor people in Hanover suffered grievously during the French occupation. Little Caroline was fully occupied in assisting her mother in household matters and in keeping her small brothers out of her sister's way, who, full of her own troubles, "could not endure to have children about her". Moreover, as her mother could neither read nor write, "my pen", she says, "was frequently in requisition for writing, not only my mother's letters to my father, but for many a poor soldier's wife to her husband in the camp".

Relief came at length. Prince Ferdinand of Brunswick had been appointed to supersede the Duke of Cumberland in command and he carried on the war with great vigour. The French were compelled to evacuate Hanover, and retired in such haste that they threw all the corn and flour which they could not carry into the river to prevent its falling into the hands of the enemy. The Hanoverian army pursued them; by July, 1759, the troops were at Hille, very near Hanover. Frau Anna Herschel was able, with other soldiers' wives, to visit her husband in camp. "She returned", says Caroline, "very comfortless, after witnessing the privations and hardships to which my poor Father, with his Asmath and broken Constitution, was exposed."

A month later Prince Ferdinand won the famous victory of Minden, and, as far as Hanover was concerned, the war was over. Isaac Herschel obtained his discharge from the band and returned to his family. Jacob also returned to Hanover from England soon after peace was concluded and resumed his place in the Court Orchestra. William preferred to remain and to follow his fortune in England.

## CHAPTER II

### 1757   1761

William and Jacob in London. Jacob returns to Hanover. William goes to Yorkshire. Lord Darlington's Militia band. Extract from Southey's book *The Doctor*. Letters to Jacob begin. William goes to Edinburgh—meets Hume—studies mathematics.

WHEN William and Jacob Herschel reached England, in November, 1757, they went straight to London. William afterwards told his son that when he arrived there he had not half a guinea in his pocket. He went at once to a music shop and asked to be allowed to write music; an opera was put into his hand to copy, he restored it with such dispatch as amazed his employer, who from that time gave him plenty to do, till he found more profitable occupation.

William Herschel's attachment to all the members of his family was sincere and unfailing, as his later conduct proved, but for his eldest brother, Jacob, he had a very special affection. They had been the closest companions from childhood; Jacob had showed himself a patient listener to all William's eager outpourings, while William on his side shut his eyes to his brother's faults and sincerely admired his musical talent. Jacob was probably the better musician of the two; some of his compositions which were published in London, have been erroneously attributed to William.[1]

Caroline, who suffered from Jacob's selfishness and fastidiousness, was more censorious in her estimate of his character, though after his death she was ready to admit that he was "on the whole an excellent and sensible creature", though "incorrigible where Luxury, Ease and Ostentation were in case". Writing in later days to Sir John Herschel, she said:

> "So it happened, my dear nephew, that your father was his brother's fag; by teaching, writing music for what he could get, to keep him from the degradation of accepting any engagement under a leader or first violin; till at the end of two years he (Jacob) was called to fill a place in the orchestra at Hanover and William was left alone to pay

[1] See Fetis, *Bibliographie des Musiciens*.

tailors' bills and parting with his last farthing for travelling expenses. With these circumstances I only became acquainted from what dropped in the course of conversations I had within the last few years of my Brother's life".

William's reason for not returning to Hanover with his brother, now that peace was concluded, was partly that, not having yet obtained his discharge, he would have had to resume his place in the regimental band till he obtained something better, and the poor pay would not have enabled him to help his family as he ardently wished to do. England held out much better prospects, as the event soon proved. His own narrative of this part of his life, as given in his Biographical Memorandums, runs as follows:

"When we arrived in London we made use of the recommendations of some families we had been known to when we were in England before. We were introduced to some private concerts, my brother attended some scholars and I copied music; by which means we contrived to live pretty comfortably in the winter; and in the summer we visited some families in and near Maidstone and Rochester, and had a concert at Tunbridge Wells.

"London was so overstocked with musicians that we had but little chance of any great success. My brother, having the promise of a place in the Hanoverian orchestra, received notice of a vacancy in September, 1759, and accordingly left me.

"Having given everything I could possibly spare to my brother when he left me I found myself involved in great difficulties. Seeing no likelihood of doing well in London, I intended to try for better success in the country. Very opportunely I had an offer of going into Yorkshire, where the Earl of Darlington wanted a good musician to be at the head of a small band for a regiment of Militia of which he was the Colonel. The engagement being upon a liberal plan and not binding for any stipulated time, I gladly accepted it. As the regiment was quartered at Richmond in Yorkshire I took a place in the stage coach, and on my arrival found that our musical band consisted only of two Hautboys and two French horns. The latter being excellent performers, I composed military music to show off our instruments".

In Southey's book, *The Doctor*, there is an interesting allusion to William Herschel at this period of his life. Dr Miller, who was organist at Doncaster, is reported in this work as describing how he had *discovered* William Herschel, then an unknown member of the Militia band; and had induced him to give up the band and to come and reside with Dr Miller, who promised to find him more remunerative employment. "This offer", Dr Miller continued, "was accepted as frankly as it was made"; but the only situation which he claimed to have found for Herschel was that of first violin at a series of private concerts which were arranged to take place at the house of a country gentleman, Mr Copley, residing near Doncaster.

The name of Miller does not appear in William Herschel's Memorandums. John Herschel noticed the paragraph in a review of Southey's book and drew his father's attention to it. He noted down the substance of William Herschel's reply:

"The paragraph in question is founded on truth but inaccurate in detail. Dr Miller, it seems, assumes a credit to himself more than his due. My Father was never living in Dr Miller's house as the Doctor would insinuate, though it is true that he was occasionally there for a night or so on a visit, and repeatedly passing through Doncaster, where Miller was organist, often called on him. Yet it is not unlikely that Miller introduced Father to Copley's concerts".

The claim which Dr Miller made that he had induced Herschel to leave the Militia to come and live with him bears its own refutation on its face. Herschel certainly did not leave the Militia before March, 1761. His frequent letters to his brother at this time are dated from Sunderland, Richmond or Pontefract, and show that he had a fixed lodging at Sunderland. Moreover, he had already made more influential friends than Dr Miller; he wrote in his notes:

"13 *August*, 1761. I spent a week at Halnaby, the seat of Sir Ralph and Lady Milbanke.[1] I had often been there before and in November, 1760, I made a long stay there to accompany Lady Milbanke who was an excellent player on the harpsichord".

Thus Dr Miller's claim to have drawn William Herschel out of obscurity seems to have been exaggerated. But like all who came in

[1] It appears from an order-book of the Richmond Militia of the North Riding of York, quoted by Sir J. Fortescue in *The Times* of 18th November, 1929, that Sir Ralph Milbanke, the ancestor of the future Lady Byron, was Colonel of the regiment in 1759. This may account for Herschel's acquaintance with him.

contact with the young violinist, Dr Miller felt the charm of his personality and talent.

William Herschel's highest ambition at this period of his life was to establish a reputation as a composer. His letters to Jacob are mainly filled with descriptions and criticisms of his own and his brother's compositions. He mentions seven symphonies as having been composed by him between June, 1760 and January, 1761, two of them while staying at Sir Ralph Milbanke's house.

In order to get from place to place to fulfil his various engagements he had been obliged to buy a horse, and he frequently rode over fifty miles a day across the moors in all weathers. He was often greatly fatigued by these long journeys on horseback, and the uncertainty of the future necessarily weighed on his spirits; but he fought these moments of depression by filling his time with work and study. He says of himself:

"During all this time, though it afforded not much leisure for study, I had not forgot my former plan, but had given all my leisure hours to the study of languages. After I had improved myself sufficiently in English, I soon acquired the Italian, which I looked upon as necessary for my business. I proceeded next to Latin, and having also made considerable progress in that language, I made an attempt of the Greek, but soon dropped the pursuit of that as leading me too far from my other favourite studies, by taking up too much of my leisure. The theory of music being connected with mathematics, had induced me very early to read in Germany all what had been written upon the subject of harmony; and when, not long after my arrival in England, the valuable book of Dr Smith's *Harmonics* came into my hands, I perceived my ignorance and had recourse to other authors for information, by which I was drawn from one branch of mathematics to another".[1]

During the two years which William and Jacob Herschel spent together in London very little correspondence must have passed between them and their family in Germany. The father was absent with the army in camp and the mother could neither read nor write. Isaac Herschel did, however, contrive to get some news of his sons. Caroline found, in a pocket-book of her father's, some notes showing that he had re-

[1] Letter to Dr Hutton.

ceived three letters from them in 1758, the year after they reached London, and that he had answered to an address given in Cavendish Street. After Jacob left him, William began a more regular correspondence through the Hanoverian agent in London, Mr Wiese. His letters were in the form of a journal, portions of which he sent to his brother from time to time. The pages are numbered, and the earlier ones are missing; those which have been preserved begin with a date in the year 1761: the packet also contains a prefatory note in William's later handwriting which runs thus:

> "A series of letters that were written in 1761, 1762 and 1763, to my eldest brother Jacob Herschel at Hanover, who was an excellent Musician and eminent Composer, and was also fond of intelligent Disquisitions. He preserved most of my letters, but chiefly those that were on musical, moral or metaphysical subjects. When I visited our family at Hanover in 1764, my brother, knowing that I was about writing a Treatise of music, allowed me, on account of their musical contents, to take these letters back again".

The packet also contains a catalogue of the letters made by William himself, in which their contents are summarized under such headings as "musical", "historical" (i.e. biographical), "moral", "metaphysical" or "characteristic". The series is not complete, several have been destroyed; when the gaps involved the omission of some event in his life which he thought worth recording, he has inserted a note of it from memory in the catalogue. Thus the letters and catalogue taken together afford a very full record of his life in Yorkshire for the years 1761 to 1763.

But these letters are also of very great interest, as they show the development of his mind and character at a critical period of his life. He was now twenty-two years of age, and was studying very carefully the few books he was able to procure on philosophy as well as mathematics and music; and as he sat in his lonely lodgings at Sunderland he solaced himself by pouring out to his brother all the reflections that passed through his mind. He recalled the demonstration of the existence of God, by the deductive method of scholastic reasoning which he must have learnt from his metaphysical French teacher at Hanover, but the freer spirit of Locke's *Essay Concerning Human Understanding* pleased him more, and he devoted more than seventy pages of manuscript to giving his brother an abstract of Locke's arguments for the Existence of God, of matter, space and time.[1]

He wrote mostly in German, occasionally in English, but when in the mood which he called in his catalogue "characteristic" or "bantering" he preferred to express himself in French. For the purpose of publication the German letters have been translated but the French ones are left untouched, as the intention of the writer would evaporate in a translation. When reading the selections given here it must be borne in mind that the writer was very young.

It was as a refuge from the depression consequent on his lonely life that he had recourse to conversing on paper with his brother. A fragment, undated, at the beginning of the series, shows how very friendless he was at this time.

### *Letters of* WILLIAM HERSCHEL *to* JACOB HERSCHEL

(In German)

(No date, probably January, 1761)

"...I have been since you left me nearly constantly *alone*. Now is it not certain that the greatest happiness is to be found in a friendly sociable life. As my sociable inclinations increase with my years, (as very often happens) I feel more and more their power. Is it not strange that during the three years which I have spent in England I have not met a single person whom I could feel worthy of my friendship. I won't assert that there is no such person in England."

After twenty pages about musical composition he concluded:

(In German)

Sunderland. 4th Feb. 1761

"I must put off these 'speculations' (in English) in order to start on my journey to Edinburgh. To-morrow I go to Newcastle and there attend the usual concert, and then continue my journey, which will altogether mean 130 or 140 miles. I hope on my return, which will be more than 16 days hence, to find a whole packet of letters from you. Adieu".

---

[1] He contributed a paper to the Bath Philosophical Society. It is printed in the *Collected Scientific Papers* under the title "Existence of Space, etc."

The letters give no account of the visit to Edinburgh, but William supplies the information in his Catalogue.

"I went from Sunderland to Newcastle, where I attended the regular subscription concert of Mr Avison, the organist, in which I was engaged as first violin and solo player. From Newcastle I set out the next day on a journey to Edinburgh, where I had been informed it was probable that there would be a vacancy in the musical department, the manager of the concerts intending to leave the place. On my arrival there I was introduced to Mr Hume, the Metaphysician; and a few days after, at one of their regular concerts, I was appointed to lead the band of musicians, while some of my Symphonies and solo Concertos were performed. Mr Hume, who patronized my performance, asked me to dine with him, and accepting his infitation (*sic*) I met a considerable company, all of whom were pleased to express their approbation of my musical talents; so that there seemed no doubt of my success had not their established manager agreed to stay."

He seems to have been sufficiently confident of getting the appointment at Edinburgh to resign his post with Lord Darlington's Militia band, and in a letter to his brother, dated 11th March, he speaks of going to Edinburgh as if it were a settled thing. The announcement therefore that the manager intended to stay must have been a great disappointment to him; it meant continuing the old uncertain rambling life. He even considered the possibility of returning to Hanover and getting an appointment in the Court Orchestra, but he added:

"By settling in Hanover, I may lose the considerable advantage in rambling about as I do now. I certainly have got in England at least three times what you had last year. But I must tell you a certain anxiety attends a vagrant life. I do daily meet with vexations and trouble and live only by hope. Many a restless night have I had; many a sigh and, I will not be ashamed to say it, many a tear, would Disappointments and Sensibility steal from me".

However, he never remained long in this despondent mood, the next letter is sufficiently cheerful.

Sunderland. 29th March, 1761

"Comme je suis d'une nature communicative je ne saurois m'empêcher de vous écrire souvent des choses qui ne vous regardent point du tout. Je viens de rire furieusement à l'occasion d'une lettre que j'ai justement reçu de Mrs D. Pauvre femme qu'elle est! Je vous raconterai quelque temps ou autre, quand je vous revois, ce qui s'est passé, mais je vous en prie...riés d'avance, riés donc.

"Comme je vous entretiens quelquefois par une narration de mes connoissances parmi les Dames, je continuerai à présent un peu à vous en parler, à cause que je suis en humeur de rire. Depuis ma lettre dernière, sur ce même chapitre, j'ai fait connoissance de deux jeunes Demoiselles; l'une en est la fille de Mrs G. et l'autre est la fille de Lady G--y. La première est la personne la plus belle du monde, la Beauté elle-même personnifiée; elle est bien jeune et son charactère est très aimable mais peu développé et tout-à-fait innocent. Pour la fille de Lady G--y, elle est déjà plus avancée, comme elle a environ 20 ou 23 ans; mais la personne est assés médiocre, et elle est assés loin d'être belle; mais *en revanche* elle joue admirablement de la guitare, und singt sehr lieblich darein. Elle aime l'hautbois à l'extrémité et m'a plusieurs fois invité à venir passer quelques semaines à la maison de campagne de My Lady sa mère pendant l'été. Elle a quelque chose de bien tractable dans sa disposition et fait *rougir* quand il en est temps; de même elle *sourit* à propos, et fait faire les yeux doux justement lorsqu'il faut. Mais tout cela avec innocence....Au diantre avec vos historiettes, dirés vous. Eh! bien, mon cher, pour vous plaire je changerai de sujet: venons en à Mademoiselle. Cessons de rire, comme il seroit injurieux de n'être pas sérieux en parlant d'une personne admirable. Comme l'on juge bien mieux des sentimens de nos amis par les lettres qu'ils nous écrivent que par les conversations, j'ai eu l'occasion de la connoître assés parfaitement puisque je reçoive de ses lettres bien souvent. Chaque lettre me surprend par sa solidité et justice; nous avons eu plusieurs conferences ensemble sur les sujets de l'amitié, des sentimens, de la connoissance du coeur humain, de la vertu et plusieurs autres, où elle fait voir un jugement au-dessus du commun. Mais ce que j'admire le plus est qu'elle sait m'écrire *en amie* et avec franchise, sans en être moins délicat ou réservé."

Writing about marriage in a letter, the commencement of which is missing, he says:

"C'est par là que nous vivrons du moins *en Compagnie*; nous ne passerons pas par la vie *en Solitude*. Nous serons du moins assuré qu'il y a *une* personne qui est notre véritable amie, qui pardonne nos fautes et nous aime. Qui pense bien de nous, qui sent comme nous-même, qui n'est point moins généreuse que nous-même, qui est de la même Religion que nous-même; enfin qui vit avec nous et qui meurt, si non avec nous, du moins comme nous-même. Adieu".

It is pleasant to think that this idea of a peaceful married life was fully realized later in the society of the gentle refined woman who became his wife. He mentions "la judicieuse Mademoiselle" in his letters, but though they seem to have corresponded for about a year he only met her once. His sentiment for her is expressed in an incomplete fragment beginning:

"que je sens pour eux. Je n'en ai principalement que deux, vous en êtes l'un, l'autre est Madlle. Il est vrai que je ne vous ai vu il y a longtemps, et pour Madlle, je ne l'ai jamais vu que trois ou quatre jours. Il est vrai que je souhaiterois de vous revoir tous les deux, du moins vous que j'estime le plus".

(In English)

11 April, 1761

"I have already several times been tempted to take a little step over to Germany for a week or two some time this summer and do not know but I may yet do so; as I do not value the trouble of travelling so much at present as I did formerly. I think I am but a very little way from you because this day week I might be with you if I choose it. But I am afraid my parents would not part with me in a fortnight's time and I could not stay longer.

. . . . . . .

"But here I must mention one thing, which is to take all manner of opportunity to convince or to let our parents know and see that as long as there is any faculties, any powers left in us they will always be employed in serving them and providing for them. I know these are your sentiments as well as mine, as you are not less grateful than I am. I am only afraid our parents should have any doubt of them. You cannot imagine how I am vexed sometimes when I am not as successful as I could wish to be, as it puts it out of my power to serve (them)."

Sunderland. 20 June, 1761

"Bon jour, Mons. Comment va-t-il donc aujourd'hui? Je viens vous voir pour prendre congé de vous pour une semaine. Il faut (avec votre permission) que j'aille à Newcastle voir les courses; il y aura un concert: et j'espère d'y voir Lady Milbanke et d'autres personnes de retour de Londres. Il y aura deux ou trois grands bals où l'on jouera quelques Menuettes que j'ai composé pour cet effet: de quoi riés-vous? donc—pour vous amuser un peu, en voilà deux. Je ne vous les aurois pas envoyées, mais comme je sais que vous aimés la composition, vous pourrés vous exercer en y mettant des Basses.... Écrivés moi bien souvent."

*Mrs Shafto's Minuet*

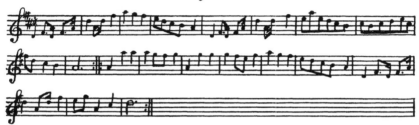

*Miss Hudson's Minuet*

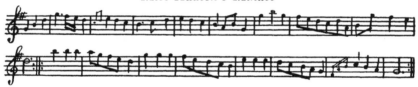

In German)

10 July, 1761

"If I found any pleasure in showing myself in a situation calling for pity, I would dilate on my experiences of the past night. However, without indulging myself so far, I will only say that at 9 o'clock, when I had still about 20 miles to ride, I was caught in an unusually heavy thunderstorm, which continued, accompanied by torrents of rain, with unbroken fury for three hours, and threatened me with sudden death. The distance from an inhabitation, the darkness and loneliness, obliged me nevertheless to ride on. I pursued my way therefore with unshaken sangfroid although I was often obliged to shut my eyes on account of the blinding lightning.

"At last the flashes all around me were so terrifying that my horse refused to go on; luckily at this moment I found myself near a house, into which, after much knocking, I was admitted. This morning, at 3 o'clock, I proceeded on my journey and arrived safely at this place."

Sunderland. August 12, 1761

"Now I must make you a little story about what I have done this week. My vanity has been a little flattered, for I had a message by a courier from Lady M. requesting my presence. I hastened thither and found at Halnaby the Duke of York, the King's brother. We made music for the whole week as his Royal Highness plays the violincello very well. I had therefore the honour not only to play several solos with his approval but to be accompanied by him. Whether this acquaintance with the Duke will be of any advantage to me I cannot yet say. You shall shortly hear more. À présent il faut me coucher, comme j'ai un grand chemin à faire demain."

Aug. 13

"I have just returned from a longish journey and am indeed so tired that I can hardly write."

Aug. 14

"We have lately arranged in Newcastle a concert in a garden after the style of Vauxhall in London. I have the direction of the music and we make up a fairly good band of about 16 persons. At present it is only once a week, but if it succeeds it will be oftener. This may help to make my things known; yesterday we played one of my symphonies."

16 Nov. 1761

"I have just received your letter No. 11, which has touched me not a little. I can hardly describe to you the present state of my feelings.... My only wish is now that I could just once see all my friends together and happy; afterwards I should care little what became of me.... I live now, as to myself, entirely without care as I have a superfluity of everything necessary. All the same I live as a hermit. No one troubles about me and I trouble about very few. Day by day I have hoped for something lucky to happen that could give me the means of helping you, as I know the sad condition of my Fatherland. Your salary cannot be sufficient. Though it cannot be sooner, yet as soon as I receive about 30 guineas which are owing to me, I will remember my obligations.

"I wish with all my heart that my father could get the situation now vacant which he well deserves, having served the King so long and grown old in his service.

"As it looks like a cold winter, I can hardly think what people in Hanover will do to protect themselves from the cold. Are coals dear in Hanover? If not, it strikes me one could make an arrangement to burn them even in the stove, by the help of some iron bars."

Jacob's letter, to which the above was the answer, must have described the troubles which afflicted the family in Hanover. There was great shortage of wood, as the French during their occupation had cut down the forests ruthlessly. Typhus fever raged in the town; poor Caroline had been struck down with it in the summer and was so weakened that as she said herself: "For several months afterwards I was obliged to mount the stairs on my hands and feet like an infant".

During the first twenty years of his life in England, William Herschel's main pursuit was music; but though in his letters to his brother he wrote pages on pages of disquisitions on harmony and the rules of composition, he never, in the opinion of good judges who have taken the trouble to read his scores, arrived at any important or original ideas. His own taste in music was simple; music must express emotion; he loved melody and hated fugues. The symphonies which he composed during these early days in Yorkshire are graceful and melodious, but do not indicate much originality. The scores, beautifully written out by himself, are preserved at Slough and have given much pleasure when performed at family gatherings.

It was only when he turned to sacred music that William Herschel's compositions rose above mediocrity. The religious elevation of his thoughts found a natural expression in praise and worship. Ozias Linley,[1] who was no mean judge, wrote to his niece, Mary Ward:

"I was on the point of adding to my Present a good number of my favourite hymns; the greater part of them the composition of the late astronomer, Sir William Herschel, formerly one of my Musick Masters. I treat my congregation (or rather myself) with them very often, not only out of respect to the memory of my good Master but because of their high *intrinsic* merit as devotional compositions".

[1] Ozias Linley was organist at Dulwich College when he wrote this letter. It is now in the British Museum; the date is only "7th August".

Unfortunately Ozias Linley's admiration for Herschel's hymns and anthems induced him to carry off to Dulwich all that had been left behind at the chapel at Bath, where he had succeeded Herschel as organist. A similar loss has overtaken the only copy of the Oratorio begun by William Herschel at Halifax. The subject and most of the words were taken from the *Paradise Lost* of Milton.

The letters which follow are not given strictly in chronological order, but they were all written in the two years 1761 and 1762, and are grouped for convenience into subjects—music and philosophy.

WILLIAM HERSCHEL *to* JACOB HERSCHEL

Sunderland. 22 Feb. 1761

"Mon cher frère; vous savés que je ne sais point de quel sujet vous écrire. Je parlerai donc de moi-même. Toute personne a quelque passion ou autre qui le gouverne principalement, ou plutôt dont il reçoit le plus de plaisir; et pourriés vous le croire, cette passion en moi est l'amour de la Musique, ou plutôt la satisfaction que je trouve en entendant ou m'imaginant une pièce de musique. Je suis jeune et d'une disposition tendre et néanmoins le plaisir que j'ai d'entendre quelque pièce favorite surpasse tout ce que l'amour, l'amitié, ou la tendresse pourroit contribuer à me plaire....Oh Dieu! Bienfaiteur qui ordonnât les règles de l'Harmonie! les Concordes, les Consonances, les Dissonances, les Modulations, le Mouvement, la Mesure! en tout cela on connoît la main de Dieu, d'un Créateur, d'un Bienfaiteur et d'un Être incompréhensible. Nous sommes des Compositeurs, vous et moi, mais que savons-nous de la musique? Rien! Rien de tout. Notre connoissance s'étend seulement à un petit nombre de faits et d'expériences; tout au plus quelque raison seconde borne notre vue. Les essences de toutes choses nous sont cachées et notre pénétration ne s'étend que jusqu'à certains accidents."

Almost as numerous as the letters on music are those in which William Herschel discoursed to his brother on metaphysics, religion and morals. He came across a book on the philosophy of Epictetus. It was written in very bad verse, but as Herschel always frankly admitted that poetry did not appeal to him, the poorness of the versification did not detract in his eyes from the value of the matter and he made long extracts for his brother's edification.

He accepted with enthusiasm the injunctions to self-discipline which he found in the *Morals of Epictetus*. Whether his character was really modified by his efforts to attain a stoic indifference so foreign to his frank, warm-hearted nature, psychologists may decide. At least the exercise may have helped him to control a certain irritability due to a sensitive pride, and impatience of contradiction, of which there are one or two examples in his life.

The predominant characteristic of his mental outlook was his love of order; he wished to array his thoughts always in a clear and logical form. As an exercise, probably, he wrote out, and afterwards sent to his brother, an abstract of Locke's demonstration of the existence of a Final cause, which we call God. But his own religious instinct was not satisfied with this cold impersonal conception of the Creator, and when writing on the subject of what constitutes happiness he concludes his letter with an expression of feeling which is not far removed from the well-known saying of St Augustine: "Thou hast made us for Thyself and our heart is restless until it rests on Thee".

WILLIAM HERSCHEL *to* JACOB HERSCHEL

(Translated from the German)

31 March, 1761

"There are two kinds of happiness or contentment for which we mortals are adapted; the first we experience in *thinking* and the other in *feeling*. The first is the purest and most unmixed. Let a man once know what sort of a being he is; how great the Being which brought him into existence, how utterly transitory is everything in the material world, and let him realize this without passion in a quiet philosophical temper, and I maintain that then he is happy; as happy indeed as it is possible for him to be.

"But as to the other kind of happiness; it is very mixed and we can only be happy when all our sensations are pleasant. Now I know by experience that, if one cannot entirely control one's feelings, we have at least some power over them. We should therefore take pains to banish all those that disturb us and to retain the pleasant ones. The most essential emotion which can help us to this end is love for our fellow-men; for this at once uproots or suppresses all those morbid emotions (irritability, jealousy, anger, envy, pride, &c.) and encourages the pleasant ones, namely kindness, generosity, pity, friendship.

"But where the attainment of our wish is accompanied by a certain melancholy, we must remind ourselves that this present life cannot be perfect; or more truly we may say that this sadness is in itself a sign that we have reached the furthest stage of happiness which our sensibilities are capable of giving us. This leads us with a kind of yearning towards our Creator. It is He Himself whom we desire and this mixed feeling of sadness and of joy comes from the thought of the perfection of the Creator and of the dependence and imperfection of the creature."

12 May, 1761

"You say that religion and laws have a great influence on manners and morals. Yes, they have; but here I would remark that the Christian religion is the best that ever has existed. I have for some time past been meaning to send you a defence of Christianity and will suggest the following points for you to consider, which I will prove or explain further if you ask me to do so.

### The Christian Religion

"(1) That in all ages and among all nations no kind of religion has prevailed so nearly without superstition.

"(2) That in all ages there have been philosophers who have had thoughts above their religion and have been true *Deists*.

"(3) That it is also impossible for a whole nation to be true Deists.

"(4) That the Christian religion is very good and one must not oppose but rather do everything to encourage it.

"Anyone therefore who wishes to be a reformer would be acting very foolishly if he should begin meddling with religion. Does one not do a thousand things in order not to be out of the common and not to offend the custom of the country? One wears 'manchetten', one frizzes one's hair, shaves one's beard, has silver candlesticks, coaches and horses, footmen and cooks, drinks tea twice a day, carries a sword in peace-time, sleeps till nine o'clock, says 'Ihr gutes Wohlsein' when one drinks, and says grace when one wishes to eat; takes off one's hat when one meets with an acquaintance, and bows when one meets an old gentleman, wherefore

then should one not &c....I leave you to make the application and will only add, that one says Good-night when one goes to bed.
Good-night."

May, 1761

"Mais dites moi un peu, n'est-ce pas naturel qu'on soit affecté si l'on est persuadé que tout ce qui nous environne est infiniment inférieur à quelque *Être Plus Parfait?*—Ajoutés-y l'ignorance dans laquelle nous vivons. Nous existons; l'universe *est*. Nous croyons donc en deviner la Cause....Tout cela est bien capable de nous affecter quelque peu, si non de nous rendre mélancolique, du moins sérieux.— Mais n'abattons pas nos esprits pour ça. Soyons philosophes et en même temps n'oublions pas que nous sommes encore hommes; n'ayons point de répugnances à faire des actions humaines; néanmoins soyons toujours prêts à quitter cette vie à l'instant.

"Et combien d'accidents n'y a-t-il qui peuvent nous l'ôter dans un moment! Quatre jours passé, venant de Richmond, comme je pensois justement à la mort et à l'ordre qu'on observe dans toutes les choses créées, je dis à moi-même: *Tout est dans l'ordre*—à l'instant, par quelque accident, je tombois de cheval avec violence. En me relevant sans être blessé aucunement, je dis: 'It would have been far better that thou hadst broke thy neck than that the laws of motion should have been altered for thy sake'."

(In German)

12 June, 1761

"As to Leibnitz, he is certainly a good philosopher, but in his *Theodicée* he goes too far and would have all actions necessary. His foreordained harmony is not the least credible nor feasible. If you can get a book entitled: *An Essay on the Origin of Evil*, by Dr W. King, you will find a much better solution of the question: 'whence comes evil?' Leibnitz does indeed reconcile it all with the goodness of God, but not nearly so reasonably as Dr King. Rollin is very pleasant to read but not convincing."

(In German)

Oct. 1761

"There are two kinds of order, the natural and the moral. Natural order is when all things are in such relationship that each can do best what it has to do. If one observes the whole natural world as one, one finds everything in the most beautiful order; it is my favourite maxim: *Tout est dans l'ordre*. But this natural order is not the one I mean; the order which concerns man is different. This leads us to consider *Moral* order, by which the former may be broken; so that if there were no moral disorder there could not be any in nature, or at least it could not have any bad results. This is no more than to say—'Sin is the source of all evil'. But I prefer to call sin disorder, and virtue order, because these words are more definite and forcible. If, for example, an inn-keeper gives his boy a box on the ear because he let a tea-cup fall, one cannot vefy well call this a sin. It is however without doubt a moral disorder, (unfitness); for neither will the broken bits of the tea-cup come together, nor will the boy be any the better for it. He will probably sit down and cry till he is either comforted with kind words and a kiss from Mama, or till he receives a sound thrashing."

William Herschel had much to say about the immortality of the soul; one extract will be sufficient to show the trend of his thoughts.

(In German)

28 July, 1761

"The immortality of the soul has always been a chief object of my thoughts. This life is so short that it scarcely merits our attention and would be much more insignificant if we had not the hope that it may be the beginning of an endless existence....

"I must acknowledge that in the 'substance' of our souls we may have a nearer connection with the unknowable, *must-exist Being* than we quite imagine. Some people consider the soul as an outflow of the Godhead, others say that we exist in God, and others that we are in the likeness of God. My feeble understanding is not capable of pushing so far into the secrets of the Almighty; and as all these propositions have something unintelligible about them, I think it better to remain content with my ignorance till it pleases the Creator of all things to call me to Himself and to draw away the thick curtain which now hangs before our eyes."

## CHAPTER III

## 1762   1766

William engaged as director of concerts at Leeds. Gets his discharge from the Guards Band. Visits Hanover. The Herschel family at Hanover. Isaac Herschel's death. William's Memorandums. Chosen organist at Halifax. Miller's story. William accepts the post of organist at Bath. The Octagon Chapel. Organ built by Snetzler.

THE unsettled life which William Herschel was obliged to lead at this time was very unsatisfactory and preyed on his spirits. He wrote to his brother:

Pontefract in Yorkshire. 22 January, 1762

"Indeed, Brother, I do not know what will become of me at last.... I am almost tired of having no home or place to be fixed in. It is true I have about three times as much as you, but I have it in such a manner as I am sure you would not take it, nor could you really do it.

"I am at present looking about for some sort or other of a place either as organist or any other fixed kind, which it is not at all unlikely I may find, as I have the good luck to make friends everywhere".

The opportunity he sought came to him a month later, when he was able to tell his family of the good prospects opening for him at Leeds. He got the appointment of director of the public concerts there and remained for four years, lodging with a Mr and Mrs Bulman, who proved themselves good and faithful friends to him. The following letter tells of his first reception at Leeds.

Leeds in Yorkshire. 25 April, 1762

"Fate has recently brought me to another spot. I have everywhere been well received, as composer and as performer. On the 16th there was a concert here, such as they have weekly, when I played for the first time; and though usually only 20 to 30 listeners in the audience, the news that I was to be there that evening brought over a hundred. I played two solos and though these were much praised I got more pleasure from finding that my symphony had such a success that I was heartily congratulated on being the composer. On the 21st there was another concert, more numerously attended.

This one was a benefit concert for a singer from York, and as he did not know that I should be there he had brought the first violin from York to lead the band. Now this violin was a celebrated man, but the Leeds directors insisted that I should have the leading part in the concert. However as the violinist from York did not like this, I thought it best to give way to him, especially as I did not know whether he were not a better performer than myself. At the same time, not to have it said that I played under him, I took, not the violin part, but the clavecin, and accompanied. In the second part he played a solo, and I declare they were unjust not to praise it as it deserved.

"Meanwhile the audience was divided in opinion as to which of us played best; those who had previously heard me play took my part and the others were for the other. The company then insisted that I should instantly play a solo, which I could not refuse to do; but as I had been playing the clavecin all the evening, I begged that we might first play a symphony, which was granted.

"The York violinist accordingly gave up his place to me and the symphony went off uncommonly well. Then I played my solo, which was received with approbation. Some went so far as to say I was the best violinist they had ever heard. The violinist from York was so courteous as to beg that he might play the second violin, to excuse himself for having taken the first place, as he had never heard me play. You know by experience that in these cases people always go too far; it was so here, for they certainly let their praise go farther than I merited. But enough of this.

"Last Friday there was another concert, when I again took the lead and I had the same approbation.

"Some of the leading and richest people in this town have interested themselves so much in me that there is talk of retaining me here; and if their intention succeeds (as I greatly hope) this town will put a pleasant end to my restless weary life.

"They talk of giving me a salary of perhaps even 4 or 5 hundred thalers, but rather than leave this delightful place I would be content to begin with 2 hundred thalers, as I am sure in a couple of years I could make it up to 600, and besides I should have a beautiful opportunity here to become known as a composer, which is always my chief ambition. But the greatest pleasure of all, without question, would be that I should be able to help my friends. I never think of Hanover without heartily wishing that this time I may succeed in doing so."

During the four years which William Herschel spent at Leeds he was fully occupied the greater part of the year with the duties of his profession. There are signs nevertheless in his correspondence with his brother that he was becoming dissatisfied with music as an intellectual pursuit. "It is a pity", he wrote, "that music is not a hundred times more difficult as a science." However, he had work enough to fill up his time and this, to his active nature, was the necessary condition for happiness. "My love for activity", he said of himself, "makes it absolutely necessary that I should be busy, for I grow sick by idleness; it kills me almost to do nothing."

In February, 1763, he heard that his official discharge had been granted and that therefore he was free to return to his Fatherland. But he delayed till his financial position had sufficiently improved to enable him to take a brief holiday and to meet the expense of the journey. In May, 1764, he was able to go, and the family party in Hanover was, as Caroline says, "transported with joy" at beholding again the son and brother from whom they had been parted for nearly seven years. It is worth noticing that William's arrival was connected in Caroline's memory with an astronomical event, to which her father had not failed to draw his children's attention.

"On the day before my Brother's arrival," she wrote, "viz. 1764, Ap. 1, 10 h.M. there happened to be an eclipse of the sun, which we were all assembled to observe in the courtyard, around a Tub of water, where my Father explained the phenomenon to us."

William must have been saddened by observing how hardly the toils and privations of these seven years had pressed on his poor father. Isaac Herschel's unremitting efforts to earn the support of his family by teaching and copying music had worn out a frame already weakened by the hardships of war. The task was not made easier for him by his eldest son's lack of consideration. It is true that Jacob, as Caroline admits, regularly brought the whole of his salary, when he received it, to his mother; but "generally", she adds, "the receipts were not enough to cover his tailor's bills, much less for the entertainments which were given to his comrades and their ladies, when he had a quartet party; where his new Overtures and other compositions were tried and praised; which my Father expected and hoped would be turned to some profit by publishing them, but no printer would bid high enough" to satisfy Jacob.

The other members of the family who welcomed William on his return were the two younger brothers, Alexander and Johann Dietrich, and the little sister, Caroline. The eldest sister, Sophia, was living with her husband and large family of boys at Coppenbrügge, a small town not far from Hanover, whither William went to see her. During the long years of his father's absence Alexander had been apprenticed as a scholar at one of the Guilds of music, living at Coppenbrügge with his sister and brother-in-law. Sophia's sisterly affection could not compensate to him for the harsh treatment which he received from Griesbach. The poor boy, naturally of a timid and listless disposition, had been completely cowed, and nothing, in his later life, but the watchful and loving care of William and Caroline saved him from the troubles to which such weak characters are liable. He atoned, in Caroline's eyes, for all his lapses, by his devotion to William.

After concluding his apprenticeship, Alexander returned to Hanover, where he obtained the easy post of "Stadt-Musicus" or Town-Musician. The chief duty was to blow the Chorale from the top of the Market-Tower at midday. He was so engaged when his brother reached home.

Dietrich, the youngest and the pet of the family, was in character the very opposite of Alexander, even as his youthful experiences had been different. He was always a great favourite with William for his happy temper. Besides being, like all the Herschels, a good musician, he struck out a line of his own and became an ardent entomologist. His life was not so intimately connected with that of William and Caroline as was Alexander's, but he also had cause over and over again to be grateful for his brother's help and sympathy in times of trouble. To Caroline, in the dreary years of her youth at Hanover, the little brother, with his sprightly, affectionate ways, was, especially after her father's death, the one bright influence in her life.

At the time of her brother's return Caroline was fourteen years old. One pictures her as very small for her age; a wistful little creature with fair hair and large blue eyes, but her childish beauty marred by the traces of small-pox. For her, William's arrival was most unfortunately ill-timed, as she was just preparing for her confirmation and was therefore obliged to attend the extra hours of instruction given by the pastor. She says in her reminiscences:

"Of the joys and pleasures which all felt at this long-wished-for meeting with my (let me say *dearest*) Brother, but a small portion could fall to my share, for my constant attendance at Church and School, besides the time I was employed in doing the drudgery of the scullery, it was but seldom I could make one in a group when the Family were assembled together.... We were in short for some days in a Tumult of joy. In the first week some of the Orchestra were invited to a concert, where some of my Brother William's Compositions, Overtures &c. and some of my eldest Brother's were performed to the great delight of my dear Father.

"Sunday the 8th was the (to me) eventful day of my confirmation and I left home not a little proud and encouraged by my dear Brother William's approbation of my appearance in my new Gown &c.[1]

"The following day my three brothers went to Coppenbrügge to visit our sister, and my Father and Mother were left with myself and Dietrich in sadness. For my dear Father this parting was felt like an anticipation of the final one, which happened on the following Sunday, April 15, when I was obliged to say adieu to this dear Brother before I went to join the assembled Confirmanten before the Altar in the Garrison Kirch to receive the Sacrament for the first time. The Church was crowded and the door open; the Hamburger Postwagen (with my Brother) passed at 11 o'clock, within a dozen yards from the open door; the Postilion giving a smettering blast in his horn. Its effect on my shattered nerves I will not attempt to describe nor what I felt for days and weeks after".

The agitation of feeling caused by the short and hurried visit of his best-loved son was too much for Isaac Herschel's failing strength. Very soon after William's departure he was one day brought home unconscious from an apoplectic stroke, from whose effects he only partially recovered. He lived on for three more years and with indomitable perseverance continued to work. Caroline wrote: "Though he never regained his former execution on the violin, every one of his scholars gladly came to him again for instruction as soon and whenever he was able to go to or receive them at home". The rest of his time, she wrote, "he passed with copying music till past midnight (for he could neither sleep nor find rest in bed), and I, as his chief nurse and companion, used

[1] The "new Gown" was of black silk, the "&c." included a bunch of artificial flowers "the same", she wrote, "which had served my sister on her bridal day".

to read to him all the while, which seemed to be no hindrance to his copying".

On the 22nd March, 1767, Isaac Herschel died, at the early age of sixty years, leaving to his sons no other legacy than an enduring example of piety and conscientious industry and of intelligent effort to use well the talents bestowed upon them.

Caroline was left for five years, the years when she was growing to womanhood, to the tender mercies of her mother and her brother, Jacob. Neither was intentionally unkind, but her mother was determined to eradicate the foolish notions of improving herself which her father had encouraged and which might lead her to wish to leave home as her brothers had done. A woman's duties, her mother thought, lay in the kitchen and parlour. Caroline was allowed to learn such refinements as washing and darning silk stockings, for Jacob's benefit, and dressmaking for herself and her mother, but nothing more. Jacob's fastidiousness added much to her troubles: "Poor I", she wrote, "got many a whipping for being awkward at supplying the place of a footman or waiter".

Alexander was always kind, but he had no influence in the home. Watching and caring for her lively little brother Dietrich was the one solace of Caroline's life till William was able to fetch her away to live with him at Bath.

William's only allusion, in his Memorandums, to his father's death is a brief entry: "*5th April*, 1767, went into mourning for the death of my father". This laconic statement might be interpreted as showing lack of feeling, if there were not abundant indications in his letters of his filial affection and respect. He constantly referred in his letters to the longing to see his father again. Moreover, though many letters passed between him and his brothers during these years, he destroyed them all, after noting the dates in his pocket-book; but the one letter he received from his father after he returned from Hanover has been carefully preserved, and, more than that, he took the trouble to copy the whole of it among his Memorandums as if afraid that the original might be lost.

The story of William Herschel's life when he returned to England can be gathered from these Memorandums. They began with the remark:

"As I at this time attended many musical families in the neighbourhood of Leeds, Halifax, Pontefract, Doncaster, &c., the memorandums of the places will partly show how I spent my time; a great deal of which was taken up with travelling on horseback".

There are no entries for the year 1764, after his visit to Hanover, nor for the whole of 1765; probably because there was nothing to record beyond the monotonous round of visits to the various places mentioned in the opening paragraph. But for the next year, 1766, which was a momentous one in his life, there are abundant notes, some of which will be given as he wrote them.

"1766, *Jan.* 1. Wheatley. This was the country seat of Sir Bryan Cook, where every fortnight I used to spend two or three days. Sir Bryan played the violin and some of his relations generally came from Doncaster to make up morning concerts. Our music was chiefly Corelli, Geminiani, &c. Lady Cook loved music and I gave her lessons on the guitar, which was then a fashionable instrument. Sir Bryan being an invalid, Lady Cook, an elderly Miss Wood, and I generally passed the evening in playing at Tredille.

"*Jan.* 7. Concert at Doncaster, at Sir Bryan's relations.

"*Jan.* 8. Having this time spent a whole week at Wheatley, my mare, standing idle in the stable and being overfed by Sir Bryan's grooms, died.

"*Feb.* 19. Wheatley. Observation of Venus.

"*Mar.* 4. Sir Bryan Cook died.

"*Mar.* 7. Halifax. The Messiah. This Oratorio was performed at a private club of chorus singers held at the Rev. Mr Bates, the Clerk of the Parish Church and father of the well-known musical Mr Joe Bates, where it was agreed to rehearse the same Oratorio every other Friday, in order to perform it in the church at the opening of a new organ erecting there. I was the leader of the orchestra and Mr Joe Bates, who played a chamber organ, directed the performance; his brother played the violincello. I was a candidate for the place of organist, which by the interest of the Messrs Bates and many musical families I attended, I had great hopes to obtain.

"About this time I was an inhabitant of Halifax. My leisure time was employed in reading mathematical books such as the works of Emerson, Maclaurin, Hodgson, Dr Smith's *Harmonics*, &c. This happened to be noticed by one of the Messrs Bates who told his brother: 'Mr Herschel reads Fluxions!'

"*Mar.* 30. I composed the first song in an intended Oratorio which I called the success of Satan against Man. The words taken from Milton's *Paradise Lost.*"

During the spring of this year (1766) he made Halifax his headquarters, though he stayed often at Leeds, having induced the organist there to let him practise on his organ with a view to the coming competition at Halifax. He gained so much proficiency that during the summer holidays in July the organists, both at Leeds and Wakefield, employed him to take services in their absence.

The movement for improving the services in churches, by substituting an organ for the old-time precentor with his tuning-fork, was spreading in England at this time. The parish church at Halifax seems to have already possessed an organ of some sort, which perhaps merited the description given in *The Doctor*: "one that might more properly be described as a box of whistles". A subscription had been raised to provide a worthier one, but there was some opposition, partly Puritanical and partly financial. The opponents went so far as to bring a lawsuit against the innovators and the organ could not be used till the dispute, which lasted a twelve-month, had been decided. But the rehearsals for the opening performance went on with undiminished confidence.

The new organ was being built by Snetzler, a workman who had considerable repute in England. Like all the English organs built at this time it had no pedals, though these had already for some time been introduced on the Continent.

After the summer holidays Herschel was back at Halifax, the lawsuit had been settled and the new organ was to be formally opened on the 28th August. The question of who was to be organist was to be decided by open competition two days later. A month earlier this would have seemed to Herschel an all-important event, but on the 9th August he had received a letter from Mrs De Chair at Bath, which opened out a different prospect, and on the 29th, the day after the first performance of the "Messiah", she wrote again, telling him that he had been definitely nominated as organist for the new Octagon Chapel. This offer must have caused him some perplexity. His friends at Halifax were evidently very anxious to have him there and he was in some sense committed to try for the post; also the Octagon Chapel at Bath was not yet ready nor was the organ built. He therefore decided to stay at Halifax for a time if he were selected for the post. On the 2nd September he answered Mrs De Chair's letter, apparently accepting the offer it contained for a later date. There is no explanation of why this correspondence was carried on by Mrs De Chair instead of by her husband. Dr De Chair was one of the proprietors of the Octagon Chapel. Mrs De Chair may have been at some time a pupil of Herschel's.

The following brief entries in William Herschel's Memorandums are all that refer to his appointment as organist at Halifax.

"1766, *Aug.* 30. Another candidate for the organist's place played first, after which I also played. The Messrs Bates and principal Gentlemen of the town were in the body of the church, and it was unanimously decided that I was to be their Organist.

"*Aug.* 31. Played the Halifax Organ during Service time.

"*Sept.* 7. Played the Organ. Took the Sacrament."

The following extract from Southey's work, *The Doctor*, gives the original version of a well-known anecdote. As we have already mentioned, John Herschel drew his father's attention to the passage and Sir William only demurred to the degree of intimacy claimed by Dr Miller. We may therefore accept the story as substantially correct.

"A new organ for the parish church at Halifax was built about this time and Herschel was one of the seven candidates for the organist's place. They drew lots how they were to perform in succession; Herschel drew the third, the second fell to Mr, afterwards Dr, Wrainwright of Manchester, whose finger was so rapid that old Snetzler, the organ-builder, ran about the church, exclaiming: 'Te Tevel, te Tevel! he run over te keys like one cat; he vill not give my piphes room for to shpeak!' 'During Mr Wrainwright's performance', says Miller, 'I was standing in the middle aisle with Herschel; "what chance have you", said I, "to follow this man?" He replied, "I don't know; I am sure fingers will not do". On which he ascended the organ-loft and produced from the organ so uncommon a fulness,—such a volume of slow, solemn harmony, that I could by no means account for the effect.

"'After this short extempore effusion, he finished with the old hundredth psalm-tune, which he played much better than his opponent. "Aye, Aye!" cried old Snetzler, "tish is very goot, very goot indeed; I vill luf tish man for he gives my piphes room for to shpeak."

"'Having afterwards asked Mr Herschel by what means in the beginning of his performance he produced so uncommon an effect, he replied, "I told you fingers would not do!" and producing two pieces of lead from his waistcoat pocket, "one of these", said he, "I placed on the lower key of the organ, and the other on the octave above; thus by accommodating the harmony, I produced the effect of four hands instead of two".'"

Having been elected organist at Halifax, William Herschel threw himself heart and soul into his new duties. Sacred music enthralled him; Handel's divine harmonies were now his model, and the brief entry in his notes referring to the first Sunday after his appointment may have been inserted with a special significance. Taking the Sacrament was a necessary formality in assuming any Church appointment, but by his noting the act in his journal we may infer that he did, at this moment, sincerely dedicate his musical talent, as later his astronomical labours, to the glory of God.

His friends, Mr and Mrs Bulman, with whom he had lived at Leeds, came to Halifax expressly to have the pleasure of hearing him conduct the Sunday service for the first time. Mr Bulman's business had failed about this time and one of Herschel's first acts, after he was settled at Bath, was to procure for him the post of Clerk to the Octagon Chapel.

During the autumn of this year (1766) he continued to take pupils at Halifax and to conduct the services, but the correspondence with Dr De Chair continued, and on the 18th October he gave notice to his employers at Halifax of his intention to resign at the end of the quarter. This announcement caused them much consternation and they endeavoured to persuade him to reconsider his decision. He wrote in his Biographical Memorandums:

"*Oct.* 22. Dined at Mr Walker's. He communicated to me a proposal from the Gentlemen at Halifax to increase my salary if it would induce me to stay".

But in spite of this substantial offer and many invitations to dine he remained fixed in his determination to leave Yorkshire. He continued to attend his scholars in different places up to the end of November; his entries in the Memorandums run thus:

"*Nov.* 24, &c. Letters to and from many preparatory to my leaving Halifax.

"*Nov.* 30. The last of playing the Organ. For the 13 Sundays of my being Organist I was paid 13 Guineas".

On the 1st December he set out for London, where he arrived on the 5th and remained three days. He mentions that he dined with Lady Cook, before starting for Bath. His notes continue:

"*Dec.* 9. Arrived at Bath, At Mrs De Chair's.

"10, &c. Was introduced to the Primate of Ireland. Letters to and from many friends. Lodged at Harpers.

"18. Concert, Linley.

"1767, *Jan.* 1. A Benefit Concert at the Rooms, the Music chiefly of my composition. I had but little company but it was select. I performed a solo concert on the violin, one on the Hautboy and a sonata on the Harpsichord.

"*Jan.* 23. Letter from Mr Derrick, the Master of the Ceremonies, offering me a situation in the established Band of Musicians that played at the public subscription Concerts, the Pumproom, the Balls, &c. This I at first refused but some time after accepted, when I found that Mr Lindley, the first musician in the place, was one of this band, and that like him I might be allowed to send a deputy when not convenient to attend personally".

The Octagon Chapel in Milsom Street had been built only the previous year. It was the first of the proprietary chapels which became a Bath institution. The invalids and fashionable visitors who crowded to Bath shrank from the gloom and chill of the Abbey Church; in these chapels they could listen to their favourite preachers and hear good music in comfort without the disagreeable presence of common people. An advertisement which appeared in several numbers of the *Bath Chronicle* in January, 1771, shows on what terms the seats were let:

Thursday, Jan. 3rd, 1771. Octagon Chapel

"As this Chapel is private property and consequently no Persons can with any reason expect that the Sexton should open the pews to them unless they have taken seats, this public Notice is given that all Strangers who come to Bath for the Seasons may have what sittings they please, for one, two, or three months, either in the Recesses (where six Fires are kept for the sake of Invalids) or any other part of the Chapel, by applying to Mr Bulman, the clerk, at

Mr Herschell's in the New King Street, or at the Vestry, an hour before Service, on Sundays, Wednesdays, and Fridays".

When William Herschel arrived in Bath the organ, which Snetzler was building, was not yet ready to be put up. Under the date February, 1767, Herschel put in his notes: "Wrote to Snetzler, the maker of the organ for the Octagon Chapel, to hasten the work".

Meanwhile the number of his pupils rapidly increased. To accommodate this increase he gave up his lodgings and took a house for six months in Beaufort Square. His friends, Mr and Mrs Bulman, were now arriving from Leeds, Mr Bulman having got the appointment at the Octagon Chapel, and Herschel had planned that the arrangement for living together, which had been so successful at Leeds, should continue at Bath, Mrs Bulman superintending the joint household.

The following entries in his Memorandums show how the rest of the year was spent.

"*May* 10 & 11. A visit to Belmont, the seat of the Marchioness of Clanricarde. I went on horseback by Salisbury, Winchester to Belmont, about 72 miles. The third day I returned to Bath. I attended Lady Amelia and Lady (Augusta) Du Bourgh in this manner once a fortnight during the summer season.

"*June* 24. My brother Jacob arrived at Bath.

"*June* 29. The organ began to be put up.

"*July* 25. Letters to Messrs Norris, Price and Mathews, to engage them for the Oratorio to be performed at the opening of the Octagon Chapel.

"*Oct.* 4. The Octagon Chapel was opened.

"*Oct.* 7. Jacob H. came to Bath again, having spent some months at the country seat of a Musical Gentleman's family.

"*Oct.* 27. Rehearsal of the Performance for the opening of the organ.

"*Oct.* 29. Oratorio. In these performances, I was the leader of the instrumental band, and my brother Jacob was at the organ; Mr Norris directed the vocal department.

"The organ being thus opened, I attended it regularly every Sunday. Dr De Chair intending to introduce Cathedral Service, I had prepared a choir of singers and composed the required music for the purpose, which on account of its simplicity was generally approved of."

A writer in a German periodical[1] for 1804, who seems to have been supplied with information by someone who had known Herschel at Bath, commenting on the energy and enthusiasm which he showed in everything which he undertook, as well as on his faculty for making use of the most unpromising materials, mentions that, on arriving at Bath, he began at once to train a choir of singers; enlisting for the purpose "young workmen, carpenters and joiners, who had no previous notion of singing, but who, under his stimulating tuition, were soon able to render the choruses of various oratorios with success". The special mention of carpenters and joiners points to his being already on friendly terms with the workmen who were to be so helpful to him later.

The notes which William Herschel made of his occupations during the next five years are very short, and record only the number and names of his scholars, the dates of concerts, the houses he lived in and the comings and goings of his brothers.

One of his most faithful patrons was Lady Lothian. He gave singing lessons to her daughter, Lady Emily Kerr, who, he remarked, had a very fine voice. Lady Lothian got up a series of fortnightly concerts during the winter season, at the houses of the most exclusive of the aristocratic visitors, which Herschel conducted up to the end of 1776. Lady Lothian's name occurs several times in the visitors' book at Slough, so she kept up her interest in Herschel even after he had abandoned music for astronomy. He mentions that she gave him a copy of Metastasio's works.

[1] *Neues Hannöverisches Magazin.*

## CHAPTER IV

### 1767  1772

William's life at Bath—Jacob joins him. Dietrich sent to Bath. Alexander settles at Bath. Caroline's life at home—William goes to Hanover—brings her to Bath—her life at Bath.

DURING the fifteen years which William Herschel passed at Bath the fashionable watering-place was at the zenith of its glory. Beau Nash was indeed no more; he died in 1761, six years before Herschel came, but the prestige he had won for the town, added to the fame of its waters, brought an ever-increasing crowd for the winter season.

It is impossible to think of Bath at this time without recalling the romantic story of Elizabeth Linley and of her elopement with Richard Brinsley Sheridan. She made her debut as a singer, in Bath, when she was only twelve years old; her brother Thomas, aged ten, playing the violin at the same concert. For six years her exquisite voice, her beauty and her charm, made her the idol of Bath, and unfortunately the mark of much ill-natured gossip. Linley, her father, was himself undoubtedly a very competent musician and conductor, yet there seems to have been something in his character for which William Herschel felt aversion. At first they worked together amicably, and young Ozias Linley took lessons on the violin from Herschel, but gradually little disagreements arose, and at last there was an open breach, which led to Herschel's resigning his post in the orchestra. He wrote a letter to the *Bath Chronicle* explaining his reasons for so doing; Linley replied, and a rather undignified and regrettable altercation ensued. It must be confessed that William Herschel on this occasion lost his temper. It is pleasant to be able to record that the quarrel was soon completely made up. Herschel and Linley not only lived on friendly terms during the remainder of the time that Linley stayed at Bath, but kept up their acquaintance after they had both quitted the town. William Herschel taught young Ozias Linley mathematics as well as the violin. When Ozias went to Oxford, he continued to write to Herschel for help in solving mathematical problems. There is a very quaint letter from Linley to Herschel written in October,

1782. After congratulating Herschel on: "ye Honours and advancement you have acquired", he proceeds thus:

"I understand you are so friendly to my son Ozias as to permit him to write to you, for which I hold myself obliged, and I am happy to find he has any sort of talent in him which you think worth your notice. For my own part, tho' I discover in particular instances Traits of a strong understanding, I have been, and still am, apprehensive of his being very deficient in common sense".

During the summer months, when all the visitors had left Bath, Herschel took pupils at schools and attended those living at country houses in the neighbourhood. To these he went on horseback, sometimes riding as far as Winchester and Salisbury for concerts. Concerted music was more cultivated in those days than in the Victorian era, perhaps on account of the absence of outdoor games to provide occupation in the country when the hunting season was over. It was quite usual for the owners of large country houses to engage a professional musician to reside during the summer months in order to conduct the musical entertainments in which the whole company took part, playing various instruments as their inclination led them; the piano had not yet monopolized the position of the only instrument for amateurs. Both William Herschel's brothers spent many summers in this way, but William himself preferred more liberty.

His financial position improved year by year. In December, 1769, he noted: "My this year's receipts amounted to 316 pounds". The following year he was able to record a total of 352 pounds; and in December, 1771, he wrote: "My receipts, cast up at the end of the year, amounted to nearly 400 pounds".

The rent of the house, No. 7, New King Street, Bath, which he took in April, 1770, was thirty guineas a year, of which Mr Bulman paid one-third. Herschel might therefore have considered himself in quite easy circumstances and with good prospects of still further improvement. Had he any thoughts of marriage? The question suggests itself, especially as there are indications of advances made to him by a rich widow, Mrs Colebrook, who lived at Bath Easton. She took music lessons from him, and, when his sister arrived in Bath, was at some pains to ingratiate herself with her. Mrs Bulman and her fellow-gossips found matter for conjecture, but William himself seems to have been quite indifferent.

His thoughts were indeed wholly preoccupied with his plans for making the rest of his family sharers in his good fortune.

Isaac Herschel, on his death-bed, had committed the charge of his youngest son's musical education to his eldest brother. This charge did not, however, hinder Jacob from getting leave of absence from the Court Orchestra, a few months after his father's death, and going off to join William at Bath.

> "The loss of so able a master", Caroline wrote, "might have been of serious consequence to Dietrich if Alex. had not fortunately at this time been a very good violin player, who could direct him in his practice, or prevent his neglecting the same."

Jacob seems to have felt some compunction for neglecting his father's wishes, and the next spring he wrote to his mother requesting her to send Dietrich over to him at Bath.

> "This proposal", Caroline wrote, "caused my Mother the greatest anxiety on account of Dietrich's not being of age to be confirmed before she parted from him, but her objections were at last removed by a promise of my brother's that it should be done immediately on his arrival at Bath."

Dietrich was accordingly sent to England and remained for one year at Bath. But it would seem that it was William chiefly who saw to his musical education and general welfare, for Jacob spent most of the year in London. When the winter had passed and the mother found, in answer to her inquiries, that no arrangements were being made for his confirmation, she insisted that he should return to her. Jacob therefore took him home with him when he went back to Hanover in July, 1769. Caroline says in her Autobiography:

> "They arrived, both in good health, and Dietrich entered school again immediately, and was no less a favourite among his young friends, after his having seen the world, than he was before he went abroad. And I suppose the violin had not been neglected, for his name was never left out in the infitations his brothers received to musical parties".

The following spring the mother had the satisfaction of herself witnessing his confirmation. He left school at the age of fourteen and was soon able to earn something as a music teacher like his brothers. Caroline had no anxiety on his account; it was otherwise with Alexander.

At the time of his father's death Alexander was a member of the Court Orchestra and added to his salary by taking pupils. During the first year after his father's death he remained quietly at home; the attention he gave to Dietrich in Jacob's absence had a steadying effect on him.

But when Dietrich was sent to England, and Alexander was freed from any responsibility for the disposal of his time, the weak tendencies of his character showed themselves. Caroline records in her Autobiography:

"I was extremely discomposed at seeing Alex. associating with young men who led him into all manner of expensive pleasures which involved him in debts for the Hire of Horses, Carioles, &c., and I was (though he knew my inability to help him) made a partaker in his fears that these scrapes should come to the knowledge of our Mother".

Caroline may have communicated her anxiety to William, or it may have been only his own desire to help Alexander to a more profitable mode of life, which led him to invite him to get leave of absence and to come to Bath. Alexander acted on this advice and accompanied Jacob the following summer to England. He must soon after have resigned his position in the Orchestra as he did not return to Hanover but continued to reside with William at Bath.

The only one of the family whom William had not so far been able to help was Caroline, and it seemed as if nothing could be done to assist her to a fuller life. She was naturally deeply concerned about her future as she grew to womanhood:

"I saw", she says in her Autobiography, "that all my exertions would not save me from becoming a burden to my brothers; and I had by this time imbibed too much pride for submitting to take a place as Ladiesmaid, and for a Governess I was not qualified for want of knowledge in languages. And I never forgot the caution my dear Father gave me; against all thoughts of marrying, saying as I was neither hansom nor rich, it was not likely that anyone would make me an offer, till perhaps, when far advanced in life, some old man might take me for my good qualities".

The first attempt she made to acquire some useful accomplishment was to beg her mother to let her go to a dressmaking school. Jacob graciously gave his consent to this, though on the condition that it should only be

"to make my own things, but positively forbidding it for any other purpose. I felt myself rather humbled on going the first time among 21 young people with an elegant woman (Mme Kuster) at their head, directing them in various works of finery. Among the groups were several young Ladies of genteel families, and as I came there on rather reduced terms I expected no other but that I would be kept back with doing nothing but the plain work of the business. But contrary to my fears I gained in the school mistress a valuable friend.

"I should not have thought it worth while to recall those trifling circumstances to my recollection if it was not for having from that period to date the favorable impression I unwittingly had made on Madame Beckedorff; who after a lapse of 35 years, when I was introduced to her at the Queen's Lodge, received me as an old acquaintance, though I could but just remember having sometimes exchanged a nod and smile with a sweet little girl about 10 or 11 years old. But I was soon sensible of having found what hitherto I had looked for in vain, a sincere friend".

Caroline was only allowed to stay a few months at Mme Kuster's school, as her mother required her help for the usual round of household drudgery. Even when her brothers were not at home and she expected to have a little time to herself:

"My Mother permitted a country Cousin to spend the summer with us, to have the advantage of my Mother's advice in making some preparation for her marriage. This young woman, full of good nature and ignorance, grew unfortunately so fond of me that she was for ever at my side, and by that means I lost that little interval of leisure I might then have had for reading, practising the violin, &c."

With Jacob's return Caroline's troubles increased.

"It was soon found", she wrote, "that at her time of life, with only my assistance, my Mother could not manage a family where so many luxurious fashions all at once were introduced, and that it was necessary to engage a servant, for which in fact we had not house room without letting her have part of my bed, to which inconvenience I was obliged to submit."

Alexander had now become settled at Bath and had given up all thought of returning to Hanover. But he had by no means forgotten the plight of his sister left behind at home. He told William that she had a good voice, and the idea occurred to the brothers that she might be trained as a public singer. William wrote home to his mother and Jacob that he would come and fetch her to make the trial of her voice, adding that if, after two years, it would not be found to answer, she should be brought back again. He also suggested that Jacob might give Caroline some lessons "by way of a beginning".

Jacob took no notice of this hint, though he seemed at first to approve of the plan; but he changed his mind later. Caroline wrote:

"At first it seemed agreeable to all parties, but by the time I had set my heart on this change in my situation, Jacob began to turn the whole scheme into ridicule and of course never heard my voice but in speaking; and yet I was left in the harrowing uncertainty, if I was to go or not!!!

"I resolved at last to prepare as far as lay in my power for both cases, by taking, in the first case, every opportunity, when all were from home, to imitate, with a gag between my teeth, the solo parts of concertos, *shake and all*, such as I had heard them play on the violin; in consequence I had gained a tolerable execution before I knew how to sing.

"I next began to net ruffles, which were intended for my brother Wm. in case I remained at home—else they were Jacob's. For my Mother and brother Dietrich I knitted as many cotton stockings as were to last two years at least. In this manner I tried to still the compunction I felt at leaving relatives who I feared would lose some of their comforts by my desertion, and nothing but the belief of returning to them full of knowledge and accomplishments could have supported me in the parting moment, which was much embittered by the absence of my eldest brother who was with the Court which attended on the Queen of Danemark at Gorde".

Having decided to bring Caroline to Bath, with or without Jacob's approval, William set out to fetch her. He took the opportunity of visiting Paris on the way and of hearing some music.

"At the Grand Opera", he notes, "I was much surprised to hear all the recitatives chanted like Cathedral music. The composition that of Lulli. There was an excellent orchestra of 60 musicians."

From Paris he went to Nancy, where he stayed a few days with Dr and Mrs De Chair, and then proceeded to Hanover, which he reached on the 2nd August, 1772. He found only Caroline and Dietrich with his mother, Jacob being absent with the Court. He stayed a fortnight, going with the rest of the family one day to Coppenbrügge to see his sister Sophia, and for Caroline to take leave of her.

His mother's objections to parting with Caroline had been overcome by his promise to contribute a yearly sum to enable her to keep a servant. They left Hanover on the 16th August and reached Bath on the 27th. Caroline's account of the journey shows what a very unpleasant experience the crossing from Germany to England could be in those days. They travelled first for six days and nights on the open seats of the mail-coach; Caroline lost her hat, which blew off as they crossed the wind-swept dykes of Holland. They encountered a severe storm at sea; the packet-boat lost her main-mast, and arrived off Yarmouth little better than a wreck. They were landed from an open boat on the backs of two English sailors and "thrown like balls", Caroline says, "on the shore". After this disagreeable beginning she seems to have been pleasantly surprised to find the inn where they breakfasted very neat, and the young women who served them clean-dressed and employed in cutting bread and butter from fine wheaten bread as fast as they could.

They had to hire a cart to take them to the place whence the coach for London started; the horse bolted and upset the cart, throwing them out: "I flying", says Caroline, "into a dry ditch". They picked themselves up, however, unhurt and were, by the assistance of some fellow-passengers, enabled to reach the coach and finally London, where they stopped at an inn in the City.

William did not allow his sister much rest. "In the evening," she says, "when the shops were lighted up, we went to see all what was to be seen in that part of London; of which I only remember the optician shops, for I do not think we stopt at any other."

Apparently they did not visit a milliner to repair the loss of Caroline's hat, for she proceeds to say:

"The next day the Mistress of the Inn lent me a hat of her daughter, and we went to see St Paul's, the Bank, &c., and we were never from our legs except for meals in our Inn. Towards evening we went to the west of the town, where after having called on Depeschen-Secretar Wiese and his lady (Mr Wiese conducted our correspondence with Hanover) we went to the Inn from whence, at 10 o'clock in the evening, we went with the night coach to Bath; where we arrived (the next day) at four in the afternoon. We were received at my Brother's house by an old lady, Mrs Bulman, my Brother's housekeeper, and her daughter, a few years younger than myself.

"After taking some tea I went immediately to bed and did not awake till next day in the afternoon; when I found my Brother had but just left his room. I, for my part, was through the privation of sleep for 11 or 12 days (not having been above twice in what they call a bed), almost annihilated".

The house which William Herschel occupied when he brought his sister to Bath was still No. 7, New King Street. The Bulman family occupied the ground floor; the first floor William kept for his own use. It was furnished, Caroline says, "in the newest and most handsome style. The front room, containing the Claficent, &c., was always in order to receive his musical friends and scholars at little private Concerts or Rehearsals".

Caroline herself and Alexander were lodged in the attics. William was evidently rather anxious lest Caroline should not get on well with Mrs Bulman, who had, up to this time, managed his domestic affairs and made him very comfortable. "On our journey", Caroline says, "my Brother had taken every opportunity of making me hope to find in Mrs Bulman a well-informed and well-meaning Friend, and in her Daughter (a few years younger than myself) an agreeable companion."

At first Mrs Bulman continued to act as housekeeper, but as soon as Caroline had acquired some facility in speaking English, she prepared to take the burden of household management on her own shoulders. Mrs Bulman was ready to instruct her in all the refinements of English cookery, but seems to have left her to face all the difficulties of marketing, engaging servants, &c., without much help. When the "hot-headed Irish servant", whom they found in the house on their arrival, had to be discharged for insolence, Mrs Bulman did indeed direct Caroline to a registry office, but without telling her to ask for a character before engaging a new servant. Consequently the Herschel family suffered for some months from a succession of dishonest maids.

"I know not when I should have got out of this scrape," Caroline wrote, "if I had not got acquainted with two women belonging to the Choir of the Octagon Chapel. One of these put me in mind of inquiring after the character of a servant at the place where they lived last, which on doing so, we hobled on a little better."

Housekeeping in all its details was ever irksome to Caroline, though she laboured at it conscientiously.

The following extracts from her Autobiography will show how she passed the years before William took up astronomy seriously.

### EXTRACTS FROM CAROLINE HERSCHEL'S AUTOBIOGRAPHY. 1772 TO 1778

"I did not find my brother Alexander at home, and he was not expected to return before the end of October from an engagement in another part of England; so I found myself all at once in a strange country and among strangers. But as the season for the arrival of visitors to the Baths does not begin till October, my Brother had leisure to try my abilities of making a useful singer for his Concerts and Oratorios of me; and being well satisfied with my voice, I had 2 or 3 lessons every day, and the hours which were not spent at the Harpsichord were employed in putting me in the way of managing the family.

"On the second morning, when meeting my Brother at breakfast, which was at 7 o'clock or before, (much too early for me, for I would rather remain up all night than be obliged to rise at so early an hour) he began immediately giving me a lesson in English[1] and arithmetic, and showing me the way of booking and keeping account of cash received and laid out. The remainder of the forenoon was chiefly spent at the harpsichord; shewing me the way how to practise singing with a gag in my mouth. And by way of relaxation we talked of Astronomy and the fine constellations with whom I had made acquaintance during the fine nights we spent on den Postwagen travelling through Holland.

[1] Recalling the many difficulties she had encountered at this time, Caroline, in writing to her niece, apologized for her "bad grammar, bad spelling, &c."—"for to the very last I must feel myself on uncertain ground, having been obliged to learn too much, without any one thing thoroughly;—my dear brother, William, was my only teacher and we generally began with what we should have ended".

"In this manner the first month or six weeks after my arrival in England were spent. Knowing then so much English as my Brother thought necessary to make myself understood by Mrs Bulman and his servant, (a hot-headed old Welsh woman) and the season approaching when my Brother's time would be taken up with attending on his scholars and public business, I began to think on how those hours I should now be left to myself might best be spent, in learning what would become necessary to know for a housekeeper of our little family. And for this purpose the first hours immediately after breakfast were spent in the Kittschen, where Mrs Bulman taught me to make all sorts of Pudings and Piys, besides many things in the Confectionary business, Pickling and preserving, &c., &c., a knowledge for which it was not likely I should ever have occasion.

"Sundays I received a sum for the weekly expenses, of which my Housekeeping book (written in English) shewed the sum laid out, and my purse the remaining cash.

"Shopping and marketing were also left to me, to which I had a particular dislike, especially the latter, having never been used to it at home. About six weeks after my being in England, I was sent alone among Fishwomen, Butchers, basket-women, &c., and brought home whatever in my fright I could pick up. My brother Alexander, who was now returned from his summer engagement, used to watch me at a distance, unknown to me, till he saw me safe on my way home.

"The three winter months passed on very heavily and my spirits were besides much depressed by receiving, in January, 1773, an account of the death of my brother-in-law, by which my sister became a Widdow with 6 children, of which the youngest was not 12 months old.

"I knew too little English for deriving any consolation from the society of those about me; so that, during dinner time excepted, I was intirely left to myself. For my brother Alex. was also much engaged with public business and giving lessons on the violincello, and if at any time he found me alone, it did me no good, for he never was of a cheerful disposition.

"My Brother (William) took great delight in a choir of singers who performed the Cathedral service at the Octagon Chapel; and as soon as I could pronounce English well enough, I was obliged to attend the rehearsals, and Sundays at morning and evening service; which, though I did not like much at first, I soon found to be both pleasant and useful."

To add to Caroline's natural depression and home-sickness, there was the consciousness of awkwardness in the society which her brother frequented. To her straightforward character the artificial manners of the polite society of Bath seemed unnatural and insincere. When the Herschels settled at Slough she felt no constraint in the company of the men of science who visited her brother and who appreciated and honoured her as she deserved. And among the members of Queen Charlotte's household she formed her closest friendships. The opinion she held of some of the ladies she met at Bath is very frankly expressed.

"Mrs —— and her daughter were very civil and the latter came sometimes to see me; but being more annoyed than entertained by her visits I did not encourage them, for I thought her little better than an idiot. The same opinion I had of Miss Bulman, for which reason I never could be sochiable with her.

"During the summer (1773) my Brother had frequent Rehearsals at home, when Miss Farinelli, an Italian singer, was met by several of the principal performers he had engaged for the winter Concerts. I was enabled (having procured a servant) to join the company in the Drawingroom, when Miss Farinelli with her Father and Mother were there, to rehearse some of her songs in the Messia or to spend the evening. But I cannot say that I was much edified in the society of these three women, (Mrs Bulman being always of the party). But my attention was mostly directed to the loud and hot conversation of Mr Farinelli and my Brother, sounding in my ears like quarrelling. But I heard afterwards from my Brother that Mr F. was a very learned man, and their dissertations were on metaphisical, &c., subjects, and that he suspected him to be one of the disbanded Jesuits of whom about that time many were, under fained names, supporting themselves by teaching languages, &c.

"About the ending of 1773, my Brother introduced me to two Ladies both very great crickers on Singers and musical performers, Mrs De Chair having spent some time at Rom with her Family. But it was long before I could get into her good graces, which was not till I had made a few of those flattering compts. laid in my mouth by my Brother, when she turned on her heel, saying—'Your sister is very much improved'.

"But to gain the good opinion of the other Lady, Mrs Colbrook, it was not quite so easy; for she was a very well-informed Woman, but wimsical to such a degree, that she seldom was of the same mind for two days together; and with this Woman it was my lot to spend six weeks in London!

"My Brother thought it a good opportunity for me to have the advantage of hearing some of the celebrated performers at the Opera. Our stay was not to have been above a fourth night, but was prolonged above six weeks on account of a deep snow which had made the roads at the Devizes impassable. This detention became in the end very embarrassing to me, for Mrs C., though learned and clever, was very capricious and ill-natured, to which I had nothing to oppose but patiens and the greatest desire to please, for I dreaded the criticism she would pass on me on our return.

"My Brother had left it to Mrs Bulman to make such additions to my Cloathing as were necessary for such an excursion and gave me 12 gueneas, in case we went to Public Places, that I might pay for my Tickets, Hairdressing, &c., and Mrs Bulman very officiously tuck care to fill my head with doubts and fears about the impossibility of pleasing this wimsical Lady, who would most likely be ere long my sister. I know not what grounds she may have had for entertaining such a supposition. My Brother never mentioned anything of the kind to me. I only knew that he visited at her villa at Bath Easton and at her town lodging in the winter as he did in the families of most of his scholars, and that I was frequently included in the infitations.

"I suffered besides much uneasiness at being obliged to accompany her five or six nights in the week, to the Opera, Pantheon or Play, at my own (or what grieved me most, at my Brother's) expenses for Tickets, Hairdresser, Chair hire, &c., &c., and to stint myself I was afraid would have a miserable appearance which I knew would have displeased my Brother. But by a deep snow the roads became totally impassable and though I wished myself heartily at home again, I was obliged, like my betters, to stay where I was. (The Duchess of Ancaster, Mistress of the Robes to the Queen, offered any sum for having a passage cut through the snow on the Hill at the Devizes, in vain, and she was prevented of being present on the 18th Jan. when the Queen's birthday was kept.)

"But I found it was no easy task to conceal my uneasiness at this detention, but remembering my Brother's instructions about going to public places, I never refused to accompany Mrs Colbrook to the Theatres of Drury Lane and Covent Garden, and am only sorry I missed seeing Garrick, who acted for the last time when we were engaged in another place. To the Opera I was particularly desired by my Brother to go, and we missed none. Millico was the first singer and I heard him between the acts perform one of his songs, (which came out afterwards in print, Millico's song) which he accompanied himself on the Harp. To the Pantheon we went twice; a wonderful and beautiful place which was not long afterwards destroyed by fire and never rebuilt. The evenings we were there above a Thousand people were assembled in the large room on the ground floor, where a Concert of the first Performers was given, and there I heard the famous Leoni (a Jue), a Counter Tenor, and Crosdel[1] perhaps his first solo on the violin play.

"The mornings were spent in doing business, going to auctions, &c., where I felt nothing but Long-while, and though my opinion was often asked on occasions, I never ventured but once to give it, and a pair of carriage horses were bought of my choosing. They were a pair of pretty Fuxe[2] with white foreheads and noses. And after having taken a ride with them about Town, they were with a new-hired coachman sent immediately to Bath. But I had the mortification to hear, on our first meeting after our return, that they were both blind!

"But I began now to be truly unhappy at our long detention, for not only all my money was spent, but I had been obliged to borrow 2 gueneas from Mrs Colbrook and having (according to my Brother's directions) called repeatedly on Mr Wiese, the Canzellist to the Hanoverian Embassy, I began to suspect that they did not trust me, for I had seen him and his lady but once two years ago in company with my Brother. But the evening before we left London I received a letter from my Brother desiring me to go to Wiese again and ask for what I wanted, as he had wrote to the same for that purpose, (and as there was full time enough to go to German Street, where Mr Wiese

[1] It may be interesting to note that Mrs Colbrook later married Crosdel.
[2] Chestnut horses.

lived and be back again by 11 o'clock when the carriage was ordered) I went as soon as we had breakfasted, and received 5 gueneas. On my way home I unluckily tuck it into my head to call at the Italian Warehouse in the Haymarket (where I had been with Mrs Colbrook before) to buy some permesan cheese for my Brother; and here I lost my way, and it was long before I found myself by chance in New Bond Street at the door of our lodging, where the carriage with the door open and all were standing and waiting for me, and in a moment I found myself with Mrs C. and her Woman on the way to Newberry where we did not arrive till late in the evening, without stopping, changing Horses without leaving the carriage or receiving any refreshment. And on getting out of the carriage all thoughts of thinking on ourselves were put to flight by finding all hands engaged to bring Mrs C's. servant (who attended the carriage on horseback) to life again from a fainting Fit, and Mrs Colbrook was in Histiricks all the evening. The next morning we met tolerably composed at breakfast and I took an opportunity to return the 2 gueneas I had borrowed, and having freed myself from obligations to the servants the evening before through Mrs Colbrook's Woman, I gave myself up intirely to the pleasing thoughts of being in the evening with my dear Brothers again.

"But I was cruelly disappointed, for on arriving some hours after dark I was received by a huge blier-eyed Woman (a new servant). My Brother Alexander not at home, and my poor Brother Wm. confined to his Bed by a sore Throuth and Feeber, where Mrs Bulman was nursing him.

"Note: Here I will remember that to her (Mrs Colbrook) and to the Marchioness of Lothian I am indebted for all I ever saw of the Fashionable world. The latter sent me tickets to the Play and Mrs Colbrook tuck me to the full dress and Cotillion Balls, to the Pump Room, &c. But I ever felt most happy on returning from these places, at the thought that I was not obliged to be every night pushed about by a gentil (genteel) crowd.

"In the latter part of the summer of 1776, I went with my Brother to Dr De Chair's; a very musical family residing at Little Risington, near Oxford, to which they were recently returned from Italy; where after having spent two or three afternoons playing and singing our best songs, I was glad at finding ourselves safe at home;

for the Jant was made in a single horse Chaise and my Brother was not famous for being a good driver.

· · · · · ·

"As I was to take part in the Oratorios, I had for a whole year two lessons a week by Miss Fleming, the celebrated dantzing Mistress, to drill me for a Gentlewoman, (God knows how she succeeded!) As the time drew near when I was to appear before the Public, it became necessary for me to provide myself with proper Cloathing for the purpose, for which my Brother gave me 10 gueneas to lay out to my own liking; and that my choice must not have been a bad one, I conclude from having been by Mr Palmer, (the then proprietor of the Bath Teater) pronounced to be an Ornament to the Stage. And as to acquitting myself in giving my Songs and Recitatives in the Messia, Judas Machabeus, &c., I had the satisfaction of being complimented by my Friends, the Marchioness of Lothian, &c., (who were present at the Rehearsals) for speaking my words like an Englishwoman.

"(Writing the above brings to my recollection the answer I gave once to a Lady, who reproved me for being my own Trumpeter by saying: 'How can I help it? I cannot afford to keep one!')"[1]

By the time Caroline was twenty-seven years of age she was fairly launched on a career which might have made her, had she wished, independent of her brother. She wrote in her Autobiography:

"I was in 1778 first singer in the Concerts and suppose I must have acquitted myself tolerably well, for before I left the Room (after a performance of the Messiah) I was offered an engagement for the Music Meeting at Birmingham. But, as I never intended to sing anywhere but where my Brother was the Conductor, I declined the offer".

She continued to sing in public for another year; but gradually the time she would have given to practising was more and more absorbed in helping her brother in his various occupations and in training the Choir. She says of the winter season of 1780:

"I was obliged to rehearse the treble of the Chorus singers for the Oratorios which were to be performed both at Bath and in Bristol;

[1] The foregoing extract is from the commencement of an Autobiography which Caroline Herschel began to write, in her old age, at Hanover.

16 or 18 Women and boys were with me on those nights when my Brother was engaged at the Theater. The solo performers he had engaged from London, and I only led the treble as I did the season before; for the interruption in my practise of the preceding months, besides accumulation of copying music, &c., left me no time to take care of myself or to stand upon nicetys".

In the last sentence we have her own, only, acknowledgment that, in order to be of use to her brother, she had to make up her mind to sacrifice her own ambition. It must have been a keen disappointment to her to relinquish the prospect of an independent career just when success seemed within her grasp. The wish to earn her own livelihood was almost an obsession with her. The hope of being fitted to do so had given her courage to leave her mother and her home and to endure the manifold discomforts of her first years in England, but her devotion to her brother stood the strain; she turned uncomplainingly to render herself useful to him in his new pursuit, and to continue the uncongenial task of managing incompetent servants. But the future had compensations for her of which she could not dream.

It is doubtful whether a purely musical career would have satisfied her. The society she would have had to frequent would have been uncongenial and her thirst for wider knowledge would have remained unquenched. Her intelligence was active and she absorbed with avidity the lessons in mathematics which her brother began to give her as soon as she came to Bath. These began playfully with the "Little Lessons for Lina", written on slips of paper during the breakfast hour; but as she showed real ability, William, who loved imparting knowledge, gave her more serious instruction. Her notebook, kept with her usual neatness, shows how carefully she worked out all the problems her brother set her. She studied algebra, geometry and trigonometry. But when she had acquired sufficient facility with spherical trigonometry to put it to practical use, she showed no inclination to proceed further. Her nephew, J. F. W. Herschel, summed up once, in conversation with his brother-in-law, his estimate of her abilities:

"She had nothing of mathematical genius, but her extraordinary powers of application in long-continued effort, with her extreme accuracy in all she did, were of as great practical value to my Father as if she had had far greater mathematical knowledge—perhaps more".

In 1866, referring especially to her self-devotion, he wrote to R. Wolf:

"She was attached during 50 years as a second self to her Brother, merging her whole existence and individuality in her desire to aid him to the entire extent and absolute devotion of her whole time and powers. There never lived a human being in whom the idea of Self ('der grosse Ich') was so utterly obliterated by a devotion to a venerated object unconnected by those strong ties of love and marriage which inspire such devotion in the female mind".

We need not deny to Caroline Herschel her place among eminent women because she was no intellectual giant. She was fortunate in being placed in a situation where all her best qualities could be used for a great purpose, and her work brought her happiness, as it was prompted by love.

## CHAPTER V

### 1773  1780

William Herschel turns to astronomy. Letter to Dr Hutton. Condition of astronomical knowledge. Ferguson. William begins to make telescopes. Caroline's autobiography. William goes to Hanover in search of Dietrich. Houses occupied by the Herschels at Bath. Meeting with Dr Watson—Maskelyne—Aubert. Papers communicated to the Royal Society.

THE year 1773, which saw Caroline settled at Bath, marked also the turning-point in her brother's career. William Herschel was now in his thirty-fifth year and in the fullness of his powers, both bodily and mental. His early privations had only served to strengthen a constitution naturally sound, and to teach him the control of all his faculties. Music had long since failed to satisfy his intellectual energy; he began to write a treatise on harmony but abandoned it as unsatisfying; his mind was ripe for a larger enterprise.

In the letter to Dr Hutton, already quoted in Chap. III, he says:

"In the year 1776 I was removed from Halifax to Bath, where I became organist of the Octagon Chapel. The great run of business, far from lessening my attachment to study, increased it, so that many times after a fatiguing day of 14 to 16 hours spent in my vocation, I retired at night with the greatest avidity to unbend the mind (if it may be so called) with a few propositions in Maclaurin's *Fluxions* or other books of that sort.

"Among other mathematical subjects, Optics and Astronomy came in turn, and when I read of the many charming discoveries that had been made by means of the telescope, I was so delighted with the subject that I wished to see the heavens and Planets with my own eyes thro' one of those instruments".

There is no mention here of any book on astronomy apart from mathematics. It has been claimed for Ferguson that it was a work of his which first attracted William Herschel's attention to this science, but we find by his own notes that he only bought Ferguson's book after he had determined to take up astronomy practically.

In the Memorandums for the year 1773, after the interesting entry

"I paid £15. 10. 0 for a handsome suit of clothes, it being then the fashion for gentlemen to be very genteelly dressed",

there occurs the following:

"*April* 7. Oratorio at Bristol.
"*April* 8. Concert Spirituale.
"*April* 19. Bought a Quadrant and Emerson's *Trigonometry*.
"*May* 10. Signor Farinelli's Concert.
"*May* 10. Bought a book of astronomy and one of astronomical tables".

In another place he wrote:

"In the spring of the year 1773 I began to provide myself with the materials for Astronomical purposes. The 19th April I bought a Hadley's quadrant and soon after Ferguson's *Astronomy*".

But though Ferguson's book may not have provided the first inspiration, it must have had a great fascination for him, and it is likely that he drew from it the material for those breakfast-table lectures of which Caroline tells. It is just possible that Herschel may have met Ferguson and have heard him lecture, as he gave popular lectures on astronomy at Bristol about this time.

Ferguson began life as a poor shepherd lad, watching the stars and trying to measure their distances apart with a knotted string as he lay on his back in the heather. Though he never acquired much knowledge of mathematics and did not himself advance the science by any original discoveries, he had earned a well-deserved reputation for the ingenuity of the mechanical devices by which he explained the motions of the planets and by his aptitude for describing astronomical phenomena in clear and simple language. It is worth recording that George III had granted Ferguson a pension of £50 from his private purse, thus already showing his genuine interest in astronomy as well as his kindness and generosity.

In Ferguson's book twenty-one chapters are devoted to the solar system and only one to the fixed stars; this division of the subject represents the attitude of astronomers at that time towards the phenomena of the heavens. Since Newton's day astronomers had been engrossed in working out the application of the law of gravitation to the motions of the planets. The new mathematical method, simultaneously discovered by Newton and Leibnitz, provided the means for solving the complicated problems involved, and observation had given place to calculation; astronomy had become almost entirely a deductive science.

As the method of Leibnitz, which was exclusively used on the Continent, was more flexible than that of Newton, it was among continental mathematicians that the greatest progress in the science was achieved in the eighteenth century. Laplace was engaged on his stupendous work, the *Mécanique Céleste*, in which the whole mechanism of the solar system is shown to be deducible from the general law of gravitation. This stupendous triumph of human reasoning laid a firm foundation for further advance in knowledge, but it confined the attention of astronomers to the solar system.

There was very little known about the fixed stars beyond their relative positions in our sky and the fact that, being at an enormous distance from our sun, they must be self-luminous. A very slight proper motion of certain stars amongst the rest had been detected and the existence of variable stars was also known. The number of stars in the British Catalogue for both hemispheres did not exceed 3000, and Ferguson considered that the number of stars "is much less than generally imagined".

A few more quotations from Ferguson's book will show the vague ideas which were current at this time about the heavenly bodies outside the orbit of the planets.

"There is a remarkable tract round the Heavens called the Milky Way from its peculiar whiteness, which was formerly thought to be owing to a vast number of very small stars therein; but the telescope shows it to be quite otherwise; and therefore its whiteness must be owing to some other cause.

"There are several little whitish spots in the Heavens, which appear magnified and more luminous when seen through a telescope; yet without any stars in them. One of these is in Andromeda's girdle.

"Cloudy stars are so called from their misty appearance. They look like dim stars to the naked eye; but through a telescope they appear broad illuminated parts of the sky; in some of which is one star, in others more. Five of these are mentioned by Ptolemy,.... The most remarkable of all the cloudy stars is that in the middle of Orion's sword. It looks like a gap in the sky, through which one may see (as it were) part of a much brighter region."

This idea of light shining through rifts in a dark envelope is a survival of the mediaeval conception of the universe as a series of concentric spheres; the outer and highest of all being the pure Empyrean of heavenly light. In spite of Newton the cobwebs of the Ptolemaic theory still hung about the minds of astronomers; it was reserved for Herschel to brush them all finally away.

When Herschel, in the words of his epitaph: "Coelorum perrupit claustra", broke through into the hidden recesses of the heavens and led the way for future exploration, he swept finally away the narrowing superstitions which clung around the Mosaic cosmogony and the mediaeval notions of perfect numbers.

Herschel could not himself have foreseen the full magnitude of the revolution in astronomy which he was initiating, but he saw clearly that no real progress was possible without a regular and systematic survey of the entire heavens. This he set before himself as the task to which he would devote the rest of his life. Every object must be scrutinized, its position and appearance carefully noted, and that not once but several times, in order that no change in either may escape detection.

Having once made up his mind to this undertaking, the first step was to equip himself with the necessary instruments. Telescopes of wide scope and high magnifying powers would be essential; these therefore were his first consideration, and for the next four or five years his attention was chiefly directed to the construction of telescopes. He constantly observed with these instruments, but rather for his own pleasure and to test their efficiency than for systematic exploration. He kept a careful record of all the processes he employed and continued to make notes in it up to his eightieth year.

The amount of manual work he got through in this task, for grinding and polishing the metal mirrors, turning wooden and brass eyepieces, shaping the tiny lenses for these, &c., is almost incredible, especially when we remember that he continued his professional duties as organist and conductor, as well as teaching on an average six to eight students a day during the Bath season. It was not till many years later that he perfected a machine for grinding and polishing specula; up till that time his work was done entirely by hand. And the number of these mirrors, some only four inches in diameter, others six or seven, up to twelve inches for the small 20-foot telescope, was astonishingly great, as he continued to cast fresh ones, to polish, test, and perhaps reject, till he was satisfied that he had got a thoroughly reliable one.

The shaping of the lenses for the eyepieces and mounting these in wooden cases, fitted with slides and ratchets for adjusting the focus, took up much time and exercised his ingenuity. The wood used was a sort called cocus, which was that used for oboes; the word has puzzled some of his biographers, who imagined that he made eyepieces out of coco-nuts.

It may not be amiss at this point to say a few words about the construction of telescopes, enough to explain Herschel's choice of the reflecting rather than the refracting model for those he made and used.

Every telescope consists essentially of two parts: (1) Either a lens or a mirror to collect more light from a distant object than can enter the unaided eye, and to form an image of the object within the tube; in the case of a lens this is called the object-glass. (2) A magnifier with which to examine this image and to reveal more of its details; this magnifier is a lens mounted in a wooden or brass tube with ratchet arrangement for adjusting the focal length; it is called the eyepiece. Eyepieces of different magnifying power can be used in the same telescope, and do not differ in principle whether employed in a reflecting or refracting telescope.

At a casual glance the main difference between a refracting and a reflecting telescope is the position of the observer. In a refractor the eyepiece and object-glass are in line with each other, the observer holds the telescope to his eye and looks in the direction of the object (though he is really gazing at its image within the tube). But when a mirror is used instead of a transparent lens, the observer must turn his back on the object in order to see its reflected image. In this position his head would obstruct some of the light. There are several devices for getting over this difficulty. Newton, the inventor of the reflecting telescope, solved the problem by placing a small flat mirror at a sloping angle within the tube so as to deflect the rays from the large speculum and bring them to a focus on the side of the tube. A hole was cut at this point and the eyepiece inserted. The observer could then examine the reflected image while standing at the side of the telescope.

Herschel made his first telescopes on the Newtonian plan, but for his larger ones—the 20-foot, which was his favourite instrument, and the gigantic 40-foot—he adopted a device of his own which enabled him to dispense with the small flat mirror in the tube.

The plan was simply to tip the large reflecting speculum very slightly to one side so that the focus of the converging rays fell, not in the centre of the aperture of the tube, but against the rim, where the eyepiece could then be placed. With a telescope of from 20 to 40 feet of focal length this sloping of the speculum produced very little distortion of the image. But the fact that there is a distortion has caused this form of construction to be abandoned by later astronomers.

The reason which prompted Herschel to make reflecting rather than refracting telescopes was principally the impossibility at that time of getting clear glass lenses made of the size that he required. Moreover, even the small glass lenses then obtainable were very expensive, while on the other hand it had been experimentally shown that metal mirrors would answer equally well and could be easily cast of much larger size than the largest glass lens yet made; nor was there any reason to suppose that the limit of size had yet been reached. Again, and this was probably to William Herschel the determining factor, the construction of a reflecting telescope is a comparatively simple matter, such as might be hopefully undertaken by an amateur of small means and little experience.

It only remains to indicate why William Herschel could not rest content with the smaller telescopes hitherto in use, but laboured continually to increase the size of his instruments; till at last, when he had completed his great 40-foot telescope, he found that the cumbersomeness of the giant outweighed the advantage in observing which he hoped to gain from its light-collecting power.

In the following letter to Sir Joseph Banks, written about July, 1785, when he was planning to build the 40-foot, he put the case for what he sometimes called the "space-penetrating power" of large telescopes:

"The great end in view is to increase what I have called 'THE POWER OF EXTENDING INTO SPACE'. Give me leave, Sir, just to mention this circumstance a little more at length. Light diminishes in a known proportion as it proceeds in its course thro' space. Hence it follows that an object, be it ever so luminous, can only be seen at a certain distance, beyond which the rays that proceed from it become so dispersed as no longer to be fit for natural vision.

We are then obliged to have recourse to artificial means: a star of the 8th magnitude, for instance, which escapes the naked eye, becomes plainly visible with an instrument that collects its scattered rays. Telescopes have this power of collecting light in proportion to their apertures, so that one with a double aperture will penetrate into space to double the distance of the other. The application of this principle to astronomy is obvious. The stars of the milky way, which cannot be seen in common telescopes, become visible in my 20 feet reflector".

William's own enthusiasm led him readily to believe that his brother and sister would engage in this new pursuit with an ardour equal to his own. Alexander indeed was ready enough to help in anything requiring mechanical ingenuity, but it required all Caroline's devotion to overcome the dismay with which she found herself swept along in such an unexpected direction. She records in her Memoirs the devastation caused in the house by her brother's new schemes.

"1773. Nothing serious could be attempted for want of time till the beginning of June, when some of my Brother's scholars were leaving Bath, and then it was to my sorrow I saw almost every room turned into a workshop. A cabinet maker making a tube and stands of all descriptions in a handsome furnished drawing-room. Alex putting up a huge turning machine (which in the Autumn he brought with him from Bristol) in a bedroom for turning patterns, grinding glasses and turning eye-pieces &c....I was to amuse myself with making the tube of pasteboard against the glasses arrived from London, for at that time no optician had settled at Bath, but when all was finished, no one besides my Brother could get a glimpse of Jupiter or Saturn, for the great length of the tube could not be kept in a straight line; this difficulty however was by substituting tin tubes soon removed."

William's own record of his experience in making telescopes is the best to follow for an account of the work he was doing in this summer of 1773. (A work in four folio volumes now at the University Observatory, Oxford.)

"In May I procured some short object glasses and had tubes made for them, beginning with a 4 feet one of the Huyghenian construction.[1]

[1] I.e. a refractor.

With this I began to look at the planets and stars. It magnified 40 times. In the next place I attempted a 12 feet one and contrived a stand for it. I saw Jupiter and its satellites with it. After this I made a 15 feet and also a 30 feet refractor and observed with them. The great trouble occasioned by such long tubes, which I found it almost impossible to manage, induced me to turn my thoughts to reflectors, and about the 8th September I hired a two feet Gregorian one. This was so much more convenient than my long glasses that I soon resolved to try whether I could not make myself such another, with the assistance of Dr Smith's popular treatise on Optics. I was, however, informed that there lived in Bath a person who amused himself with repolishing and making reflecting mirrors. Having found him out he offered to let me have all his tools and some half-finished mirrors, as he did not intend to do any more work of that kind. The 22nd of September, when I bought his apparatus, it was agreed that he should also show me the manner in which he had proceeded with grinding and polishing his mirrors, and going to work with these tools I found no difficulty to do in a few days all what he could show me, his knowledge indeed being very confined. About the 21st October I had some mirrors cast for a two feet reflector, the mixture of the metal was according to a receipt I had obtained with the tools. It was at the rate of 32 copper, 13 tin, and one of Regulus of Antimony, and I found it very good, sound white metal. In the beginning of November I had other mirrors cast, among them was one intended for a $5\frac{1}{2}$ feet Gregorian reflector, and as soon as they were ground and figured as well as I could do them, I proceeded to the work of polishing. About the middle of December I got so far as to give a tolerable gloss to some metals, and having advanced considerably in this work it became necessary to think of mounting these mirrors."

During the remaining nine years which William Herschel, his brother and his sister spent at Bath, their life flowed on in the same incessant round of work. William's own memoranda are very brief, the names of the ladies who came to him for music lessons are curiously intermingled with lists of astronomical instruments bought and work done in making telescopes; a few extracts will suffice.

"1774, *Jan.* Gave 6, 7 and 8 private lessons every day.
"*Feb.* The same. Caroline was in London with Mrs Colebrooke.

"*March.* Nearly the same number of scholars. *Astronomical Journal* begun.

"*April.* Specula grinding. Mirror cast. Tools bought.

"*Augt.* Maps, glasses, putty &c. &c. Astronomical timepiece.

*Midsummer. House at Walcot*

"*Sept.* Lady Priscilla and Lady Charlotte Bertie became my scholars. The Marchioness of Lothian presented me with a set of the works of Metastasio.

"*Dec.* Attended 6, 7 or 8 scholars every day. At night I made Astronomical observations with telescopes of my own construction.

"1776, *May.* I observed Saturn with a new 7 feet reflector.

"*July.* I observed Saturn with a new 20 feet reflector I had erected in my garden."

In Caroline's Memoirs we get a more intimate picture of their daily life.

"1774. During the season for public amusements he had of course not much time to spare; for on undertaking to be the conductor of all the public concerts after his predecessor Mr Linly; (who left Bath with his son and daughters for London), it was not likely soon to find a singer who would make a favourable impression after Miss Linly (Mrs Sheridan). Therefore my Brother composed Glees, Catches, &c. for voices such as he could find. Besides keeping up his practice for giving sometimes, in the absence of Fisher, a Concerto on the Oboe or a Sonata on the Harpsichord. And the solos on the Violoncello of my Brother Alexander were divine!

"1775. But every leisure moment was eagerly snatched at for resuming some work which was in progress, without taking time for changing dress, and many a lace ruffle (which were then worn) was torn or bespattered by molten pitch, &c.; besides the danger to which he continually exposed himself by this uncommon precipitancy which accompanied all his actions; of which we had a melancholy sample one Saturday evening when both Brothers returned from a Concert between 11 and 12 o'clock, my eldest Brother pleasing himself all the way home with being at liberty to spend the next day (except a few hours attendance at Chapel) at the turning bench, and recollecting at the same time that the tools wanted sharpening, they ran with the Lanthorn and tools to our Landlord's grinding stone in a public yard where they did not wish to be seen on a Sunday morning. But they were hardly gone when my Brother William was brought back fainting by Alex. with the loss of the nail of one of his fingers. This happened in the winter of 1775, at a house situated near Walcot Turnpike, which at that time was without the district of the Mayre and free from Buildings till near Bath Easton.

"Here my Brother had moved to at midsummer 1774 and on a grass plot behind the house preparation immediately was made for erecting a twenty foot telescope for which, among 7 and 10 feet mirrors then in hand, one of 12 inch was preparing; and the house also afforded more room for workshops and a place on the roof for observing.

"During this summer I lost the only female acquaintances (not friends) I ever had the opportunity of being very intimate with, by the Bulmans' Family returning again to Leeds. For my time was so much taken up with copying Music and practising, besides attendance on my Brother when polishing, that by way of keeping him alife I was even obliged to feed him by putting the Vitals by bits into his mouth;—this was once the case when at the finishing of a 7 feet mirror he had not left his hands from it for 16 hours together. And in general he was never unemployed at meals, but always at the same time contriving or making drawings of whatever came into his mind. And generally I was obliged to read to him when at some work which required no thinking; and sometimes lending a hand, I became in time as useful a member of the workshop as a boy might be to his master in the first year of his apprenticeship."

She adds as a note here:

"My only reason for saying so much of myself is to shew with what miserable assistance your Father made shift for obtaining the means for exploring the heavens".

1777. In March of this year three Oratorios were performed at Bath under William Herschel's conductorship, for which, as already related, Caroline had copied the scores and trained the treble chorus singers, and in these Concerts she for the first time appeared as a soloist. With these performances, she says:

"public business ended, and my Brother had now a little more time for perfecting his Instruments and looking at the heavens in fine nights. But on the 30th July 1777 our quiet was cruelly interrupted by a letter from our Mother informing us that our jungest Brother had secreatly gone away with the intention of going to the East Indies in company with a young idler not older than himself. The letter arrived when my Brother was busy at the Lathe, turning an eye-piece in Cocos of which at that time he made many of various construction, and of which a whole boxful stood about his workroom for years after. The piece he was about was left in the Lathe unfinished and a few hours after he was on the road to go over to Holland where he hoped to find Dietrich. And not meeting him there went on to Hanover where he arrived on the 11th Augt."

William kept his sister constantly informed of the progress of his journey and of his hopes of finding the runaway. He reached London on the 31st July and wrote (in German) to Caroline.

"Dear Sister,

"I arrived punctually in London yesterday after a pleasant journey, and have arranged all my affairs here so that I can start this evening and probably shall reach the sea to-morrow at noon; therefore when you receive this letter I shall already be in Holland. But if I am detained at Harwich I will write to you again.

"I have already learnt that no ship will sail for the East Indies before December or January, so that John D. can quite certainly not have gone there. Very likely he has gone back to Hanover or he may even have come to England.

"I am going to-day to buy a pair of handsome spectacles for Mama.

Greetings to Alex. &c. Farewell."

(No signature.)

Friday morning, Aug. 1, 1777

Helvotsluis. Monday morning, Aug. 4, 1777

"Here I am, having arrived yesterday evening after a very good passage. I slept well and breakfasted well and want for nothing, *Praise God*. It is only a pity that my lathe is not here so that I might finish my *Eye-piece* or a polisher that I might do a little polishing; the Dutchmen know nothing about these things.

"Adieu. The 'Buller' coach is ready to fetch me; this evening I shall be in Rotterdam and to-morrow in Amsterdam. John D. shall certainly not escape me, they tell me that nothing can be easier than to find him if only he has not changed his name; even then he will soon be traced. Once more, adieu.

"Greet my dear Alex. I hope he visits you as often as possible. Once more adieu."

Hannover. Friday, Augt. 22, 1777

"Dear Sister,

"To Day Mama has seen Luther who is returned to Hannover; he has left Joh. Die. at Amsterdam who is gone to London and I suppose will soon be with you at Bath. I shall set out from Hannover the 1st, 2d or 3d Septr. and go by Amsterdam and London. I have already wrote to Mr Wiese to let Joh. Di. have what money he may want in London, and if he is not come to Bath before me I shall bring him with me; be so good as to write me a line or two that I may find in London and let me know if Joh. is with you or where he is to be found in London. Inclose the letter for me to Mr Wiese as you did your last letter to Mama and direct the letter to me as follows:

To Mr Wm. Herschel<br>'to be kept till called for';

as soon as I come to Town I will call at Mr Wiese for it.

"We are all very well here, your letter of the 31st only arrived yesterday. Mama is extremely well and as I have represented things gives her consent to your staying in England as long as you and I please. I wish very much to see my own home again, and conclude at present—remaining your affectionate Brother

WM. HERSCHEL.

"Complim. to all Friends. I hope to be in Bath about the 14, 15 or 16 of Sept."

CAROLINE'S REMINISCENCES (continued)

"None of the letters I received brought any tidings of the run-a-way, till at last, on a Saturday, I received one dated from a little Inn near the Tower of London, informing me of Dietrich's laying there very ill. Alex was engaged at Bristol but used to walk over to see me on Sundays. I went to meet him and he no sooner saw the letter than we hastened home and whilst taking a hasty breakfast, post horses were ordered; and he was with D. next morning and had him removed to a lodging, where Dr Smith and a fourthnight's nursing brought him so far that by easy stages he could be removed to Bath.

"Poor Alex was rejoysing at having saved the life of a brother, though it was at the loss of his business and all his stock of cash (£20 I had in keeping for him); for the poor fellow was never famous of being an economist; especially where he thought he could assist a fellow creature in want."

William, believing Dietrich to be safely with his brother and sister, lingered a little while in Hanover with his mother, and it was not till a fortnight after Dietrich had been conveyed to Bath that his elder brother arrived. Caroline finishes this part of her Autobiography with these consoling words:

"In about another fourthnight our Brother returned from Hanover, finding poor Dietrich much reduced and feeble, and seeing him walk about the room whilst we were seated at our substantial dinner, insisted on his joining us and I was released from nursing him (according to the Doctor's direction) with roasted apples and Barley-water".

It is noticeable that none of the family uttered a word of blame for Dietrich's bid for freedom from the monotonous life at Hanover. As soon as his health was restored William procured musical engagements for him in Bath and London. He returned to Hanover towards the end of the summer of 1779, and shortly afterwards married his landlord's daughter Miss Reif, and settled down there for good; occasionally coming over for short visits to his relations in England.

Many of his letters to William (dating from 1780 to 1807) have been preserved, and they reveal a very lively and warm-hearted disposition. He addresses William as "dearest" or "most dear brother" and concludes one letter with the words: "Love me only half as much as your devoted J. D. Herschel does you". Caroline and "good old Alexander" are also warmly remembered and on more than one occasion he refers to the debt of gratitude which he owes to Alexander for coming to his rescue when he lay ill and forlorn at Wapping. Dietrich was a keen entomologist. William shared his interest and continued to send him specimens of moths, after his return to Germany.

The Herschel family changed their abode several times while they lived at Bath. When Caroline joined her brother, in the summer of

1772, he was living at No. 7, New King Street, a house on the north side of the street, the Bulman family sharing it with him. Two years later they all moved to a house higher up, the exact position of which cannot be exactly determined. We have only Caroline's description of it as being "situated near Walcot turnpike, which at that time was without the district of the Mayre and free from buildings till near Bath-Easton".

In the following year Mr Bulman left them, to Caroline's great relief, for she had never been able to make real friends of Mrs Bulman and her daughter, and she lived in dread of the effect on Alexander of association with the empty-headed company whom they invited to the house. Alexander did in fact become engaged to one of Miss Bulman's friends, but soon found reason to regret his imprudence and turned to his brother to get him out of the scrape. The engagement was broken off, and this unfortunate episode may have been one reason for Mr Bulman to accept the offer of employment at Leeds and to leave his old friend.

The house at Walcot, almost in the open country, with plenty of space for carpentry work on the telescope mountings, was very pleasant in the summer, but when the winter season commenced and the brothers and sister had to be constantly at the concert rooms in the lower part of the town, the distance proved very inconvenient; and so, after spending three years at Walcot, the Herschels moved back again to New King Street in September, 1777, just after Dietrich had come to join them. The house which William had chosen was No. 19; it lay on the south side of the street with an unimpeded view of the sky and a fair-sized garden, which afforded sufficient space for erecting the 20-foot telescope, and several sheds which could be used as workshops. This house seemed in every way suitable, yet in two years' time they moved again to the upper part of the town; this time to No. 5, Rivers Street. In the same street a few doors off were the rooms occupied by the newly formed "Literary and Philosophical Society of Bath", and Caroline alleges this to have been the attraction which made her brother move to Rivers Street, though he was not yet a member. The Society, Caroline tells us,

> "was provided with all sorts of instruments for making experiments in various branches of Natural Philosophy, among which was an excellent electrifying machine put there by Mr Bryan, a member".

The house in Rivers Street must have been very inconvenient. It had no garden, so William was obliged to carry his telescope out into the open space before the house; it was this circumstance which led to the meeting which was to have such an important bearing on his subsequent life. We find the following account in his Biographical Memorandums:

"1779, *December*. About the latter end of this month I happened to be engaged in a series of observations on the lunar mountains, and the moon being in front of my house, late in the evening I brought my seven feet reflector into the street, and directed it to the object of my observations. Whilst I was looking into the telescope, a gentleman coming by the place where I was stationed, stopped to look at the instrument. When I took my eye off the telescope he very politely asked if he might be permitted to look in, and this being immediately conceded, he expressed great satisfaction at the view. Next morning the gentleman, who proved to be Dr Watson, jun. (now Sir William), called at my house to thank me for my civility in showing him the moon, and told me that there was a Literary Society then forming at Bath, and invited me to become a member of it, to which I readily consented".

Herschel's pursuit of star-gazing had already attracted some notice at Bath; people began to talk about this eccentric musician who made his own telescopes and spent his nights observing with them. The actor Bernard tells in his *Retrospections* that he took lessons in music from William Herschel and that, one January evening, in the middle of the lesson, the sky which had been cloudy began to clear. "There it is at last", cried Herschel to the amazement of his pupil, and dropping his violin, rushed to a telescope and apostrophized a certain star, saying how enraptured he was to see it again.

Bernard took his lessons in Herschel's rooms, and tells us that:

"His lodgings resembled an astronomer's much more than a musician's, being heaped up with globes, maps, telescopes, reflectors, &c., under which his piano was hid, and the violoncello, like a discarded favourite, skulked away in a corner".

William himself says that some of his scholars "made me give them astronomical instead of music lessons".

Early in this year, 1779, he made his first appearance in public as a writer. He says in his Memorandums:

"1779, *February*. A musical prize question being proposed in *The*

*Ladies' Diary* for this year, I sent a solution of it directed as required. My solution of the question was published in *The Ladies' Diary* for 1780".

Dr Dreyer states that the question was one proposed by Landen:

*The Prize question, by Peter Puzzlem*

"The length tension, and weight of a musical string being given, it is required to find how many vibrations it will make in a given time when a small given weight is fastened to its middle and vibrates with it".

Dr Dreyer adds that the result was only a very rough approximation; but Herschel finished his short note with the reservation that "the above solution is not to be considered as mathematically true but as a practical solution approaching near the truth".

In the *Correspondance Astron.* for 1829, vol. IV, p. 626, is an amusing note by Von Zach.

"Il y a des pays où on moquera de vous si vous faites des Almanachs,.... Le célebre M. Herschel a débuté en Angleterre dans un Almanach, et qui plus est dans un Almanach des Dames; 'The Ladies' Diary'. Il y a résolu plusieurs problèmes de mathématique, surtout sur la théorie de la musique, qu'on proposait tous les ans dans cet almanach. Nous en avons reçu de lui plusieurs exemplaires en présent, pendant notre séjour en Angleterre en 1783, 1784, que nous conservons encore comme un monument curieux et comme un souvenir précieux de l'incomparable donateur."

As soon as he joined the Bath Philosophical Society he began to contribute papers to be read at their meetings. There are thirty-one of these papers in all, written between January, 1780 and March, 1781. The first, "On the growth of Corallines", reminds us of the interest in Natural History with which his brother Dietrich had inspired him. There are some ten or eleven on physical experiments, electricity, light and heat, &c.; one, "On Della Torre's method of making Crystal Globules", suggests the manner in which he may have made his tiny lenses; and three papers on astronomical subjects which were communicated by Dr W. Watson to the Royal Society.

There are besides several papers on more general subjects, especially on metaphysics, which seems always to have had a strong attraction for him.

Even while living at Walcot he had been visited by some of those whom Caroline calls "Philosophical gentlemen". She says that up to the time of his joining the Bath Philosophical Society:

"I do not remember any living Astronomer with whom my brother could have any communication except every now and then a stranger introduced by some of his scholars to have a look through his telescopes or otherwise satisfy their curiosity; among whom was Dr Maskelyne, who I remember came with Dr Lysons when he lived at Walcot in 1777; but not being introduced as Astronomer Royal, my Brother only pronounced him (after having had several hours' spirited conversation with him) to be 'a devil of a fellow',—after he had seen him to the door".

Dr Maskelyne seems to have carried away a favourable impression of William Herschel, in spite of his vehemence in conversation, for in all their subsequent relations he proved a warm and steadfast friend. About the same time Herschel met Dr Hornsby, the Director of the Oxford Observatory, and had some correspondence with him on the subjects of the eclipses of Jupiter's satellites and the rings of Saturn. William's letters are lost, but it is clear from Dr Hornsby's replies that he already recognized in William Herschel an amateur observer whose work was worthy of attention.

In a letter dated "Oxford, July 26, 1779", Dr Hornsby wrote:

"If you have any observations made either in the present year or before, with which you yourself are satisfied, (on the eclipses of the satellites of Jupiter) and will send them to me, you shall have in return those I have made either corresponding or at the distance of one or two revolutions: and if I have none myself I will procure Dr Maskelyne's for you".

In concluding this letter Dr Hornsby adds:

"I took the liberty to desire two of my particular friends, Admiral Campbell and Mr Aubert, to call upon you when at Bath to see your telescope".

This passage is interesting, as it marks the beginning of Herschel's acquaintance with Aubert; an acquaintance which later developed into a warm friendship. Alexander Aubert was a well-known amateur astronomer, who possessed in his private observatory at Loam-pit Hill,

Deptford, a 2-foot Cassegrain reflector, by Short, reputed to be one of the best in England; the model was known as "Short's Dumpy".

1780. On the interesting occasion when Dr William Watson first made Herschel's acquaintance, the latter was engaged on observations for ascertaining the height of the lunar mountains by measuring the shadows which they cast. He wrote a paper describing these observations and giving his inferences from them. This paper Dr Watson communicated to the Royal Society; it was read there on the 11th May, 1780. However, before the paper was printed in the *Transactions* of the Society, Dr Maskelyne wrote to Dr Watson pointing out some alterations and additions which he thought desirable. Dr Watson sent the letter on to Herschel with this covering note.

DR WILLIAM WATSON, JUNIOR *to* WILLIAM HERSCHEL

June 5th, 1780

"Dear Sir,

"I have enclosed to you a letter to me from the Rev. Dr Maskelyne, Astronomer Royal, in which he desires some further elucidations with respect to your paper on the 'heights of the Moon' Mountains. You will find that it has been read at the Meeting of the Royal Society and canvassed at the Committee, and that the publication of it is deferred until something more be added, as you will find by the enclosed. I take it extreamely obliging in Dr Maskelyne to take the pains of giving an account of what was wanting to the further improvement of your paper, and am very certain that the motive for pointing out further illustrations is from a desire he has of having the subject thoroughly handled by so accurate and ingenious a Philosopher.

"I think you would do right (pardon my giving you advice) either to add the desired improvements, or to write over again the Paper, and to send it to Dr Maskelyne, who, as he is Astronomer Royal, will be pleased, I believe, with the compliment paid him, and he will present it anew to the Society.

"I hope that Mr Bryant and you are pursuing the quarry started before I left you in electricity; the alternate attractions and repulsions are likely to lead to the knowledge of very interesting Principles....

I am, Dear Sir,
Your very humble servt. and friend,
WM. WATSON."

It appears from a memorandum concerning this letter of Dr Maskelyne's, which Herschel communicated to the Bath Society, that in his original paper he had advanced some speculations or assertions of his own belief in the probability of the moon being inhabited. These speculations Dr Maskelyne rightly considered out of place, and he also desired fuller details of Herschel's procedure in taking his measurements. The engaging frankness with which Herschel was ever ready to impart his ideas and speculations in conversation was one part of his charm in social intercourse, but he had yet to learn that such frankness might be prejudicial to his reputation, in communications to the Royal Society. He withdrew the passage and added the description of his method of making the measurements which Dr Maskelyne had asked for; but he could not resist, in his private letter to Dr Maskelyne, restating his belief and endeavouring to give reasons for it. He wrote:

"I beg leave to observe, Sir, that my saying there is an almost absolute certainty of the Moon's being inhabited, may perhaps be ascribed to a certain Enthusiasm which an observer, but young in the Science of Astronomy, can hardly divest himself of when he sees such wonders before him. And if you will promise not to call me a Lunatic I will transcribe a Passage...which will shew my real sentiments on the subject".

The argument is from analogy—that the moon being very similar to the earth is equally habitable and therefore probably is inhabited. Herschel apparently did not realize that the moon has no atmosphere; for a long time he clung to the conviction that he had observed active volcanoes in the moon.

## CHAPTER VI

## 1781

Discovery of Uranus. Letters from Hornsby, Maskelyne. Paper "Account of a Comet" read at the Royal Society. Letter from Messier. William's attempt to cast a mirror for a 30-foot telescope. Experiments on metals.

THE Herschels did not stay much more than a year in Rivers Street. Whatever the inducement may have been which made them leave New King Street, it was evident that it did not compensate for the inconvenience to William of having his telescopes at a distance from the house and workshops. Early in March, 1781, he arranged to transfer all his instruments and furniture back to No. 19, New King Street. Caroline, for once, was not present when the telescopes were again erected in the garden there, for she had remained behind in Rivers Street to wind up some business matters.

Thus it happened that she was not at her brother's side when he made the discovery which was to bring him so much fame and to change the whole course of his life. Yet the moment when the new planet sailed into his view was not in truth one of unusual excitement to him. The quest on which he was then engaged was part of his settled plan for investigating the whole structure of the heavens. His first efforts were directed to determining the distances of some of the fixed stars. For this purpose he had begun a systematic review of the entire visible sky with the special object of finding suitable pairs of stars. If in the course of this survey a hitherto unknown planet lay in his way he was bound to discover it, as his admirable telescope at once revealed to him its difference from a fixed star.

William Herschel had brought his telescopes down to No. 19, New King Street, and had set up the 7-foot instrument in the garden before the rest of the household moved in. He was observing there on the night of the 13th March, 1781, when he noticed a "curious either nebulous star or perhaps a comet", as he described it in his Astronomical Journal. In the account of his life which he wrote to Dr Hutton, for publication, he says:

"It has generally been supposed that it was a lucky accident that

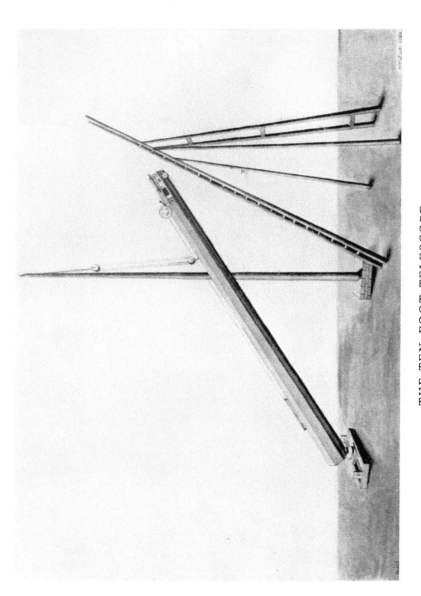

THE TEN-FOOT TELESCOPE
as mounted for use at No. 19 New King's Street, Bath

*From a drawing made by Sir W. Watson*

brought this star to my view; this is an evident mistake. In the regular manner I examined every star of the heavens, not only of that magnitude but many far inferior, it was that night *its turn* to be discovered. I had gradually perused the great Volume of the Author of Nature and was now come to the page which contained a seventh Planet. Had business prevented me that evening, I must have found it the next, and the goodness of my telescope was such that I perceived its visible planetary disc as soon as I looked at it; and by the application of my micrometer, determined its motion in a few hours".

He announced his discovery at once, both to Dr Hornsby at Oxford directly, and to Dr Maskelyne through Dr Watson. Dr Hornsby replied on the 24th March as follows:

"Sir,

"I am much obliged to you for your intelligence respecting the comet which you have discovered and am sorry I have not yet been able to find it. I rec$^d$ your letter on Wednesday, at which time the air was not very clear. I looked for it with my Finder & a night glass & thought I had seen it but it proved to be a Cluster of small stars which in a telescope of that sort looked like a faint comet or a cloudy star".

On the 31st March, he wrote again, still unable to find the "comet" and asking for "an eye draft of the stars at a small distance from it".

Dr Maskelyne meanwhile had forestalled Dr Hornsby. On the 4th April he wrote to Dr Watson from the Royal Observatory:

"Dear Sir,

"I thank you for your favour of the 29th of last month (in answer to mine). The three last nights I observed stars near the position pointed out by Mr Herschell, whereby I was enabled last night to discern a motion in one of them, which, as well as from its agreeing with the position pointed out by him, convinces me it is a comet or a new planet, but very different from any comet I ever read any description of or saw. This seems a Comet of a new species, very like a fixed star; but perhaps there may be more of them.

. . . . . . .

"Astronomers are obliged to Mr Herschell for this discovery and

I among the rest. I desire you will present my compliments to him. I shall be glad to hear of any farther observations he may make of it, and shall be ready to communicate to him any of mine.

I am &c.

N. MASKELYNE.

"P.S. I think he should give an account of his telescope, and micrometers".

Dr Hornsby now also succeeded in locating the "comet". He wrote from Oxford on the 14th April, 1781:

"Sir,

"In consequence of the information in your last letter I immediately found the comet & perceived that I had unknowingly observed it on the 29th & 30th of last month; having while I waited for your letter observed some stars near the Parallel of H Geminorum.

. . . . . . .

"I should be glad to know how the comet appears to you &, if you have measured it lately, what its apparent diameter is?

"I do not in the least question but this is the comet of 1770, whether it has passed its perihelion or has not yet come to it, is more than I can say at present. I will very soon try to construct its orbit.

I am Sir,

THOS. HORNSBY".

Dr Maskelyne wrote, this time to Herschel himself:

Greenwich. April 23, 1781

"Sir,

"I am to acknowledge my obligation to you for the communication of your discovery of the present Comet, or planet, I don't know which to call it. It is as likely to be a regular planet moving in an orbit nearly circular round the sun as a Comet moving in a very eccentric ellipsis. I have not yet seen any coma or tail to it.

. . . . . . .

"I shall be obliged to you for the relative positions of the three small stars which you compared the comet with, that I may be able to make use of your observations, which if they do not err more than you suppose, are very good ones. I have been trying to find them but find them very small; I should be glad to know what aperture your telescope had when you compared the Comet with them, and whether you illuminated the wires. They will not bear the light in my telescope.

. . . . . .

"On the 6th of April I viewed the Comet with my 6 ft. reflecting telescope and greatest power 270, and saw it of a very sensible size but not well-defined. This however showed it to be a planet and not a fixt star, or of the same kind as the fixt stars as to possessing native light with an insensible diameter.

I am Sir, &c. &c.

"To W. H."
N. MASKELYNE.

(This letter is addressed to "Mr William Herschel, Musician, near the Crescent, Bath".)

It is clear that Dr Maskelyne was prepared from the first to accept the new comet as a planet, while Dr Hornsby had quite made up his mind that it was a comet.

Herschel continued to note down the position and to measure the diameter of the "comet" during his nightly observations up to the middle of April, and then embodied them in a paper which he communicated to the Royal Society through Dr Watson. It was read on the 26th April and later printed, with some alterations, in the *Philosophical Transactions*. It was simply entitled, "Account of a Comet", and begins thus:

"On Tuesday the 13th March, between ten and eleven in the evening, while I was examining the small stars in the neighbourhood of H Geminorum, I perceived one that appeared visibly larger than the rest; being struck with its uncommon magnitude, I compared it to H Geminorum and the small star in the quartile between Auriga and Gemini, and finding it so much larger than either of them, suspected it to be a comet.

"I was then engaged in a series of observations on the parallax of fixed stars which I hope soon to have the honour of laying before the Royal Society; and those observations requiring very high powers, I had ready at hand the several magnifiers of 227, 460, 932, 1536, 2010, &c., all of which I successfully used upon that occasion. The power I had on when I first saw the comet was 227. From experience I knew that the diameters of the fixed stars are not proportionally magnified with higher powers, as the planets are; therefore I now put on the powers 460 and 932, and found the diameter of the comet increased in proportion to the power, as it ought to be, on the supposition of its not being a fixed star, while the diameters of the stars with which I compared it were not increased in the same ratio. Moreover the comet, being magnified much beyond what its light would admit of, appeared hazy and ill-defined with these great powers, while the stars preserved that lustre and distinctness which from many thousand observations I knew they would retain. The sequel has shown that my Surmises were well founded, this proving to be the comet we have lately observed".

He then gave tables and diagrams representing the "comet's" position up to the 19th April, and measurements of its diameter.

When this paper was read at the Meeting of the Royal Society, its contents excited great interest and astonishment and some dissent. Astronomers were puzzled to account for his having noticed so small a star and detected a proper motion in it in so short a time, while his calm assertion that he employed magnifying powers of 932, 1536, 2010, *et cetera*, seemed to them wild exaggeration; a power of 270 being at that time considered very good indeed. Herschel's telescopes were in fact immensely superior to any possessed even by the Royal Observatory, for not only had he brought the shaping and polishing of his mirrors to the highest degree of perfection, but the eyepieces, which he had made himself, were marvels of delicate workmanship.[1] The figures and diagrams which Herschel had given in his paper, showing the position of the "comet" from his own observations, showed some curious irregularities. He thought these might be explained by supposing that the star was sufficiently near to the earth to give a parallax. Writing to Dr Hornsby, on the 22nd May, he said:

"My observations also have shown that the Comet had a very visible daily parallax which was sufficient to prove it to be on our

[1] These eyepieces, with their lenses, have been preserved and are now exhibited on loan at the Science Museum, South Kensington. They have been carefully tested by Dr Steavenson and fully bear out all Herschel's claims; indeed the smallest lens, which is considerably smaller than an ordinary pin's head, would give a magnification of about 8000; it was probably made by him as a *tour de force*, as it could not have been put to any practical use.

side of the Sun; but on account of small unavoidable errors that may be in such nice measures, I only suppose (without having entered into calculations) that a parallax of not less than 10″ nor much more than 20″ would follow from my observations".

It was unfortunate that Herschel should have alluded to this hypothesis in his paper to the Royal Society, as further observations soon showed that the course of the comet or planet lay wholly outside and very far beyond the earth's orbit. Herschel himself in revising his observations found that the fault had lain in the measurements taken with the micrometer he had been using. It was modelled on one made by Nairne and Blunt, and, in a paper communicated by him to the Royal Society in December, 1784, he mentions that it had a defect which he subsequently remedied by an improvement of his own requiring a special precaution.

"I found the absolute necessity for this precaution", he wrote, "when I came critically to examine the positions of the Georgium Sidus as they are given in Table III, *Phil. Trans.* vol. LXXI. The measures were affected by a small and pretty regular error, which I was at a loss to account for; and the distance of this star being then totally unknown I looked for the cause of the deviation in a diurnal parallax of that heavenly body."

The passage referring to parallax was not printed, but it had been read, and supplied additional ground for his detractors to throw doubt on the efficiency which he claimed for his instruments, and to consider him as a rather hasty and superficial reasoner.

At the end of the paper, as printed for the *Phil. Trans.* vol. LXXI, Herschel added a paragraph, headed "Remarks on the path of the Comet", in which he offered a kind of apology, and at the same time a defence of his unconventional method of determining the position of a star.

"We may observe, that the method of tracing out the path of a celestial body by taking its distance from certain stars, and the angle of position with regard to them, cannot be expected to give us a compleatly just representation of the tract it describes, since even the most careful observations are liable to little errors, both from the remaining imperfections of instruments, though they should be the most accurate that can be had, and from the difficulty of taking angles, and positions of objects in motion. Add to this a third cause of error, namely, the obscurity of very small telescopic stars that will not permit the field of view so well to be enlightened as we could wish, in order to see the threads of the micrometer perfectly distinct.

"This will account for the apparent distortions to be observed in my figures of the Comet's path.

. . . . . . .

"But though the method may be liable to great inconveniences... yet the advantages are not less remarkable. Thus we see that it enabled me to distinguish the quantity and direction of the motion of this Comet in a single day (from the 18th to the 19th of March) to a much greater degree of exactness than could have been done in so short a time by a sector or transit instrument; nay even an hour or two, we see, were intervals long enough to show that it was a moving body, and consequently, had its size not pointed it out as a Comet, the change of place, though so trifling as $2\frac{1}{4}$ seconds *per* hour, would have been sufficient to occasion the discovery. A gentleman very well known for his remarkable success in detecting Comets seems to be well aware of the difficulty to discover a motion in a heavenly body by the common methods when it is so very small; for in a letter he favoured me with, he says: 'Rien n'étoit plus difficile que de la reconnoître et je ne puis pas concevoir comment vous avés pu revenir plusieurs fois sur cette étoile ou comète; car absolument il a fallu l'observer plusieurs jours de suite pour s'appercevoir qu'elle avoit un mouvement'.

"I need not say that I merely point this out as a temporary advantage in the method I have taken; for as soon as we can have regular, constant and long continued observations by fixed instruments, the excellence of them is too well known to say anything more upon that subject; for which reason I failed not to give immediate notice of this moving star, and was happy to surrender it to the care of the Astronomer Royal and others, as soon as I found they had begun their observations upon it."

Having thus formally handed over his "comet" to the professional astronomers, Herschel's chief desire was to be left to prosecute the work he had planned out for himself. In comparison with the wonderful vision which was ever before his mind, that of probing into the hitherto invisible depths of space and unravelling the problem of the structure of the heavens, the discovery of a strange comet, added to the solar system, seemed a minor matter. He intended, as soon as the 20-foot telescope was properly erected in the garden at No. 19, New King Street, to prepare for casting an even larger mirror to suit a 30-foot telescope. But before starting on this we learn, from his own Memorandums and from the letter just quoted to Dr Hornsby, that he went up to London and was received at the Royal Society on the 3rd May. It must have been a brief visit, and apparently his friends, among whom Sir Joseph Banks, the President, could be counted, desired that he should make a more formal appearance as the question of bestowing the Society's gold medal on him was being considered.

Dr Watson wrote to him:

London. May 25th, 1781

"Dear Sir,

"I received your paper on the Comet or rather star, in due time, and carried it in my hand to Sir Joseph Banks. He had received your letter containing further particulars of the *great* discovery and told me they were extreamely satisfactory. Your election did not come on last Thursday; Sir Joseph had given my father reason to believe so by mistake; it will not come on till next Thursday Se'en-night.... You will therefore, I hope, be here on Thursday night or Friday morning, time enough to dine with the Society, Sir Joseph thinking it very proper in your case that you should dine with us on account of the medal's being to be conferred on you, tho' it will happen that you will not be a member at that time.

"I breakfasted with Dr Maskelyne at Sir Joseph's who enquired very kindly after you. He says he is satisfied the star has not the diameter you have given it and supposes your instruments to give the figure indistinct and of greater visible diameter.

I am, Dear Sir, your very sincere Friend,

W. WATSON".

This projected dinner seems never to have taken place. The Copley medal was not conferred on William Herschel till November, 1781, and he was only elected a Fellow of the Royal Society on the 6th December.

Considerable interest was aroused among continental astronomers by the accounts which reached them of the strange new star. Messier, the

French astronomer, who had made the pursuit of comets his speciality, wrote as early as the 29th April, to Herschel, to ask for particulars. This is the letter from which Herschel quoted in his paper for the Royal Society; the following is a translation:

MONSIEUR HERTSTHEL at *Bath*

Paris. 29th April, 1781

"Having learnt from Mr Maskelyne, Astronomer Royal of England, in a letter which he has done me the honour to write to me, that he was observing a new comet, of a very singular character, having no atmosphere, no coma nor tail, resembling a little planet with a diameter of 4 to 5 seconds, a whitish light like that of Jupiter, and having the appearance when seen with glasses of a star of the 6th magnitude, I sought it and as it resembles the stars I had some little trouble in recognizing it, and I found that it had the characters which Mr Maskelyne noticed.

"I am constantly astonished at this comet, which has none of the distinctive characters of comets, as it does not resemble any one of those I have observed, whose number is eighteen. I will add here the observations I have already made....

"I have since learnt by a letter from London that it is to you, Sir, that we owe this discovery. It does you the more honour, as nothing could be more difficult than to recognize it, and I cannot conceive how you were able to return several times to this star—or comet—as it was absolutely necessary to observe it several days in succession to perceive that it had motion, since it had none of the usual characters of a comet....

"For the rest this discovery does you much honour; allow me to compliment you for it. I should be very curious, Sir, to learn the details of this discovery, and you will oblige me if you will be so good as to inform me of them.

With equal consideration and respect,

(Signed) MESSIER,
Astronomer to the Navy of France,
of the Academy of Sciences, Paris".

There was now a lull in the astronomical world, as the new planet was temporarily out of sight; the short summer nights were unsuitable for observing. At Bath the season practically came to an end with the performances of the Oratorios in Passion Week, and William Herschel was therefore free to devote his time to telescope construction comparatively undisturbed.

We turn again to Caroline's narrative.

"As soon as the public amusements were ended and no scholars except a few residents of Bath remaining besides the two principal Boarding schools, my Brother applied himself to perfect his mirrors, erecting in his garden a stand for his 20-ft. Telescope; which caused (I remember) many trials to be made before the required motions for such an unwieldy machine could be contrived; of which many (I suppose) were made by way of experiment against a mirror for an intended 30-ft. Teles. should be completed, for which between whiles, (not interrupting the observations with 7, 10, and 20 feet, and writing papers for both the Royal and Bath Ph. Society's), Gages, Shape, weight &c. of the mirror were calculated, and trials of the composition of the metal were made. In short I saw nothing else and heard nothing else talk but about those things when my Brothers were together. Alexander was always very alert and assisting when anything new was going forward, but he wanted perseverance and never liked to confine himself at home for many hours together; and so it happened that my Brother was obliged to make trial of my abilities in copying for him Catalogues, Tables, &c. and sometimes whole papers which were lent him for his perusal; of which among others was one of Mr Michel and a Cat. of Christian Mayer in Latin which kept me employed when my Brother was at the telescope at night; for when I found that a hand sometimes was wanted when any particular measures were to be made with the Lamp micrometer &c. and a fire kept in, and a dish of Coffe necessary during a long night's watching; I undertook with pleasure what others might have thought a hardship."

Among William Herschel's manuscripts there are four volumes of Records of work done in making telescopes, and in addition a large bundle of folio slips entitled: "Results of Experiments on the Construction of Mirrors". From these and from Caroline's graphic narrative of what she saw and herself participated in, we learn to appreciate the courage and perseverance with which her brother attempted, with the crudest appliances, tasks of such great difficulty.

With the object always in view of casting a mirror of 3 or 4 feet diameter, the first proceeding was to settle upon the best material, which would fulfil the two requirements of taking a good polish and of being strong enough to bear rather severe handling, as those large mirrors had to be lifted by a crane and dropped into the tube. The first entry in the Records for 1781 is under date 26th Feb. In this Herschel says that he intends to cast a number of small mirrors as tests for different compositions.

As soon as he got to New King Street, he set about building a furnace and melting-oven in his own house. No. 19 is built on sloping ground, the garden at the back being lower than the street. There is a basement room having an entrance from the garden; it was here that he set up his furnace; a fortunate choice as it turned out, as the room had two exits, the one to the garden and the other to the house. Caroline describes the eagerness with which her brother watched the progress of the work.

> "Though being at times much harassed with business, the mirror for the 30 feet Refl. was never out of his mind and if a minute could be spared in going from one scholar to another or giving one the slip, he called at home to see how the men went on with the furnace which was built in a room below even with the garden."

As soon as the furnace was ready he started his experiments, trying a great number of different mixtures. Copper and tin were always the principal ingredients, other metals being added in small quantities with the view of increasing the lustre. After numberless trials he came to the conclusion that there was no advantage to be gained by these additions, and that the main thing was to ascertain what proportion of tin could safely be used without rendering the metal too brittle. In his subsequent work he used for small mirrors 29·4 per cent. of tin; for the second mirror of the 40-foot telescope he reduced this proportion to 25 per cent. of tin. He also experimented on other metals such as wrought and cast iron and hammered steel; the latter, he recorded, took a fine polish but did not reflect much light.

At the same time that he was occupied with these experiments and with making some mirrors for some of his friends, he was, as his Records show, very busy in planning the mode of erecting the 20-foot telescope as well as the bigger one that was to be. The weight of copper and tin for the great mirror had also to be estimated.

By the beginning of August all was ready for the casting. The moulds which were used for small mirrors were made of "loom", annealed by having charcoal burnt in them. The preparation of the mould for the great mirror must have been a more unpleasant task, but the devoted sister shrank from nothing to aid her brother and was helped by Alexander and their kind friend, Dr Watson.

"The mirror", says Caroline, "was to be cast in a mould prepared from horse dung of which an immense quantity was to be pounded in a morter and sifted through a fine seaf; it was an endless piece of work and served me for many hours' exercise and Alex frequently took his turn at it, for we were all eager to do something towards the great undertaking; even Sir Wm. would sometimes take the pestel from me when he found me in the workroom where he expected to find his friend;—in whose concerns he took so much interest that he felt much disappointed at not being allowed to pay for the metal; but I do not think my Brother ever accepted pecuniary assistance from any one of his friends, and on this occasion declined the offer by saying that it was paid for already."

The day fixed for the casting arrived. William Herschel notes briefly in his Biographical Memorandums:

"*August* 11. I cast the great mirror".

But in his Records he gives a fuller account of what must have been to him a sore disappointment:

"The metal ran very quietly into the mould so that we had the greatest hopes of success, but when nearly full a small crack near the bottom of the mould began to let some of the metal out which caused a great deficiency in one side of the mirror. The disappointment", he says, "did not last long, as in cooling the mirror cracked in two or three places, which proved our mixture too brittle.

"We began immediately to make up another mould and as this is the work of some days, we altered in the mean time our composition of the metal by an addition of copper.

"When everything was in readiness we put our 537·9 pounds of metal into the melting oven and gradually heated it; before it was sufficiently fluid for casting we perceived that some small quantity began to drop through the bottom of the furnace into the fire. The crack soon increased and the metal came out so fast that it ran out of the ash hole which was not lower than the stone floor of the room. When it came upon the pavement the flags began to crack and some of them to blow up, so that we found it necessary to keep a proper distance and suffer the metal to take its own course".

Caroline, we can imagine, was waiting in the garden, eager to hear that the metal had been successfully poured into the mould, when, as she tells:

"both my Brothers, and the caster and his men were obliged to run out at opposite doors, for the stone flooring (which ought to have been taken up) flew about in all directions as high as the ceiling. My poor Brother fell exhausted by heat and exertion on a heap of brickbatts".

William continued his account thus:

"When all was quiet enough to permit us to come near we removed the fire and gathered up the metal of which we picked up $408\frac{1}{4}$ pounds that were pure, the rest being mixed with dross and ashes".

The original weight of metal employed had been 537 pounds; it is worth noting that the first mirror which Herschel made for his 40-foot telescope weighed 1023 pounds, and was found to be too thin to keep its shape in the tube; the second mirror, which was the one he always used, weighed 2118 pounds. These were 4 feet in diameter; the one he tried and failed to cast in August, 1781, had a diameter of only 3 feet; still it would have been much too frail to keep a good figure when in use.

This second failure put an end for the time to Herschel's endeavours to make a telescope larger than his 20-foot instrument. No further entries were made in his Records till August, 1782, when he was settled at Datchet, and these only refer to repolishing the mirrors of the 20-foot telescope.

There were at this time two other amateur makers of reflecting telescopes in England, both of them clergymen. With one of them, the Rev. J. Michell, Herschel had tried to get into touch in the summer of 1780, through Dr Priestley, as Michell's experiments were mentioned in Priestley's *History of Optics*. Priestley wrote to Michell at Herschel's request but got no reply. As Michell was reported to be making a telescope with a mirror of 30 inches diameter, Herschel was really desirous of hearing from him and in January, 1781, he got Dr Priestley to write again to Michell, who replied with a very long disquisition on the making of specula. The letter began:

REV. J. MICHELL *to* DR PRIESTLEY

21st Jan. 1781

"Dear Sir,

"I look upon myself as very much obliged to you for your favour from Bath and particularly for the very interesting acc$^t$ both of what Mr Herschel has done and what he has seen, both which seem very important. I shall be very happy if I should be able to succeed as well, as from your account he seems to have done, and I shall be very glad of the favour of his correspondence; at the same time I think it very probable that I may be more likely to learn from him what may be useful to myself than he to learn anything from me".

He then described his own aim as follows:

"The object I am seeking after, is to grind and polish my speculums to the form of conic sections in what degree I chuse,—in order to make telescopes of short focus and large aperture".

He then dilates on the great trouble and expense to which he had been put in carrying on his experiments and concluded his letter with sending "compliments to Mr Herschel, though I have not the pleasure of a personal acquaintance with him".

Herschel then wrote himself to Mr Michell and received in reply a voluminous communication written in a rather superior tone, recommending him to read Smith's *Optics* and Dr Priestley's *History of Optics*, and giving at great length his reasons for preferring the Gregorian form of telescope to the Cassegrain, or to the Newtonian, which he dismissed as "much the worst of all". At the end of his letter he wrote:

"Having said thus much, you must excuse me from entering more minutely into the methods of computation which I have used and of which, if you are sufficiently acquainted with these subjects, as I daresay you are, I have said enough to show you how to investigate yourself; to pursue them more at length would be in some degree writing a treatise upon the subject. For the same reason I must decline giving you any detail concerning the several tools and contrivances I mean to make use of or have made use of already".

Michell certainly deserves much credit for his self-sacrificing efforts to improve the construction of telescopes; but it is evident that Herschel did not learn anything useful from him. Herschel did not think that the advantage of the short focus compensated for the trouble of figuring the mirror to a greater curvature. He did, however, make one telescope on this plan himself, in 1700 (it was marked in his catalogue as the "large 10-feet", or "X-feet", and was sold to Lucien Bonaparte in 1814). To shape the mirror for this telescope he used Michell's own polishing tool which, together with the great telescope, he had bought after Michell's death from his son-in-law. The mirror of that telescope was cracked, so Herschel never used it. Dr Dreyer thinks that Herschel made the 10-foot telescope for himself when his advancing age made it difficult for him to mount the scaffolding of his own long-focussed instruments.

Herschel was more interested in the work of another amateur, the Rev. J. Edwards, who had been for some time experimenting on the composition of a good metal for specula. Dr Maskelyne was particularly anxious that Herschel should use Edwards' mixture. It was very complex, consisting of copper, tin, brass, silver and arsenic. Herschel had tried it and readily admitted that the brilliance of the metal was excellent, but he found it very brittle. Writing in 1783, to Dr Wilson of Glasgow, he said that his large mirrors had to be strong enough to stand the jars and shocks of being lifted and dropped into the tube.

## CHAPTER VII

### 1781    1782

Reappearance of the "comet". Letters from Hornsby, Gerstner. William Herschel receives the Copley medal. Paper "On the Parallax of the Fixed Stars"—its reception by the Royal Society. Dr Watson's letter. Correspondence with Watson, Hornsby, Maskelyne, Sir Joseph Banks.

THE time had now come when the "comet" was again visible and the telescopes of all the observatories were turned upon it. Dr Watson wrote to William Herschel:

Dawlish. Aug. 31st, 1781

"Dear Sir,

"I am happy to hear your Comet—if it is to be called a comet—has reappeared.—Your success, you perceive, has set both the English and French astronomers on exerting themselves to the utmost; a comet which is likely to be visible for at least 12 years longer is a very extraordinary phenomenon".

In October Hornsby wrote from Oxford:

Oct. 14, 1781

"Sir,

"I cannot allow my very excellent friend Sir James Stonehouse to come to Bath without troubling him with this letter to you, in order to give you some account of the last obsns. of the comet which I made upon the meridian.

. . . . . . .

"When you first discovered this comet, or planet as some are disposed to call it, it had not long been progressive; its distance from the Earth was increasing, and was the greatest in June when in conjunction, and since that time the Earth has been moving towards it; so that about this time its distance from the Earth is nearly the same as when first discovered. It will now be retrograde for many (5) months. Mr Lexell, a very celebrated mathematician, has determined its elements from the observations made before its conjunction with the sun, and found its perihelion distance to be 16 times the distance of the Earth from the sun.

. . . . . . .

"In December and January next we shall be as near to it as we can yet have been in this part of its orbit; and therefore we shall be able to see it to the greatest advantage when upon the meridian at midnight, and consequently you may determine how much brighter it then is than when you first discovered it. I shall be glad to hear your opinion of its appearance, and am Sir,

With great regard<br>
THOS. HORNSBY".

The burning question of whether the new body was to be considered a comet or a new planet could not be absolutely decided till many more observations had been made; though after Lexell's announcement the balance of opinion inclined on the side of its planetary nature. "From this time", wrote Lalande, the eminent French astronomer, in the *Mémoires* of the French Académie des Sciences,[1] "it appeared to me that the body ought to be called the new planet". But it required months of careful observations and elaborate calculations to establish this belief as indisputable. The continental mathematicians revelled in the problem set before them. One of them, Gerstner of Prague, wrote to Herschel some years afterwards in these words:

"And as the whole astronomical world is indebted to you for the interesting discovery of this planet, so am I in particular, as you have given me the opportunity of making a new application of that sublime analysis, by means of which the greatest geometers of our century have exhausted the theory of the other planets".[2]

During the autumn of this year Herschel was diligently working at the materials for the paper "On the Parallax of the Fixed Stars", which was communicated to the Royal Society in November and read on the 6th December. It was followed by a detailed "Catalogue of Double Stars", read on the 10th January of the following year, 1782.

In November, 1781, he went up to London to receive the Copley

---

[1] "Dès lors il me parut qu'on devait lui donner le nom de nouvelle planète."

[2] "Et comme toute l'astronomie vous est redevable de la découverte intéressante de cette planète je le vous en suis en particulier, parce que vous m'en avez donné l'occasion de pouvoir faire une nouvelle application de la sublime analyse, avec laquelle les plus grands géomètres de notre siècle ont épuisé la théorie des autres planètes."

medal; Sir Joseph Banks, the President of the Royal Society, wrote to him on the 15th November:

"Sir,

"The Council of the Royal Society having ordered their Annual Prize Medal to be presented to you in reward for your discovery of a new star, I must request that (as it is usual for me on that occasion to say something in commendation of the discovery) you will furnish me with such anecdotes of the difficulties you may have experienced in the discovery, &c.,...as you may think proper to assist me in giving due praise to your industry and ability.

. . . . . .

"Some of our astronomers here incline to the opinion that it is a planet and not a comet; if you are of that opinion it should forthwith be provided with a name and our nimble neighbours, the French, will certainly save us the trouble of Baptizing it.

Your faithful servant,

JOS. BANKS".

In presenting the medal Sir Joseph Banks said:

"In the name of the Royal Society I present to you this Gold Medal, the reward which they have assigned to your successful labours, and I exhort you to continue diligently to cultivate those fields of science which have produced to you a harvest of so much honour.

"Your attention to the improvement of telescopes has already amply repaid the labour which you have bestowed upon them; but the treasures of the heavens are well known to be inexhaustible. Who can say but your new star, which exceeds Saturn in its distance from the sun, may exceed him as much in magnificence of attendance? Who can say what new rings, new satellites, or what other nameless and numberless phenomena remain behind, waiting to reward future industry?"

Herschel remained in town three or four days as the guest of Sir William Watson, the father of his devoted friend, Dr Wm. Watson. The impression which he made on those Fellows of the Royal Society whom he met, as well by the frankness and courtesy of his manner as by his great ability, increased their desire that he should in his next contribution justify their high opinion. They kept him informed of all that was published in continental scientific journals about the new star, as well as anything relating to the double stars on which they knew he was engaged.

Dr Watson wrote on the 7th December to William Herschel to announce that "You are just now unanimously elected a fellow".

The next day he added:

"Since then I heard that Christian Mayer of Mannheim had published his observations in which he gave an account of a great many double stars. I breakfasted at Sir Joseph Banks' on purpose to know more of the matter, and he was so kind as to permit me to send you the 4th vol. of the Mannheim Acts which contains the paper (I have marked the place). I am sure the perusal of this will give you much pleasure. I was (pleased) however when I counted the catalogue at the end of the memoir to find that the double stars do not amount in number to yours.... Sir Joseph whispered to me that possibly the Council of the Royal Society would, on application being made, excuse you from paying the 30 guineas, in case you should have no objection. The application is not meant to be made as from yourself, you are only to signify whether you chuse another should make the application for you. This you will inform me of. As I was informed that some nebulous stars are like-wise inserted in the *Connaissance des Temps* for 1783, I bought it and have sent it likewise, begging your acceptance of it. I have also sent a quire of ruled paper.

"I am just going to Dr Heberden to dinner. I wish you was to be of the party.

Your sincere friend and humble servant

W. WATSON".

Though it is not the intention of this work to give a detailed account of William Herschel's scientific work, yet his life story would be very imperfect without some allusion to those researches by which he advanced so immensely the science of astronomy, as these show the great originality of his mind as well as the sagacity and perseverance with which he conducted his observations. The paper "On the Parallax of the Fixed Stars", which was read before the Royal Society on the 6th December, 1781, affords an excellent example of the manner in which he attacked the problems which he set himself to solve.

The first step towards a knowledge of the sidereal heavens was to ascertain, if possible, the distance of the fixed stars. This problem had already proved too hard for the most patient investigators; as indeed it did for Herschel himself.

Bradley had previously attempted to find a parallax for some stars supposed to be nearest to us, by observing them from opposite points in the earth's orbit when they crossed the meridian, say, in March and September. If any difference could be detected in the direction of the star from these two points of view this would indicate, not that the star had moved, but that the motion of the earth had produced a change in its apparent direction. By measuring this difference it would be easy for a mathematician, knowing the diameter of the earth's orbit, to calculate the distance of the star. This method had failed, as the stars are too far off to give an angle then measurable.

Herschel approached the problem in a totally different manner. He shunned meridian observations which require a fixed telescope. He was by nature adventurous and he chose a method which, if not entirely original, had never been seriously attempted before. It offered unlimited scope for accumulated observations, in which, as an amateur himself, he invited the co-operation of other lovers of astronomy who possessed good telescopes. It had been known for some time that there were some bright stars which, on being viewed with powerful telescopes, appeared to be double, generally one bright star accompanied by a much fainter one. Herschel saw in these pairs of stars the very objects for his purpose. Assuming that the smaller star was very much farther off than the larger one, he hoped, by carefully repeated observations, taken at six months' intervals, to find evidence of a change in the distance between them, which would give him the long-sought parallax. With this object in view he began searching the heavens with his powerful telescope, to find examples of such pairs of stars. He did not succeed in his primary object, but he was gradually led by these observations to a very important discovery: namely, that most of the pairs of stars he examined were really double stars, linked in close association with each other, and revolving round a common centre; thus proving that gravitational force acts even in those remote regions.

The assumption with which Herschel set out, following a generally received opinion at the time, that the difference in the size of the fixed stars is due to the smaller being farther off, is erroneous as a general proposition; but the method he proposed to employ (which had originally been suggested by Galileo) is quite suitable for obtaining a parallax, in cases where the two stars chosen for measurement are really independent.

The Jesuit astronomer, Father Christian Mayer of Mannheim, had already, in 1776, collected some examples of double stars, and deduced from their appearance the conclusion that they were intimately connected; the smaller he considered to be satellites of the principal ones. This theory received some confirmation from the Rev. J. Michell, the maker of telescopes, who demonstrated by mathematical reasoning that it was extremely improbable that among scattered stars many should be found in the same line of sight. In a paper printed in the *Philosophical Transactions* for the year 1784, he suggested "that some of the great number of double, triple, &c., stars which have been observed by Mr Herschel, are systems of bodies revolving about each other".

However fascinating to the imagination such a theory might be, Herschel was cautious in adopting it. Though he mentions it in a Postscript to his first paper "On Double Stars",[1] it was only to say: "in my opinion it is much too soon to form theories of small stars revolving round larger ones".

He went over his observations again and again with the greatest care, but it was not till nearly twenty years later that he considered himself able to prove:

"by a series of observations on double stars, comprehending a period of about 25 years,—that many of them are not merely double in appearance, but must be allowed to be real binary combinations of two stars, intimately held together by the bond of mutual attraction".[2]

The reception which the paper "On the Parallax of the Fixed Stars" met with, when read at the meeting of the Royal Society, is described by Dr Watson in the following letters:

DR W. WATSON *to* WILLIAM HERSCHEL

London. Dec. 18th, 1781

"Dear Sir,

"Your paper was read last meeting; it gained great applause from the generality of those who were present. Yet I have reason to think that the astronomers were put to a stand for to think of it. Dr Maskelyne

---

[1] Printed in *Phil. Trans.* vol. LXXII, 1782.
[2] *Phil Trans.* 1803.

is, from what I could find, ready to admit its merit, tho' he said, he did not think the method of computing the interval between two stars by guessing how many diameters of either star it would subtend, a good one: tho' he was ready to own he might change his opinion when he came to read over the paper. He said likewise some of the postula were very disputable, especially that by which the stars of equal size were supposed to be at equal distance from us. Mr Russel, who has not like the rest had an opportunity of knowing more of your merit, told me frankly he thought the paper a flighty one, so that your prognosis that some would think you fit for Bedlam when you talked of a power of 5400 has been verified. I dined with him yesterday in company with Aubert, Nairne and others. I found it was the high power you use which prevented him from giving you your due merit. He said, however, that he would suspend his judgement. I told him that was all I required, and that (to use an expression I have heard made use of by Dr Priestley) you desired only a clean stage and no quarter. Even Nairne doubted of your power and Aubert could not help being suspicious of some mistake, and so, my good friend, you have arrived at such perfection in your Instrument, or at least have dared to apply it to such uses and so have overleaped the timid bounds which restrain modern astronomers, that they stand aghast and are more inclined to disbelieve than to admit such unusual excellence. Though I contrived to dine with these gentlemen on purpose I had no opportunity of knowing their sentiments more fully.

"Pray inform me (for I find many different opinions prevail) what are your thoughts with respect to the diameters of the fixt stars? I find you can see them round with an uniform edge like a shilling on a piece of black paper, but I believe you do not suppose you can get at their true diameters. Many have asked me about your opinion on that subject and I should be glad to give a better answer than I can at present".

. . . . . . .

London. Dec. 25th, 1781

"I have now had a better opportunity of knowing the opinions of astronomers and opticians than I had before and I will now give you the result of my enquiries. Those who are not very much conversant in these studies have in general united their suffrages in your favour. I wish I could say the same with respect to those who are well acquainted with these matters. Of these, I am sorry to say, few are inclined to give you credit for your assertions. Some, however, I have found among these, who from what you have already done and from their personal knowledge are much inclined in your favour. What! say your opposers, opticians think it no small matter if they sell a telescope which will magnify 60 or 100 times, and here comes one who pretends to have made some which will magnify above 6000 times! is this credible?

"So that by what I can learn, the *trade*, as well as astronomers, oppose your pretentions. Yet, my dear Sir, be not in the least dejected, your facts will be verified and the greater will be your glory. In the meantime it will be necessary for you to take such measures as may as much as possible counteract the effect of this opposition. In your case it may even be permitted to those who do not know you as well as I do, to remain in suspense. Hitherto I have met with no words or actions that really had indicated envy; all that has passed may be placed to the account of the greatness of your pretentions with too little knowledge of you to give you all the credit you deserve. You have however many warm friends, who I dare say will be ready on every occasion to espouse your cause."

London. Jan. 4th, 1782
"Dear Sir,

"I received your Catalogue of the fix'd stars which I hope will greatly tend to alter the present doubts entertained by several concerning your observations. Besides the doubts entertained concerning the high powers you use, I find they hesitate with respect to your seeing the fixed stars round and well defined in very fine nights. Maskelyne and Aubert say they never saw the fixt stars without aberration. I was surprized to hear Mr Aubert mention yesterday that he had never seen your *Star* tolerably well defined. I then took an opportunity of telling him, that tho' I could not pretend I ever had the fortune to see any fixt star round, having very seldom looked at them, yet I took upon me to say that I had seen your star like a planet round and well defined....

"The only method to verify your experiments is just the method you are pursuing, which is to get some scientific persons to observe with your instruments and by the testimony of others to authenticate your facts; particularly as to seeing the stars without aberration, as well as the seeing any one of them magnified 6000 times."

WILLIAM HERSCHEL *to* DR W. WATSON, JUNIOR

Jan. 7, 1782

"Sir,

"From the contents of your letter I begin to have a much better opinion of my own observations than I had before. I thought what I had seen had been within the reach of many a good telescope and am surprized that neither Mr Aubert nor Dr Maskelyne have seen the stars *round and well-defined*. I do not say without the least aberration; for so far I will not go even with Jupiter or Saturn; but that I have a thousand and a thousand times seen them (with 460) as well defined as ever I saw Jupiter (with 227) I am very well convinced of to myself. And I believe till these gentlemen can see them so, they will not be able to find that ζ Cancri, ♄ Draconis, &c. are double; (I mean the preceding of the two stars of ζ Cancri) for the aberration of one of the stars will efface the other star, or make them appear as one. I make no doubt of soon being able to mention a good list of respectable names to ascertain a great many of the facts I have reported; but there will always be a few that cannot be otherwise than so difficult to prove to another person that we must not be surprized to see them doubted, if my own observations are not sufficient to make them believed, I do not suppose there are many persons who could even find a star with my power of 6450, much less keep it, if they had found it. Seeing is in some respect an art, which must be learnt. To make a person see with such a power is nearly the same as if I were asked to make him play one of Handel's fugues upon the organ. Many a night have I been practising to see, and it would be strange if one did not acquire a certain dexterity by such constant practice.

"I promise myself great happiness in spending a few days or a week with Mr Aubert and shall certainly carry my own instrument with me that it may be compared with his own noble apparatus.

. . . . . . .

Your most obedt. Humble servant

WM. HERSCHEL."

It is remarkable how much public curiosity was aroused by Herschel's statement that he saw the fixed stars round without rays. This was simply due to the excellent polish of his mirrors. When the pencil of light from a star passing through the aperture of a telescope is gathered to a focus, there must be a certain amount of aberration even with the best mirrors and lenses. But if the instrument is well made this aberration is symmetrical and the image will appear round and well defined in proportion to the care bestowed on the construction.

Sir John Herschel was fond of recalling an anecdote which he had heard from his father in this connection. At a dinner given by Mr Aubert in the year 1786, William Herschel was seated next to Mr Cavendish, who was reputed to be the most taciturn of men. Some time passed without his uttering a word, then he suddenly turned to his neighbour and said: "I am told that you see the stars round, Dr Herschel". "Round as a button", was the reply. A long silence ensued till, towards the end of the dinner, Cavendish again opened his lips to say in a doubtful voice: "Round as a button?" "Exactly, round as a button", repeated Herschel, and so the conversation ended.

### WILLIAM HERSCHEL *to* ALEX. AUBERT

Jan. 9, 1782

"Sir,

"I have sent two maps with the double stars of the 1st, 2nd, 3rd, 4th and 5th Classes marked upon them with red ink. The figure annexed denotes the class to which they belong. I have also marked some that I could particularly wish you to try your telescopes upon, with an additional margin of red ink, by way of pointing them out. If you succeed in some of the most difficult you will receive a very considerable pleasure.

"I find by my very excellent friend, Dr Watson, that my observations will stand much in need of the protection of some kind gentleman well known in the astronomical line, who may give credit to them by his own observations; give me leave to beg of you (for the love of Astronomy's sake) to lend your assistance, that such facts as I have pointed out may not be discredited merely because they are uncommon. It would be hard to be condemned because I have tried to improve telescopes and practised continually to see with them. These instruments have played me so many tricks that I have at last found them out in many of their humours and have made them confess to me what they would have concealed, if I had not with such perseverance and patience courted them. I have tortured them with powers, flattered them with attendance to find out the critical moments when they would act, tried them with Specula of a short and of a long focus, a large aperture and a narrow one; it would be hard if they had not proved kind to me at last.

"I find the powers of my telescopes are doubted. Let but the facts appear, the powers will follow. A telescope that will show $\eta$ Coronae, $\hbar$ Draconis, $\zeta$ Cancri and the star near Procyon as I have seen them, must possess such a degree of distinctness that no person will doubt its being capable of a very great power.

"You have in your possession a treasure beyond value;[1] such instruments as yours, I should think, will enable you to verify most of my observations. I could wish you would have a set of single eye glasses adapted to your $3\frac{1}{2}$ ft. achromatic to magnify 200, 250, 300, 500 times.

. . . . . . .

"If I say too much or am too particular, pray excuse it, Sir, but, judging from myself I know that no person could give me more real pleasure than by entering into such Minutiae as I know to be generally of considerable consequence when we come to refinements and want *to screw an instrument up to its utmost pitch.* (As you are an *Harmonist* you will pardon the *musical phrase.*)

"You will soon hear the description I am now drawing up of the last new micrometer, which occasion'd you obligingly to exclaim that 'we would go to Bedlam together'. I shall have no objection to any place in such astronomical company; but pray, Sir, do not let me be sent there by myself, which I fear will be the case, unless you procure me a reprieve by verifying some of the facts I have mentioned, and by that means make it appear probable that they may *all* be true.

"I am much obliged for the fine silver wire which you have been so kind as to send me and remain

Sir, your most obedt. humble servt.

WM. HERSCHEL."

To this letter of Herschel's Aubert replied:

London, ye 22nd January, 1782

"Dear Sir,

"Although I have not returned you thanks for the valuable present you conveyed to me by our worthy friend Dr Watson, give me leave to assure you that I am not the less grateful for your kind attention

---

[1] A very fine Cassegrain reflector by Short.

and trouble. The marking all the stars you have found double, treble, &c. must have given you a deal of trouble, but trouble is nothing to you; and the least thing that we can do in return is to bestow our labour under your advice and assistance in order to convince the world that though your discoveries are wonderful, they are not imaginary. I am firmly of opinion that we shall do you justice; none will contribute towards doing so more than I shall, for I go to work with a full confidence in everything you have asserted.

"The Doctor will have mentioned to you some hints I took the liberty of giving him concerning your paper and you may be assured that I will stand by you in everything that I can with some degree of experience make myself reconciled to. Your great power of 6450 continues to astonish, your micrometer also; in regard to the first, our President gave me Dr Watson's letter to read and it has convinced many that what you asserted about the time a star took to pass over the field of this power is true. Go on, my dear Sir, with courage, mind not a few barking, jealous puppies; a little time will clear up the matter and if it lays in my power you shall not be sent to Bedlam alone, for I incline much to be of the party.

Your most humble &c.<br>ALEX. AUBERT".

WILLIAM HERSCHEL *to* MR AUBERT

King Street, Bath. Jan. 28, 1782

"Dear Sir,

"The favour of your letter has given me great pleasure and I am very glad to find you intend to explore the powers of your telescopes.

"I shall beg leave to give a hint or two that experience has taught me in the use of high powers. One of no small consequence is that when you have looked a considerable time with a small power and change it for a higher, not to be too ready to lay it aside if you should find it not so distinct as you could wish. The eye is one of the most extraordinary Organs. I remember the time when I could not see with a power above 200, with the same instrument which now gives me 460 so distinct that in fine weather I can wish for nothing more so. When you want to practise seeing (for believe me Sir,—to use a musical phrase—you must not expect to *see at sight* or a *livre ouvert*) apply a power something higher than what you can see well with, and go on encreasing it after you have used it some time. These practices I have repeated upon Castor, $\epsilon$ Bootis, &c.: encreasing the power by degrees till I fairly could not distinguish the least appearance of the two stars, and they had run perfectly into each other. The consequence of this was that every time I tried again, my eyes acquired more practice and I can now see with powers that I used to reject for a long time.

. . . . . . .

"I promise myself much pleasure in the expectation of spending a few days next summer in observing with your instruments; my intention is to bring also one of mine along with me that we may compare them together.

I remain sir &c.

WM. HERSCHEL."

DR W. WATSON *to* WILLIAM HERSCHEL

London. Jan. 11th, 1782

"My dear Sir,

"I have the satisfaction to inform you that at a council of the Royal Society held yesterday, Sir Joseph Banks made a motion which was unanimously agreed to, that you should be complimented with an exemption of the usual payment, and it was put upon the footing, that the money which was withheld from the Society would still be expended on such objects which would best answer the view of the Society. This gives me infinite pleasure as this shews the great respect they have for you, as well as for other considerations. Your catalogue was read yesterday in the manner you directed, and I am sure it had a great effect. Let us have a little patience, and all will end in the best manner. Dr Maskelyne told me yesterday after the Society was over, that tho' he did not in the least doubt your veracity, he could not help thinking that you was mistaken as to your high powers.

"My Father, who attended the Council meeting yesterday, likewise got permission for you to stay as long as you pleased before you came to Town to be admitted, as the usual form makes the election void for any member, who should not come to be admitted within one month after election."

Herschel took the opportunity at the same time of writing to Dr Hornsby in the hope of dispelling some of his prejudices, but without any great success, as appears from Hornsby's reply:

REV. THOS. HORNSBY *to* WILLIAM HERSCHEL

Oxford. Feb. 26, 1782

"Sir,

"I am much obliged to you for your very long letter, and cannot help expressing my surprise at your extreme diligence in observing so many stars that have circumstances, &c., peculiar to them.

. . . . . . .

"In your paper sent to the Royal Society have you given a description of your two micrometers? I confess I can have no idea of your being able to measure with great certainty with them, and I the rather think so because your observations led you to think that the Comet, which you discovered, was below the orbit of Mars, when he has hitherto been at a distance more than double that of Saturn.

. . . . . . .

"It is the fashion I think now to call it a new star or a planet, but I cannot help thinking that it will prove to be a comet; but it does not necessarily follow that it should be either a Planet or a Comet.

. . . . . . .

"I presume you have been informed that Mr Cavendish has computed several obs$^{ns}$ of the comet made by Dr Maskelyne & myself, as also your first obs$^n$ & has now determined its mean distance to be not more than 19 times the Earth's distance, and not less than 17 or $17\frac{1}{2}$, & that its least distance in all probability will be 13 more or less; but in all these comparisons your obs$^n$ differs 2 minutes and more in defect, when others do not differ more than $\frac{1}{2}$ a minute, almost always much less; & their differences generally in excess; there must therefore I think be some considerable error in your Obs$^n$.

I am, Sir, with great regard,

THOS. HORNSBY".

In William Herschel's "Letter Book", preserved at the Royal Astronomical Society, there is a copy of a "Memorandum for Mr Cavendish", which was probably the only reply which he made to Hornsby's aspersion on his accuracy. It runs thus:

"By a letter from the Rev. Mr Hornsby I find that Mr Cavendish has computed several observations of the Comet made by Dr Maskelyne and Mr Hornsby and that in all these computations my observations differ 2 min. and more *in defect*, when others do not differ more than

½ minute, almost always much less, and their differences generally *in excess*. From this Mr Hornsby thinks there must be some considerable error in my observations. I could wish Mr Cavendish to be acquainted that there is in my opinion no possibility of an error that can amount to more than 5″ in any of my observations except I have marked it as such, and that in general the error is much less than 5″. The reason of the disagreement I imagine is owing to a mistake in the stars. For as I could not determine the situation of them otherwise than by an eye-draught I suppose either Mr Maskelyne or Mr Hornsby must have taken the wrong stars...."

William Herschel was quite justified in maintaining the correctness of his observations. They have been verified, and confirm his assertion that there could be no error greater than a few seconds of arc. Hornsby's remark had no reference to the slight irregularities which Herschel had attributed at first to parallax; yet this mistake no doubt contributed to his ill opinion of Herschel's accuracy. It must be admitted that the indications of the positions of the star which Herschel had given in his paper for the Royal Society were strangely lacking in the precision customary among professional astronomers. The diagram resembled a sketch which an untrained explorer might make of rocks and islands with no reference to scale or to longitude or latitude. He merely mentioned that it represented a portion of the sky, "in the neighbourhood of H Geminorum", and the little stars which he took as guides were not included in any catalogue existing at that time; they are difficult to find even now with the aid of a photographic map. We must remember that up to this time Herschel had been working entirely independently; he had not acquired what one might call the etiquette of the profession. To one in Hornsby's position there appeared a sort of impertinence in the casual manner in which Herschel had presented his observations to the Royal Society, and he was not sorry to think he had proof that they were seriously wrong. Had he been as broad-minded as Dr Maskelyne or Sir Joseph Banks he would have recognized Herschel's genius. But being blinded by prejudice he did not see that it was much more likely that the orbit computed by Cavendish was in fault. Considering that Uranus takes eighty-four years to complete the circuit of its orbit and that Cavendish had to rely on observations only nine months apart for his calculations, it would have been wonderful indeed if they had been so reliable as to discredit Herschel's observations. The incident is only of interest as showing Hornsby's attitude. As he occupied a position second only to that of the Astronomer Royal, Herschel's friends, particularly Dr Watson, were anxious that he should be conciliated, but Herschel himself took no further steps to do so. In later years, when Herschel had achieved a world-wide reputation, they exchanged amicable visits.

### SIR JOSEPH BANKS *to* WILLIAM HERSCHEL

Soho Square. March 15, 1782

"Dear Sir,

"I have much pleasure in informing you that Mr Aubert, our Friend, has succeeded in verifying your observation of the Pole star being double. He, with a magnifying power of 1,000, saw the star itself almost if not totally deprived of scintillation, flat, & with an edge not ill defined, & saw its small companion very plain.

"Believe me, I had great pleasure to hear him inform me of it & more when he told me that he would take pains to verify as many of your observations as his instruments & powers will allow him to do.

"Our astronomical Friends however still hesitate at the immense powers which you tell us of in your papers & wait with anxiety for the explanation of them which you propose to give us. 1,000 is to them a power of great difficulty & how you can arrive at 6,000 of course to them unintelligible.

"The Censors of Papers have postponed printing your last papers, not out of any disrespect of them, but only till that matter, so little understood by them, shall be cleared up; they are unwilling to believe at this moment that you have not made somehow a mistake in calculating the quantity, but in that case would not the less respect the discoveries you have made; they will always speak for themselves & the other will of course be cleared up in time for the papers to be printed in the present volume.

Jos. Banks."

### WILLIAM HERSCHEL *to* SIR JOSEPH BANKS, BART.

28 March, 1782

"Sir,

"I have the honour of laying before you the results of a set of measures I have taken in order to ascertain once more the powers of my Newtonian 7 ft. reflector...._

"To prevent any mistakes I wish to mention again that I have all along proceeded *experimentally* in the use of my powers, and that I do not mean to say I have used 6450 (or 5786) upon the planets or even upon double stars; every power I have mentioned is to be understood as having been used just as it is related; but further inference ought not to be drawn; for instance my observations on $\epsilon$ Bootis mention that I have viewed that star with 2010 (or as in the above table with 2175) extremely distinct; but upon several other celestial objects I have found this power of no service. Many plausible suggestions have already occurred to account for these appearances, but I wait till further experiments shall have furnished me with more material to reason upon. The use of high powers is a new and untrodden path and in this attempt variety of new phenomena may be expected, therefore I wish not to be in haste to make general conclusions; I shall not fail to pursue this subject and hope soon to be able to attack the celestial Bodies with a still stronger Armament, which is now preparing.

Your most obedt. and most humble servant,

WM. HERSCHEL."

DR MASKELYNE *to* WILLIAM HERSCHEL

Greenwich. April 19, 1782

"Sir,

"Last night at a Committee of papers at the Royal Society your valuable paper on the parallax of the fixed stars came under consideration, and was highly applauded for its ingenuity and the new lights it throws on that difficult subject; and would have been immediately ordered to be printed, but on account of some corrections which were then suggested as requisite to be applied to it in order to obviate future cavils and objections and to make the paper do you honor, which the whole and every member of the Committee are desirous of, I was therefore desired by the Committee to write to you and explain to you those passages which it was suggested might be improved....

"The principal correction that seems requisite is in the last 6 pages but one, where you have founded all your calculations on a supposition that the proportional distances of the fixt stars are as the numbers expressing their magnitudes.... I should also mention to you that the principle that all the fixt stars are equal in real magnitude and lustre to one another and to our sun, seems a very hard hypothesis and not agreeable to the very great variety observable in the works of nature...."

Herschel's reply to Sir Joseph Banks' letter was printed in the *Philosophical Transactions* of the Royal Society. He also wrote a very long categorical reply to Dr Maskelyne, from which an extract will be given here, as it is a good example of his style of reasoning.

WILLIAM HERSCHEL *to* DR MASKELYNE

April 28, 1782

"Rev.<sup>d</sup> Sir,

"It is not a little difficult for me to express how much I think myself honour'd by the kind attention of the Gentlemen of the Committee of Papers, and also in particular by your friendly communication of those passages in my Paper on the Parallax that may be improved by proper alterations.

"The Respect and Deference I have for the opinion of the Gentlemen who favour'd my paper with their examination is of such a nature that I submit with pleasure to their judgement to omit any part of it that may appear to them liable to objections; and if I offer a few remarks on those passages you have pointed out to me, I beg it may be looked upon only as a further exposition of my Sentiments upon those subjects, in order that it may the better appear how far they really are liable to such objections, as might be made to them.

"I beg leave, Sir, to follow the order of your letter; therefore to begin with the Theory of the Parallax of Double Stars; when I say 'Let the stars be supposed *one with another* to be *about* the size of the Sun', I only mean this in the same extensive signification in which we affirm that *one with another* Men are of such or such a particular height. This does neither exclude the Dwarf, nor the Giant. An Oak-tree also is of a certain general size tho' it admits of very great variety. And when we consider the vast extent of relative magnitudes we shall soon allow that by mentioning the size of a Man or of an Oak-tree we speak not without some real limits; for, tho' one man may differ much in size from another, yet are they both sufficiently distinguished from much larger objects such as a mountain, or much smaller as a particle of dust. If we see such conformity in the whole animal and vegetable kingdoms, that we can without injury to truth affix a certain general Idea to the sizes of the species, it appears to me highly probable and analogous to Nature, that the same regularity will hold good with regard to the fixt stars. It may be said that, if I admit such latitude there can be but little dependence on calculations founded on such Postulata, to which I answer that a single observation indeed might mislead us considerably, but when a number of stars are taken we may expect the result of the whole will be very near the truth. If a person desirous to know the common stature of a man were to measure the first person he chanced to meet with and no other it is possible he might be deceived, tho' even there a certain limit would take place; yet if he were but to measure one score of them promiscuously, we should find that he would not be far from the mark."

. . . . . . .

DR MASKELYNE *to* WILLIAM HERSCHEL

Greenwich. May 7, 1782

"Sir,

"I received your answer which I laid before the Committee of papers, who approve of the corrections and alterations you have sent and I have accordingly marked them upon your papers, as you will find by the proofs which are ordered to be sent to you....I think your papers will do very well now. They are all ordered to be printed....

"I could have wished you had said something of the method you use to find a star in your telescope without a polar axis; this would have lessened the difficulty attached to the use of the large magnifying powers you mention. It has cost me some time and pains to conceive how you do it; I believe I have now hit upon it.

Your sincere friend and humble servant,

N. MASKELYNE."

## CHAPTER VIII

### 1782

William Herschel's prospect of becoming private astronomer to the King. Letters. William goes to London. His letters to Caroline and Alexander. The King's offer accepted. Congratulatory letters. Name for the new planet. Letter from Bode. Sign for Uranus. Herschel dedicates it to the King as "Georgium Sidus". Letters from Lalande, C. Mayer, Magellan, Lichtenberg, Schroeter.

HERSCHEL'S friends at the Royal Society were very desirous that he should be freed from the trammels of his musical profession in order to devote himself wholly to astronomy. King George III, whose interest in astronomy was very genuine, had before this time established a private observatory at Kew. The director of this observatory, Demainbray, was at this time an old man, and it occurred to Sir Joseph Banks and others that this position was exactly suitable for Herschel, and they wished that the King's attention could be drawn to his qualifications.

Sir Joseph Banks wrote in February, 1782, to Dr Watson:

Soho Square. February 23, 1782

"Dear Sir,

"For our friend Mr Herschel, his discoveries are now I think by all hands acknowledged and no more said of course about them which even tends to controversy....

"For the name half my objection is unfortunately past. I wished the new star, so remarkable a phenomenon, to have been sacrificed somehow to the King. I thought how snug a place his Majesty's astronomer at Richmond is and have frequently talked to the King of Mr Herschel's extraordinary abilities. I knew Demainbray was old but as the Devil will have it he died last night. I was at the Levy this morning but did not receive any hopes. I fear this (the time) has passed by which a well timed compliment might have helped if the old gentleman had chose to live long enough to have allowed us to have paid it.

Adieu my dear Sir, yours faithfully,<br>Jos. Banks.

"My best compliments to Mr Herschel with wishes that for the sake of science his nights may be as sleepless as he can wish them himself".

It turned out that the King had promised the reversion of this appointment to Demainbray's son, and would not go back on his word, but Dr Watson, who did not know of this, continued to cherish the hope that he might see his friend established comfortably at Kew.

Meantime the Easter season at Bath was at its height. William Herschel spent his time in the usual routine of music lessons and concerts; and as Easter drew near the labours of both brother and sister were much increased by preparations for the performances of the Oratorios which always took place in Passion Week, both at Bath and Bristol.

In the beginning of May Herschel received a letter from his friend Colonel Walsh, which ran as follows:

"Dear Sir, Chesterfield Street. 10th May, 1782

"In a conversation I had the Honour to hold with His Majesty the 30th ult° concerning You and Your memorable Discovery of a new Planet, I took occasion to mention that You had a twofold claim, as a native of Hanover and a Resident of Great Britain, where the Discovery was made, to be permitted to name the Planet from His Majesty.

"His Majesty has since been pleased to ask me when You would be in Town, to which I could not certainly answer. I am now setting out for Taunton, where the Worcester Regiment at present is quartered and shall endeavour to see You in my way. I yesterday discoursed with Sir Joseph Banks on this subject, who has the same sentiments with me on the matter.

I remain, Dear Sir, your very obedient humble servant

JOHN WALSH".

But before this letter reached him Herschel had received other intimations that it would be desirable for him to go to London. In his Memorandums he wrote:

"1782, *March* 20*th*. In passion week we had Four oratorios, two at Bath and two at Bristol.

"*April*. Attending scholars by day and astronomical observations at night. I was informed by several gentlemen that the King expected to see me, and by my journal it appears that about the end of April I made out a list of celestial objects that I might show the King.

"*May* 19*th*. This being Whit Sunday one of my Anthems was sung at St Margaret's Chapel when for the last time I performed on the Organ.

"*May 20th.* Having packed up my 7 feet telescope, I went the 20th May to London on a visit to Dr Watson, sen., who lived in Lincoln's Inn Fields, and on the Sunday following I had an audience of His Majesty, who received me very graciously. The King said that my telescope in three weeks was to go to Richmond and meanwhile to be put up at Greenwich. Some letters to my sister at Bath which are still in her possession contain an account of what passed at this and other audiences I had of the King.

"*May 29th.* My telescope having been carried to Greenwich, I prepared it for Dr Maskelyne's observations.

"*June.* Not only Dr Maskelyne, but every gentleman acquainted with astronomical telescopes who observed with mine, during the course of all this month that it remained at Greenwich, declared it to exceed in distinctness and magnifying power all they had seen before. My telescope having undergone this kind of examination, the King desired it to be brought to the Queen's Lodge at Windsor.

"*July 2nd.* I had the honour of showing the King and Queen and Royal Family the planets Jupiter and Saturn and other objects".

The following are the letters to which Herschel refers as having been written by him at this time. (The first is undated; as is also the one to his brother Alexander.)

"Dear Lina,

"I have had an audience of His Majesty this morning and met with a very gracious reception. I presented him with a drawing of the solar system, and had the honour of explaining it to him and the Queen. My telescope is in three weeks time to go to Richmond, and meanwhile to be put up at Greenwich, where I shall accordingly carry it to-day. So you see, Lina, that you must not think of seeing me in less than a month. I shall write to Miss Lee myself; and other scholars who inquire for me you may tell that I cannot wait on them till His Majesty shall be pleased to give me leave to return, or rather to dismiss me, for till then I must attend. I will also write to Mr Palmer and acquaint him with it.

"I am in a great hurry, therefore can write no more at present. Tell Alexander that everything looks very likely as if I were to stay here. The King inquired after him, and after my great speculum.

He also gave me leave to come and hear the Griesbachs play at the private concert which he has every evening. My having seen the King need not be kept secret, but about my staying here it will be best not to say anything but only I must remain here till His Majesty has observed the planets with my telescope.

"Yesterday I dined with Colonel Walsh, who inquired after you. There were Mr Aubert and Dr Maskelyne. Dr Maskelyne in public declared his obligations to me for having introduced to them the high powers, for Mr Aubert has so much succeeded with them that he says he looks down upon 200, 300 or 400 with contempt, and immediately begins with 800. He has used 2500 very completely, and seen my fine double stars with them. All my papers are printing, with the postscript and all, and allowed to be very valuable. You see, Lina, I tell you all these things. You know vanity is not my foible, therefore I need not fear your censure. Farewell.

I am, your affectionate brother,
WM. HERSCHEL."

Monday evening, June 3rd, 1782

"Dear Lina,

"I pass my time between Greenwich and London agreeably enough, but am rather at a loss for work that I like. Company is not always pleasing, and I would much rather be polishing a speculum. Last Friday I was at the King's concert to hear George play. The King spoke to me as soon as he saw me, and kept me in conversation for half an hour. He asked George to play a solo-concerto on purpose that I might hear him; and George plays extremely well, is very much improved, and the King likes him very much. These two last nights I have been star-gazing at Greenwich with Dr Maskelyne and Mr Aubert. We have compared our telescopes together, and mine was found very superior to any of the Royal Observatory. Double stars which they could not see with their instruments I had the pleasure to show them very plainly, and my mechanism is so much approved of that Dr Maskelyne has already ordered a model to be taken from mine and a stand made by it to his reflector. He is, however, now so much out of love with his instrument that he begins to doubt whether it *deserves* a new stand. I have had the influenza but am now quite well again. It lasted five or six days and I never was confined with it. Dr Watson has it now, but will probably be well in about 2 or 3 days. There is hardly one single person but what has had it.

"I am introduced to the best company. To-morrow I dine at Lord Palmerston's, and the next day with Sir Joseph Banks, &c. &c.

"Among opticians and astronomers nothing now is talked of but *what they call* my Great discoveries. Alas! this shows how far they are behind, when such trifles as I have seen and done are called *great*. Let me but get at it again! I will make such telescopes, and see such things... that is, I will endeavour to do so."

June 5, 1782

"Dear Lina,

"There is to be a Court Mourning which I believe is to begin on Sunday; be so good as to send me my black clothes, what there is of them; when they come I shall be able to see what will do; if not I can then get anything made here that I may want. I believe you may put them up in brown paper with a towel sewed about as there will be no occasion for a box. Let them come by the Coach that goes off either at three in the afternoon (which will come sooner) or, if you cannot have them ready so soon, by the evening Coach. Direct them to Mr Herschel at Dr Watson's, Lincoln's Inn Fields, London.

"Dr Watson is returned to Bath, he set off this morning. I believe some time next week the King will see my telescope, and if nothing happens to prevent my return I shall be at home again by Sunday sev'n night. My telescope has been tried at Greenwich and is allowed to be a better instrument than any they have there. This evening I shall go to Court and may perhaps have time when I return to write something about what has happened; therefore I will leave room for that purpose. Adieu."

Wednesday Morning, June 6, 1782

"(As a postscript) I have received your letter."

WILLIAM HERSCHEL *to* ALEXANDER HERSCHEL

(No date)

"Dear Brother,

"I received your letter just now at my return from Mr Aubert where I have spent a day or two; we have tried his instruments upon the double stars and they would not at all perform what I had expected, so that I have no doubt that mine is better than any Mr Aubert has; and if that is the case I can now say that I absolutely have the best telescopes that were ever made. However we intend to try them both together at the same time, for perhaps mine would not have acted that evening tho' it was in all appearance very fine weather.

"You guess very rightly that my present situation is not very agreeable. I have no regular business. But I see a good deal of the world. I got acquainted with Arnold the time-piece maker. He is a very great genius and his watches go to such a degree of exactness that you can hardly have any idea of it. He maintains that the principle of watch-work is so compleat that he thinks they will in a little time be superior to the best pendulum clocks in Observatories. You may roast them before a fire, put them in an Ice cellar, lay them down on the face or on the back, hand them one way or another way, ride on Horseback or do any other kind of exercise, they always go equally well.

"Last Thursday evening I was at the Concert again. Mr Fisher was there and I heard them play. The Queen and the two Princesses and one Prince were present in the Music Room. As soon as the King saw me he came and spoke to me about my telescope but he has not yet fixed a time when he will see it.

"I have received my clothes and shall soon write again to Lina but the Postman goes about with his Bell, therefore I must finish at present in a hurry.—Farewell, dear Brother, I am constantly yours

WM. HERSCHEL."

Monday evening.

On the 3rd July, Herschel, who had now been six weeks absent from Bath, wrote to his sister the following account of the evening spent in showing his telescopes to the King and the Royal Family. He had been summoned to Windsor and was lodging with his nephew, George Griesbach.

Written July 3rd, 1782<br>(the date in Caroline's writing)

"Dear Carolina,

"I have been so much employed that you will not wonder at my not writing sooner. The letter you sent me last Monday came very safe to me. As Dr Watson has been so good as to acquaint you and

Alexander with my situation I was still more easy in my silence to you. Last night the King, the Queen, the Prince of Wales, the Princess Royal, Princess Sophia, Princess Augusta, &c., Duke of Montague, Dr Hebberdon, Mons Luc &c. &c., saw my telescope and it was a very fine evening. My Instrument gave a general satisfaction; the King has very good eyes and enjoys Observations with the Telescopes exceedingly.

"This evening as the King and Queen are gone to Kew, the Princesses were desirous of seeing my Telescope, but wanted to know if it was possible to see without going out on the grass; and were much pleased when they heard that my telescope could be carried into any place they liked best to have it. About 8 o'clock it was moved into the Queen's Apartments and we waited some time in the hopes of seeing Jupiter or Saturn. Meanwhile I showed the Princesses and several other ladies that were present, the Speculum, the Micrometers, the movements of the Telescope, and other things that seemed to excite their curiosity. When the evening appeared to be totally unpromising, I proposed an artificial Saturn as an object since we could not have the real one. I had beforehand prepared this little piece, as I guessed by the appearance of the weather in the afternoon we should have no stars to look at. This being accepted with great pleasure, I had the lamps lighted up which illuminated the picture of a Saturn (cut out in pasteboard) at the bottom of the garden wall.

"The effect was fine and so natural that the best astronomer might have been deceived. Their Royal Highnesses and other Ladies seemed to be much pleased with the artifice. I remained in the Queen's Apartments with the Ladies till about half after ten, when in conversation with them I found them extremely well instructed in every subject that was introduced and they seem to be most amiable characters. To-morrow evening they hope to have better luck and nothing will give me greater happiness than to be able to show them some of those beautiful objects with which the Heavens are so gloriously ornamented."

This letter ends abruptly and is the last of this series except a fragment without date:

"Tell Alex. I have got a Micrometer made by Mr Nairne for my Cross-wire. I have had a great deal of trouble to make them do it well, and have been no less than ten or a dozen times backwards and forwards 2 or three miles to look after it; the last time I saw it there were still 4 capital blunders in it but I am determined not to say that it is to my liking till it really is so. Last Thursday I had a description of a method with a string such as turners use to make the piece which is turning upon two centers always go the same way round. Pray let Alex. try if he can find out a way to do it. It is very simple.

"I must conclude my letter, farewell Lina, or as you understand greek, αδιευ".

Saturday.

Dr Watson had returned to Bath early in June and was impatient to hear that his friend's affairs were all happily settled. He was evidently convinced that the King intended to make some provision for Herschel, but thought that the latter was too diffident in advancing his own cause. As William Herschel's own letters have been lost we have only Dr Watson's to guide us; from these we gather that Herschel thought it best to wait patiently on the King's pleasure, confident that he had his goodwill, but that after he had been summoned to Windsor, and had had so many marks of the King's favour granted to him, he yielded to his friend's prompting and did make some sort of application. This led to the King's sending General Freitag to see Herschel and to explain what the King was able to offer him.

### DR W. WATSON *to* WILLIAM HERSCHEL

Addressed: at Mr Griesbach's, Musician to Her Majesty, Windsor.

Bath. July 14th, 1782

"My dear Friend,

"I need not tell you with what satisfaction I received the account of the King's offer to make you independent of music, and was in hopes before now of hearing further the mode in which it is to be put in execution. You are now upon the eve of entering into a new course of life, to take upon you a new character, and to be in a situation which will at the same time command respect and what is still more desirable, enable you wholly to give up yourself to an employment which is attended with the highest gratification.

"I presume General Freitag is a German, you would do well to cultivate his acquaintance, for as the King thought proper to make him the messenger of his offer, he will be the most proper person thro' whom you may communicate to the King anything you may wish to have him informed of relative to this business....

"Sir Joseph Banks sent me an answer to mine in which he mentions the anxious wishes he had of serving you and how often he had weighed in his mind how it was to be effected and how much he was perplexed with a thousand delicacies, and that he had actually come to Town on Thursday morning in order to inquire after you and to consult with you what was to be done, being by no means satisfied with himself. He then heard of the offer the King made and expresses the pleasure he received at the news in the warmest manner. I thought it proper to tell you this, that you might see that tho' it was not in his power to speak to the King of late, he nevertheless felt great zeal for your service, and has a very great regard for you.

"Permit me, my dear Friend, once more to congratulate you on this great event of your life, and to assure you that my satisfaction on this occasion can fall very little short of what you yourself must feel.

I am, Dear Sir,
Your most sincere friend,
W. WATSON."

ALEX. AUBERT *to* WILLIAM HERSCHEL

London, y$^e$ 31st August, 1782

"Dear Sir,

"I can assure you from the bottom of my heart that no one can have felt more pleasure than I have done at the good news of his Majesty's having provided for you in such a manner as to put it in your power to give up all your time to your favourite pursuits. May you enjoy health and happiness during a number of years and make all the progress you can possibly expect! How great! How generous! How kingly! this encouragement of science is, in his Majesty; may God bless him for it and make every matter contribute to his happiness; there never was a more deserving, a more virtuous, a more benevolent sovereign. If the blessings and the greatest respect of a single individual could procure him long life, health and peace of mind my sentiments would procure them all to him and to his family.

"My friend De Luc had informed me that the matter was in agitation. I thank you for informing me of it's being certain. As soon as you are settled a little I will call upon you and shall be happy to give you the exact place of the telescopic star from which you made your first comparison of the distance of the *Herschel*; my Transit Instrument and mural Quadrant will do it very well. I need not say, my good Sir, that I shall ever be glad to serve you and to cultivate your friendship. I can assure you that independent of your knowledge your behaviour is such as would make you deserving of my regard. I shall say no more upon the subject for fear of displeasing you; when we meet we will talk of many things. I thank you for your kind intention about a speculum for my 6 feet reflector. I am persuaded you will improve our telescopes much and that the astronomical world will have great obligations to you.

"I have been lately...at Ramsgate in Kent where I took with me Mr Smeaton in order to consult him about the Harbour; it empties every day and is now of great use after being many years almost useless and by his good advice it will be much improved soon.

I am with the sincerest regard

Dear Sir, your most humble and obedient servant,

ALEX[R] AUBERT."

## SIR JOSEPH BANKS *to* WILLIAM HERSCHEL

Soho Square. Aug. 28, 1782

"Dear Sir,

"That you have received a Pleasure and that you feel a sensation of gratitude beyond the expression of language in being enabled to pursue without interruption the studies for which beyond a doubt the great Creator originally intended you I can easily believe.

"I trust that you will believe that I feel on this occasion a pleasure almost as great. The pleasure I feel for having contributed my mite for the promotion of science and at the same time of a deserving man. Allow me however to remonstrate against one part of your letter; no gratitude whatever is due to me in mentioning your name to his Majesty; I did but my duty and that could have answered, God knows, no part of the purpose of recommending you to his Royal protection had not that love of science which has ever distinguished his reign, and has not been damped even by the remembrance of the improper channels into which it has formerly been guided, led him to inquire and examine into your merits, of which he was personally a judge before he made the least step towards reward. Was every kingdom blessed with a sovereign as capable of distinguishing and as ready to reward merit as ours is, Philosophy would indeed be a Fashionable study, we should no longer hear 'Laudatur et alget'.

"That you will perform your duty to science and to his Majesty by prosecuting with vigor the studys for which he patronises you and Nature has so eminently fitted you I cannot doubt. Wishing then that you may be able to repay your royal Protector by Future as you attracted his notice by your Former discoveries, believe me,

Your faithfull servant<br>
and assured Friend,<br>
Jos. Banks."

The idea which had already occurred to Sir Joseph Banks that the new planet should in some way be dedicated to the King appealed to Herschel as an appropriate way of showing his appreciation of the King's favours to him. He wrote a letter to Sir Joseph Banks, as President of the Royal Society, recommending the name of "Georgium Sidus" for the new planet, and prepared a drawing of the star, showing its position with regard to neighbouring stars at the time when it was discovered. This drawing he proposed to have engraved with an inscription dedicating it to King George. He applied to his friends in Bath to help him with the wording of this inscription. Dr Watson wrote in reply:

Bath. July 20th, 1782

"I will now tell you the result of my consultation with our Friends Mr Collings and Mr Webb. In the first place we think the star should be called not 'Georginum Sidus' but 'Georgium Sidus' in the same manner as Horace, Liber I, Ode XII:

> 'Micat inter omnes
> Julium Sidus'.

"Mr Webb recommends that either in the print or at the bottom with some mark referring to the star, the words 'Georgium Sidus' should be written and under it these words 'jam nunc assuesce vocari'. Thus:

> Georgium Sidus
> jam nunc assuesce vocari.
> Virg.

"The quotation is taken from the first Book of the Georgics, line 42, where Virgil, after invoking Caesar as a future God, among other things tells him he must now accustom himself to be called upon with vows.

"You see that the words as we apply them simply imply 'to be called' but you know that upon such occasions it is a common and allowed of practice to make a quotation and apply it in a different sense....

"I confess the quotation from Virgil pleases me and is as good a one as could be expected on the subject.

"I wrote this according to the date since which I had the pleasure of receiving yours of the 18th and am sorry to find that you are not likely to be settled at Kew. I am likewise a little fearful that your appointment may not be so considerable as your friends could have wished it. On the other hand if we had known that Demainbray's son had succeeded his father our expectations would not have reached even this arrangement, especially if we consider the King's present situation".

The name which Herschel proposed for the new planet was for a short time used in England, but, on the Continent, that suggested by the Prussian astronomer, Bode, was considered more appropriate, and was soon universally adopted. Bode wrote in July, 1783, to Herschel:

(In German)

"Very highly respected Sir,

"I have the honour to congratulate you heartily on the fortunate discovery of the new planet and take this opportunity with much pleasure of commencing an interchange of letters on the subjects which interest us both. You know perhaps that I was the first person in Germany to see the new star; this was on the 1st August, 1781,—since when I have observed it as often as opportunity offered, and have published my observations in my *Astronomisches Jahrbuch*.

"I have proposed the name of Uranus for the new planet as I thought that we had better stick to mythology and for several other reasons. However I assure you that had I been in your situation I should have felt it proper to do as you have done.

"I have further proposed the sign ⛢ for our new planet.

J. E. BODE,
Astronomer to the Royal
Prussian Academy of Science".

The following is the text of the letter which Herschel wrote to Sir Joseph Banks and which was printed in the *Philosophical Transactions*, vol. LXII, 1783, of the Royal Society.

*A Letter from* WILLIAM HERSCHEL, ESQ., F.R.S.
*to* SIR JOSEPH BANKS, BART., P.R.S.

"Sir—By the observations of the most eminent Astronomers in Europe it appears, that the new star, which I had the honour of pointing out to them in March, 1781, is a Primary Planet of our Solar System. A body so nearly related to us by its similar condition and situation, in the unbounded expanse of the starry heavens, must often be the subject of conversation, not only of astronomers, but of every lover of science in general. This consideration makes it necessary to give it a name, whereby it may be distinguished from the rest of the planets and fixed stars.

"In the fabulous ages of ancient times the appellations of Mercury, Venus, Mars, Jupiter, and Saturn, were given to the Planets, as being the names of their principal heroes and divinities. In the present more philosophical era, it would hardly be allowable to have recourse to the same method, and call on Juno, Pallas, Apollo or Minerva, for a name for our new heavenly body. The first consideration in any particular event or remarkable incident, seems to be its chronology; if in any future age it should be asked, *when* this last found Planet was discovered? It would be a very satisfactory answer to say, 'In the Reign of King George the Third'. As a philosopher then, the name of GEORGIUM SIDUS presents itself to me, as an appellation which will conveniently convey the information of the time and country where and when it was brought to view. But as a subject of the best of Kings, who is the liberal protector of every art and science;—as a native of the country from whence this Illustrious Family was called to the British throne;—as a member of that Society which flourishes by the distinguished liberality of its Royal Patron;—and last of all, as a person now more immediately under the protection of this excellent Monarch, and owing everything to His unlimited bounty;—I cannot but wish to take this opportunity of expressing my sense of gratitude, by giving the name Georgium Sidus,

Georgium Sidus
——jam nunc assuesce vocari. Virg. *Georg.*

to a star, which (with respect to us) first began to shine under His auspicious reign.

"By addressing this letter to you, SIR, as President of the Royal Society, I take the most effectual method of communicating that name to the Literati of Europe, which I hope they will receive with pleasure. I have the honour to be, with the greatest respect,

SIR, Your most humble and obedient servant,

W. HERSCHEL."

If some of the expressions in this letter appear to us rather exaggerated and even fulsome, we may remember that such was the conventional expression of loyalty at the time; if we hesitate to give the King credit for "unlimited bounty" in according a salary of £200 a year to Herschel, we may also remember that that of the Astronomer Royal was no more than £300, and that the appointment carried with it no more onerous duties than to afford facilities for the Royal Family to look through his telescopes whenever they had a mind to do so. Doubtless Herschel had been given to understand that the King's bounty would not end with the payment of the quarterly salary; and in fact the King did almost immediately order five 10-foot telescopes to be made for him by Herschel, and later gave £4000 for the construction of the great 40-foot instrument.

At the time of writing this letter William Herschel was in buoyant spirits, full of schemes for future work and disposed to make light of all temporary inconveniences. He saw that he would have to reside near Windsor, and being then on the spot, began at once to look out for a suitable house. He found one at Datchet, and this he took without further consideration, as it satisfied the one condition necessary in his eyes, that of possessing a large garden and plenty of sheds for workshops. Having secured the house, he returned to Bath to superintend the removal of all his telescopes, tools and household goods. Poor Caroline had no time to reflect on her own feelings as she hurriedly packed all her own and her brother's goods into the vans. By the end of July the move was accomplished, and William and Alexander were busily employed in erecting the telescopes in the garden of the new house.

The letters which follow will show the estimation in which William Herschel and his discoveries were held in France and Germany. Two of his most diligent correspondents, Lichtenberg and Schroeter, were Hanoverians, and hailed him as a fellow-countryman, feeling a personal interest in his reputation. They naturally wrote to him in German, but though he had certainly not forgotten his native tongue, in which he seems to have habitually spoken to his sister, he never used it in correspondence, even with that sister herself, after he had fairly settled in England. He seems to have carried his objection to the language so far as to try and induce Schroeter to write to him either in English or French. The objection may have been merely the difficulty of writing and reading the German script, but underlying this there would seem to have been a definite intention to maintain an attitude of aloofness. Both Lichtenberg and Bode constantly besought him to supply them with some original information regarding his discoveries which they might publish in their periodical journals, but he passed over all such appeals in silence, and expected them to be content with the copies of his papers, printed for the Royal Society, or with some brief abstracts of such papers, which he sent them.

In May, 1782, Herschel received a letter from the French astronomer, Lalande, of which the following is a translation. (The calculations have been omitted.)

Paris. 1st May, 1782

"Sir,

"You are doubtless interested in the work of those astronomers who have observed and calculated the singular planet which you have had the good fortune and the glory of discovering. I have occupied myself specially with it and I owe you an account of my results.

"The observations made in Paris by M. Messier, M. Mechain, M. Dagelet and myself agree with the hypothesis of a circular orbit with a distance of 18·931, and a revolution of 82 sidereal years....

"At the end of this month we shall lose sight of it,—we shall see it again on the 21st of July; on the 11th October it will be stationary at $97° 57'$; the 26th December in opposition at $95° 26'$, and on 1st January 1783 at $95° 24'$ of right ascension.

"Having rendered account to you of my calculations, permit me to ask what are the dimensions of the telescope with which you discovered it? Was it one of those which you mentioned in the *Philos. Transactions* in connection with the spots on Jupiter? What diameter do you find in seconds for this planet? We find about 6" to 7".

"And as your name given to this planet will make astronomers curious about everything which concerns you, I beg that you will send me the day and place of your birth, how long a time you have spent in England, and when you began to occupy yourself with astronomy or optics? in order that I may complete with these particulars the historical memoirs which I am writing on this science.

"I await with impatience the honour of your reply and beg you to receive the assurance, &c. &c.

DE LA LANDE,[1]
of the Académie des Sciences."

Christian Mayer of Mannheim, the author of the Catalogue of fixed stars, also wrote to Herschel in August of this year. The letter is addressed: "À Monsieur, Monsieur Herschell, Gentilhomme et Astronome très célèbre, à Bath—en Angleterre".

Mayer wrote in French—the following is a translation:

Manheim. 23 August, 1782

"Monsieur,

"I do not believe that among all astronomers there is one who experienced a more lively and sincere joy than I at the news of your discovery of a new planet, of which Mr Maskelyne informed me last year. Your discovery confirms that which I myself made in my Observatory at Manheim, of which I gave a full account in my work, printed at Manheim in 1779, 'De novis in coelo sidereo phoenomenis'; and which you may see at Mr Maskelyne's.

"Having the good fortune to possess a mural circle of 8 feet diameter made by the celebrated Bird, and which I have had for six years, in my new Observatory very solidly rebuilt, it is astonishing how many new celestial objects I have perceived among the fixed stars. To give you an example; Mr Maskelyne has observed Regulus

---

[1] The true name of this French astronomer was "Joseph Jérôme le François." He adopted the surname of "De La Lande" when he was sent, in 1751, on an astronomical mission to the Court of Frederick the Great. When the Revolution broke out in France he dropped the aristocratic "De" and signed himself, at first "La Lande," and afterwards simply "Lalande," by which name he is generally known. While Napoleon was Emperor he resumed the "De". See p. 239.

99 times from 1765 up to 1775, without a companion, but I, in 1776, found it always accompanied by a little star which always preceded the principal star by 9½ seconds of time, more north than that one by one minute forty seconds. You will find, Sir, in the above-mentioned work, several similar observations; together with my clear and precise declaration that I hold it for certain that among the fixed stars there are a large number of planets, which we have hitherto taken for fixed stars. I congratulate you with so much the more pleasure and sincerity that you are the author of a truth which will make your name as immortal among Astronomers as the tribute of their well-earned gratitude....

"Mr Maskelyne...praises much your telescopes of 7, 10 and 20 feet which you make yourself after the manner of Newton; in his letter of August 2nd last he tells me that your telescopes surpass all that have hitherto been made, and that with their assistance you have discovered 400 double, or triple stars; and that also you are contributing an article to the *Philosophical Transactions* on this subject, which he promises to send me, and which I await with great impatience.

"Might I be permitted to ask you two things:

"1. Could you send me, at a price fixed by yourself, one of your telescopes?

"2. Your later observations on the new planet?

"You would oblige me by stating the price of your telescopes of different lengths.

"If you do me the honour to reply I beg that you will send your answer to M. le Comte de Haslang, the envoy of my Elector Palatine in London, who will forward your letter to Manheim without your having to pay postage. We can carry on our correspondence in future in this manner. You may write to me in French or in Latin, or in English or in German. They say that you are of German origin; which would be a cause of much pride for our nation.

"You will find at the end of the above-mentioned work a Table of double stars to the number of 70.

"Believe that I will make good use of your observations and that my gratitude unites with the respect with which I am, Sir,

Your very humble and obedient Servant,

CHRISTIAN MAYER,
Astronomer to His Electoral Highness Palatine and Bavarian".

J. H. DE MAGELLAN *to* WILLIAM HERSCHEL

London. 27 Aug. '82

"Dear Sir,

"The following article relating to your new Planet I received lately from our friend Mr Lexell of Petersburg, and as it may prove agreeable to you, I send its copy here, which runs so:

"'Quand vous verrez Mr Herschell je vous prie de lui présenter mes compliments. Voici une circonstance singulière qui regarde son nouvel Astre. Mr Bode en faisant la revision du catalogue des étoiles fixes de Mayer, a trouvé que Mayer avoit observé en 1757 une étoile dans les Poissons avec 16° 52' de long. et 48° de lat. Austr. qui ne s'y trouve plus. Or, d'après les éléments pour l'Astre nouveau il auroit dû se trouver dans ce lieu l'année 1757, et même avec une latitude telle que Mayer l'a observé; ce qui donne une preuve, si non tout à fait décisive, au moins très probable, que cet Astre est une Planète, dont la révolution est de 82 ans. Il y a outre cela une étoile observée par Ticho, qui est dans le même cas; mais pour celle-ci la chose est plus incertaine'.

"I have yesterday seen a letter from Mr Delalande of Paris directed to the Rev. Dr Shepherd, in which he says that he observed the same Planet on y$^e$ 21 July, last.—You see by this, my good friend, that you have set the European Astronomers at work on a new train of investigation; and I hope they will henceforth be more attentive on y$^e$ respective positions of y$^e$ visible stars.

"I am now about sending abroad parcels of books to different friends, viz. to *St Petersburg*, to *Berlin*, *Paris*, *Italy* and *Portugal*. Therefore, if you have a wish of sending copies of your last Memoir, printed in the present volume of y$^e$ *Philosophical Transactions*, to any Astronomer in any of those places, please to send them as soon as possible to me.

"I want very much to have a little conversation with you about y$^r$ astronomical pursuits; and beg therefore, that when you come to town, you may be pleased to take a dish of tea, on any day you may appoint, with me. Now I'll conclude this by assuring you that I am very happy to hear of your new situation being such, that you may be able to pursue y$^r$ favourite applications without any hindrance and with those conveniences of life as shall enable you the most to the same purpose.

"I ever am with y<sup>e</sup> utmost regard and true hearty friendship

Dear Sir, y<sup>r</sup> most h<sup>ble</sup> Ser<sup>t</sup><br>J. H. DE MAGELLAN."

Herschel sent a copy of his paper "On the Parallax of the Fixed Stars" to Lichtenberg, astronomer at Göttingen, and received a long letter in reply thanking him for the gift. Lichtenberg wrote (in German):

Göttingen. 12 June, 1783

"The accuracy of your observations is hitherto unheard of in astronomy; and you may believe that we in Germany are proud of your name. It has given me special pleasure to see the courage with which you undertake to examine afresh things which we had thought finished and done with.

. . . . . . .

"I have observed your new Georgium Sidus since the 6th Sept. as often as the weather permitted, not at the Observatory here, in which I have no part, but at my own house, and I have the satisfaction of knowing that I am the only person, within at least 50 to 60 German miles of this spot, who has found it. Herr Bode, a young and diligent astronomer in Berlin, has given the name of Uranus to this Star, on not bad grounds. Mayer of Manheim, whom you mention in your treatise, considers it to be about the size of Saturn; I should be curious therefore to know what diameter you assign to it.

"Now I have a request to make and I hope from your kindness you will not refuse it. Young Forster, who travelled with Capt. Cook, and I, together, are issuing a Journal entitled: 'The Göttingen Magazine of Science and Literature', which has a large circulation in Germany and some of our most learned writers have contributed to it. As all eyes are now turned upon you, you would gratify the whole German public if you would let me publish some account of the chief events of your life. I beg for this, not on my own account alone but for all those who care for the history of Astronomy. Something of the kind is sure to be brought out, and I should count it a great proof of friendship if you would allow me to be the first to make the true facts public. For I fear lest false and misleading accounts should get about.

"My God! if I had only known, when I was for a few days in Bath in October 1775, that such a man lived there! As I am no friend of Tea-rooms, nor of cards or balls, I was much ennuyé, and at last spent some of my time at the top of the tower with a field glass.

"I remember always with amusement the answer which I got from a boy who was with me on the tower; when I asked him if he knew of any one in Bath who cared about books. I said it as a joke and was rewarded with his answer: 'He knew of no one but his schoolmaster'.

"Have you got the drawings of stars and the Charts which Bode has published in Berlin? In my opinion they are the best and most complete we have.

"I have taken the liberty of enclosing a letter for Mr Magellan.

And remain,

Your obed$^t$ serv$^t$,

LICHTENBERG".

### KÖNIGLICHER OBERAMTMANN SCHROETER *to* WILLIAM HERSCHEL

(Translated from the German)

Lilienthal. Jan. 1783

"Well-born and highly honoured Sir,

"As I have had for several years the honour, as a practical admirer of musical art, of the acquaintance of your brothers in Hannover I hope that I am not making a mistake if I introduce myself to your friendly notice as an even more eager admirer of astronomy, which has been for many years my favourite science; and venture to make a request which you will surely not reject both out of common humanity but still more out of sympathy with a fellow-astronomer in difficulties."

Schroeter goes on to explain at great length that he possesses a good Newtonian reflector of 4 feet focal length, in very good order except that the mirror is unfortunately broken. He begs Herschel to tell him of an instrument-maker in London who could make him a new one. Also on behalf of Professor Bode as well as himself if a 7-foot telescope, similar to that with which Herschel discovered the new planet, is procurable; and Herschel's paper on double stars, is it printed, can it be bought? He concludes by asking:

"Is it true that you have given the name of George, or Sidus Georgium as the newspapers have it, to the new planet, which in Germany from the first has been called Uranus? It is to be hoped that his Majesty has presented you entirely to astronomy and provided you with a handsome pension. With all lovers of astronomy I wish this with my whole heart".

NOTE. The true name of the French astronomer who corresponded so diligently with William Herschel was Joseph Jérôme le François. He adopted the surname of *De la Lande* when he was sent, in 1751, on an astronomical errand to the Court of Frederick the Great. When the Revolution broke out in France, in 1789, he dropped the aristocratic *De* and signed himself henceforth *La Lande* or *Lalande*.

## CHAPTER IX

### 1782   1786

William and Caroline settle at Datchet. Caroline's Autobiography. Damp at Datchet. William has ague. Making the large 20-foot telescope. Move to Clay Hall. The King's gift of £2000 for the 40-foot telescope. Move to Slough. Mrs Papendiek's Journal.

IN the summer of the year 1782 William Herschel and his sister were definitely settled near Windsor and preparing to take up their new life. Music was entirely set aside, to be considered only as a recreation, and every waking moment was to be dedicated to astronomy.

William Herschel was now in his forty-third year, and though actually nearly as long a span of life lay before him, he could not be aware of this, and felt that no instant must be lost if he would carry out his ambitious projects of research. The most immediate need was for a larger and better telescope.

Caroline gives a full account of the removal from Bath in her reminiscences:

"1782. In the last week of July he (William) came home and immediately prepared for removing to Datchet, where he had taken a house with a garden and grass plot annexed quite suitable for the purpose of an observing place. Sir W$^m$ Watson spent nearly the whole time at our house, and he was not the only friend who truly greaved at my Brother's going from Bath, or feared his having perhaps agreed to no very advantageous offers; their fears were in fact not without reason, for my B.'s desire for returning to the quiet enjoyment of observing and his optical works, had, by having been separated from them so long, been raised to such a pitch that the prospect of entering again on the toils of teaching &c. which awaited him at home (for the months of leisure were now almost gone by) appeared to him an intolerable waste of time, and by way of alternative he chose to be Royal Astronomer with a salary of 200 pounds a year. Sir W$^m$ was the only one to whom the sum was mentioned, who exclaimed, 'Never bought Monarch honour so cheap!' To every other inquirer my Brother's answer was that the King had provided for him.

"In the course of 4 or 5 days all the household goods, Telescopes and most of the contents of the work-rooms stud ready loaded in the evening of the 30th of July for going off at two in the morning; my eldest Brother left Bath with the 11 o'clock coach which goes through Slough, where Alex. and I joined him on the 1st of Aug$^t$, from which after having taken our dinner with the rest of the passengers we walked over to Datchet where we were obliged to sleep at the public house near the Church, and were glad to see on our first awaking the wagon with all the Tubes &c., &c., safely arrived, and we left our Inn to see the wagon unloaded and the furniture to be put in our *new house*; or rather the ruins of a place which had once served for a hunting seat to some Gentleman and not been inhabited for years. It was more than two months that we were obliged to move about amongst brick and morter surrounded by labourers. The garden wanted trenching 4 feet deep to destroy the weeds.

"My Brothers were at first quite delighted with all the conveniences of the place. The stables were to serve as a place for ruff grinding mirrors. A very roomy Laundry was to be converted into what we called the Library of which one door opened on a large grass-plot where the small 20 ft. was to be erected and in looking about for a proper situation poor Alex. narrowly escaped going down a Well which was grown over with weeds, and they found it necessary to call in the assistance of the Gardener to mow down the whole that they might see what ground they had to step upon.

"The greatest hindrance towards making ourselves comfortable was our having brought no servant with us, for my Brother on the recommendation of the King's Upholsterer had engaged one to be ready to receive us, but on inquiring the reason of her absence, it was found that she was in prison for teft, and it was a fortnight before we could get one with a good Character, the whole intermediate time I could not get sight of any woman but the wife of the Gardener whom my Brother had then only on trial and was yet a stranger but afterwards remained (till he died of the smallpox) and proved a very steady attendant at night at the working handle of the telescope. This old woman could be of no further service to me than shewing me the shops, and the first time I went to make markets I was astonished at the dearness of every article and saw at once that my Brother's scheme of living cheap in the country (as jokingly he said) on Eggs and Bacon would come to nothing; for at Bath I had the week before bought from 16 to 20 Eggs for 6d, here I could get no more than five for 4d, Butcher's meat was 2½d and 3d per pound dearer; and the only Butcher at Datchet would besides not give honest weight and we were obliged to deal, all the time we lived there, at Windsor. Coals were more than double the price. O dear! thought I, what shall we do with 200 a year! after Rent and Taxes are deducted; the latter were in consequence of the overcharged rent (and there being upwards of 30 Windows on the premises) enormous, and notwithstanding all the labour and expense bestowed on it; it remained an incomfortable habitation.

"The beginning of October Alexander was obliged to return to Bath; the separation was truly painful to us all, and I was particularly affected by it; for till now I had not had time to consider the consequence of giving up the prospect of making myself independent by becoming (with a little more uninterrupted application) a useful member to the musical profession. But besides that my Brother would have been very much at a loss for my assistance, I had not impudence enough for throwing myself on the public after losing his protection.

"Poor Alexander we had hoped at first to persuade to change Bath for London, where he had the offer of the most profitable engagements and we should then have had him near us. My eldest Brother at Hannover too, feared he would not be happy at Bath and wished him to accept a place in the Orchestra at Hannover at that time vacant; but he refused all, and before we saw him again the next year he was married (wretchedly) and we saw him never otherwise but discontented after this our separation.

"Much of my Brother's time was taken up with going when the evenings were clear to the Queen's Lodge to shew the King, &c. objects through the 7 feet. But this was soon (when the days began to shorten) found impossible to be continued for the telescope was oftens (at no small expense and risk of damage) obliged to be transported in the dark to Datchet for the purpose of spending the rest of the night with observations on Double stars for a second Catalogue, and my Brother was besides obliged to absent himself for a week or 10 days for the purpose of bringing home the metal of the cracked 30 feet mirror and the remaining materials of his workroom; but before the furnace was taken down at Bath a second 20 feet mirror 12 inch dia$^r$ was cast which happened to be very fortunate, for on the 1st of Jan.y 1783, a very fine one cracked by frost in the tube. (I remember to have seen the Therm. $1\frac{1}{2}$ degrees below Zero for several nights in the same year.)

"I had the comfort to see that my Brother was satisfied with my endeavours in assisting him when he wanted another person, either to run to the Clocks, writing down a memorandum, fetching and carrying instruments, or measuring the ground with poles, &c., &c., of which something of the kind every moment would occur. For the eagerness with which the measurements of the G. Sidus, and observations on Mars, D. Stars, &c., &c., were made was incredible, which may be seen by the various papers that were given to the Royl. Soc. in 1783; which papers were written in the daytime or when cloudy nights interfered; besides which the 12 inch speculum was perfected before the spring. And many hours were spent at the turning bench, as not a night clear enough for observing ever passed but that some improvements were planned for perfecting the mounting and motions of the various instruments then in use, and some trials of new constructed eye-pieces to be made which mostly all were to be made by my Brother's own hands.

"Through wishing to save his time he began to have some work of that kind executed by a watchmaker who had retired from business and lived on Datchet Common; but the work was so bad and the charges so unreasonable that he could not be employed; and it was not till some time after that in his frequent visits to the meetings of the R. Society (made in the moonlight nights) he had an opportunity of looking about for Mathematical Workmen, Opticians and Foundries, &c. But all what was done in Town seldom answered expectation, and was reserved to be executed with improvements by Alex, during the few months he spent with us.

"1783. The sommer months passed in the utmost activity for getting the large 20 feet ready against the next winter. The Carpenters and Smiths of Datchet were in daily requisition, and as soon as patterns for tools and mirrors were ready my Brother went to Town to have them cast, and during the three or four months that Alex could be absent from Bath, the mirrors and optical parts were nearly completed.

"My Brother began his series of Sweeps when the Instrument was yet in a very unfinished state, and my feelings were not very comfortable when every moment I was alarmed by a crash or fall;

THE TWENTY-FOOT TELESCOPE
*From a water-colour sketch made probably at Datchet*

knowing him to be elevated 15 or 16 feet on a temporary crossbeam instead of a safe gallery. The ladders had not even their braces at the bottom and one night in a very high wind he hardly had touched the ground before the whole apparatus came down; some neighbouring men were called up to help extricating the mirror which was fortunately uninjured; but much work for Carpenters was cut out for next day.

"That my fears of danger and accidents were not wholly imaginary I had an unlucky sample of on the night of the 31st of Dec$^r$. The evening had been cloudy but about 10 o'clock a few stars became visible, and in the greatest hurry all was got ready for observing. My Brother at the front of the Telescope directing me to make some alteration in the lateral motion which was done by a machinery in which the point of support of the tube and mirror rested (it is described somewhere as belonging to the small 20 feet). At each end of the machine or trough was an iron hook such as butchers use for hanging their joints upon, and having to run in the dark on ground covered a foot deep with melting snow, I fell on one of these hooks which entered my right leg about 6 inches above the knee; my Brother's call 'make haste' I could only answer by a pittiful cry 'I am hooked'. He and the workman were instantly with me, but they could not lift me without leaving near 2 oz. of my flesh behind. The workman's wife was called but was afraid to do anything, and I was obliged to be my own surgeon by applying aquabaseda and tying a kerchief about it for some days; till Dr Lind hearing of my accident brought me ointment and lint and told me how to use it. But at the end of six weeks I began to have some fears about my poor limb and had Dr Lind's opinion, who on seeing the wound found it going on well; but said, if a soldier had met with such a hurt he would have been entitled to 6 weeks' nursing in a hospital.

"I had however the comfort to know that my Brother was no loser through this accident, for the remainder of the night was cloudy and several nights afterwards afforded only a few short intervals favourable for sweeping until the 16th of Jan$^y$ before there was any necessity for exposing myself for a whole night to the severity of the season.

"I could give a pretty long list of accidents of which my Brother as well as myself narrowly escaped of proving fatal; for observing with such large machineries, where all round is in darkness, is not unattended with danger; especially when personal safety is the last thing with which the mind is occupied at such times; even poor Piazi did not go home without getting broken shins by falling over the rack-bar which projects in high altitudes in front of the telescope when in the hurry the cap had been forgotten to be put over it.

"In the long days of the sommer months many 10 and 7 feet mirrors were finished, there was nothing but grinding and polishing to be seen. For 10 feet several had been cast with ribbed backs by way of experiment of reducing the weight in large mirrors. In my leisure hours I ground 7 feet and plain mirrors from ruff to fining down and was indulged with polishing and the last finishing of a very beautiful mirror for Sir W$^m$ Watson".

In Zach's *Monatliche Correspondenz* for 1802 there is a letter from Von Magellan to Bode, quoted in Holden's *Life* of William Herschel, describing a visit to Datchet, of which the following extract is a translation:

"I spent the night of the 6th January at Herschel's, in Datchet, near Windsor, and had the good luck to hit on a fine evening. He has his twenty foot Newtonian telescope in the open air and mounted in his garden very simply and conveniently. It is moved by an assistant who stands below it.... Near the instrument is a clock regulated to sidereal time.... In the room near it sits Herschel's sister and she has Flamsteed's Atlas open before her. As he gives her the word, she writes down the declination and right ascension and the other circumstances of the observation.... I went to bed about one o'clock, and up to that time he had found that night four or five new nebulae. The thermometer in the garden stood at 13° Fahrenheit; but in spite of this, Herschel observes the whole night through, except that he stops every three or four hours and goes into the room for a few moments. For some years Herschel has observed the heavens every hour when the weather is clear, and this always in the open air, because he says that the telescope only performs well when it is at the same temperature as the air. He protects himself against the weather by putting on more clothing. He has an excellent constitution and thinks about nothing else in the world but the celestial bodies. He has promised me in the most cordial way, entirely in the service of astronomy and without thinking of his own interest, to see to the telescopes I have ordered for European observatories, and he himself will attend to the preparation of the mirrors".

It added much to the hardships endured by William and Caroline Herschel that the first winter at Datchet should have been so exceptionally cold. The situation of the house on the marshy banks of the Thames was the most unhealthy imaginable for one whose occupation kept him out at night. William Herschel found that damp air, and even fog, did not interfere with telescopic observations. In a paper which he contributed to the *Philosophical Transactions* some years later, he gave some particulars from his own experience of the effect of moisture in the air, of fog, frost, etc. on telescopic vision; these particulars unintentionally reveal with what extraordinary hardihood he went on gazing through his telescope under the most trying conditions of weather, especially while he lived at Datchet.

Two extracts from this paper will be sufficient:

"*Sept.* 7, 1782. I viewed the double star preceding 12 Camelopardis with 932. In this, and several other fine nights which I have lately had, the condensing moisture on the tube of my telescope has been running down in streams; which proves that damp air is no enemy to good vision.

"*Dec.* 28, 1782. 17$^h$ 30'. The water condensing on my tube keeps running down; yet I have seen very well all night. I was obliged to wipe the object glass of my finder almost continually. The Specula however are not in the least affected with the damp. The ground was so wet that, in the morning, several people believed there had been much rain in the night, and were surprised when I assured them there had not been a drop".

Sir John Herschel has recorded that his father, when observing at Datchet, "when the waters were out round his garden, used to rub himself all over, face and hands &c., with a raw onion, to keep off the infection of the ague, which was then prevalent; however he caught it at last".

Even William Herschel's robust constitution could not stand such exposure. After three winters at Datchet he succumbed to the attacks of ague, and was seriously ill. He went to London to consult the father of his friend, Dr Watson. (Sir William Watson, senior, was Physician to the King and highly reputed.) A letter referring to his prescription for the treatment of Herschel's ague has been preserved.

SIR WILLIAM WATSON, SENIOR, *to* WILLIAM HERSCHEL

28 Mar. '85

"Dr Watson presents his compliments to Mr Herschell. He called this morning, as he mentioned he would, in Leicester fields, but found Mr H. removed, to whose house Mr Campbell did not know.

"He is pleased that Mr H. has missed his ague fit. He would have him proceed to take the second two ounces of *Red bark* as he did the former.

"During the taking the bark, in such frequent doses, he should abstain almost from solid food; and is advised to take as nourishment mutton broth and bread; and between whiles, wine negus and toast.

"Dr Watson proposes to call in Jermyn Street about eleven to-morrow."

Dr Watson doubtless advised a change of residence, and another, to him more important consideration, now decided Herschel to move from Datchet to a drier spot and one more suitable for the erection of the giant telescope which was the goal of his ambition.

Amongst the mass of apparatus and material which he had brought from Bath were three completed telescopes for his own use. These were the 7-foot one, with which he had discovered Uranus; one of 10 feet; and one of 20 feet focal length; the latter with a mirror 12 inches in diameter. This one was known in the family as the *small* 20-foot, to distinguish it from the large 20-foot which he immediately proceeded to make, and which was finished and in use by the end of the year 1783. The mirror of the large 20-foot measured 18·8 inches in diameter.

This telescope turned out to be much the most efficient of all those made and employed by Herschel; the mirror, which weighed 128 lb., was not too heavy for easy movement, and the length of the tube was manageable with the kind of mounting which he devised for it. Also the observer's seat was not at quite such a dangerous elevation as was the case with the huge 40-foot. It was with this instrument that William Herschel made most of his regular observations at Slough, and with its successor his son carried on his father's work, first at Slough and afterwards at the Cape of Good Hope.

Besides constructing this telescope, William Herschel was busily employed at Datchet in making telescopes for the King, and in fulfilling other orders which began to pour in when to possess one of Herschel's telescopes became a fashion. He and his brother continued to do all the delicate work for the telescopes themselves. Alexander, being especially expert in brass-work, probably undertook all the turning of eyepieces, screws, etc., on the lathe which they had brought from Bath (and which is still preserved at Slough). William attended to the shaping and polishing of the mirrors. Every mirror which he sold with a telescope was polished by his own hand. By much practice at Bath he had acquired an almost instinctive skill in producing the necessary curvature. This had not been done without exhaustive experiments, following other men's methods as well as those suggesting themselves to him in practice.

"Every possible combination of strokes was tried; the 'glory stroke', directing the strokes over the centre of the polisher to different points of the compass like the strokes of a wheel; the 'eccentric stroke', avoiding the middle of the polisher; &c.... It is not to be wondered at that he did not succeed in laying down any hard and fast rule for polishing mirrors by hand; every speculum seemed to require different treatment."[1]

In the introduction to the "Description of a Forty-feet Reflecting Telescope", read before the Royal Society in 1795, William Herschel said:

"When I resided at Bath...in my leisure hours, by way of amusement, I made for myself several 2 feet, 5 feet, 7 feet, 19 feet, and 20 feet Newtonian telescopes; besides others of the Gregorian form, of 8 inches, 12 inches, 2 feet, 3 feet, 5 feet, and 10 feet focal length. My way of doing these instruments at that time, when the direct method of giving the figure of any of the conic sections was unknown to me, was to have many mirrors of each sort cast, and then to finish them all as well as I could; then to select by trial the best of them, which I preserved; the rest were put by to be repolished. In this manner I made not less than 200, 7 feet; 150, 10 feet; and about 80, 20 feet mirrors".

Dr Dreyer does not think that Herschel intended to say that all these mirrors were made before he left Bath. It is rather the enumeration of

[1] See Dr Dreyer's Introduction to the *Collected Scientific Papers of Sir William Herschel*, p. xxvi.

all the mirrors shaped by himself by hand. He left an autograph list of some sixty telescopes made by himself for sale, but no dates are given; the years when the most important ones were made can be traced through letters, but we cannot assign dates to all the 7 and 10-foot ones, many of which were made during the three years when he lived at Datchet. Besides the five 7-foot ones ordered by the King, he made a similar one for the Royal Observatory at Greenwich and a mirror for the German astronomer Schroeter, who owned a private observatory at Lilienthal, near Hanover. Schroeter was already acquainted with William Herschel's two musical brothers, Jacob and Dietrich, and exhibits in his letters, which are very numerous and effusive, an unbounded admiration for William himself. He earnestly desired to have a telescope like the one with which Herschel had discovered Uranus. It is characteristic of William's solicitude for his brothers that he suggested to Schroeter that, as a 7-foot telescope of his own making would be very expensive, he should get the mirrors and eyepieces made under his own direction by his brother, whom he had trained to follow his methods; the cost would then be twenty-two guineas. Schroeter accepted the offer, and was so well pleased with his purchase that he procured an order for a similar telescope from Bode at Berlin, and several commissions for mirrors and eyepieces for William's brother. But Alexander did not take to telescope-making seriously, and after his marriage does not seem to have given much regular help.

Besides all this manual work, which must have been sufficiently exhausting, William Herschel's active brain was continually devising improvements in the mounting of his telescopes. The ingenuity and knowledge of mechanics displayed in these is very remarkable. The stand upon which the small 20-foot had been mounted at Bath was, as Herschel himself described it,

> "singular but very simple. It would enable me, in a long narrow garden behind my house, the south view of which was clear of buildings, to observe the heavens in the meridian from about 10° above the horizon up to the zenith. With my 20-feet telescope I use a long pole, to the top of which is fastened a very short arm holding a set of pulleys, and when the telescope is elevated by them I use a movable ladder with a back, to mount up to the eye-piece".

From this very simple beginning we can, if we wish, follow the gradual evolution of the gigantic structure of spars and ladders which supported the tube of the great 40-foot.

With every finished telescope which William Herschel sold he sent also an elaborate list of all the parts, with directions for assembling these together. This pamphlet was copied out for each purchaser, no doubt by the indefatigable Caroline in her neat round hand; but however legible and clear the directions, it is to be feared that they were not always read through. Dr Burney supplies an amusing example of this negligence in a letter to his daughter quoted in his Memoirs.

15th Sept. 1799

"Now I must give you a little episode. Canning told me Mr Pitt had gotten a telescope, constructed under the superintendence of Herschel, which cost 100 guineas; but that they could make no use of it, as no one of the party had knowledge enough that way to put it together; and knowing of my astronomical poem, Canning took it for granted that I could help them. The first day I went to Walmer Castle I saw the instrument, and Canning put a paper in my hand of instructions; or rather, a book, for it consisted of twelve to fourteen pages; but before I had read six lines company poured in, and I replaced it in the drawer whence Canning had taken it; and to say truth, without much reluctance, for I doubted my competence. I therefore was very cautious not to start the subject; but when I got to Dover I wrote upon it to Herschel, and received his answer just in time to meet the Dover visit of Mr Pitt. It was very friendly and satisfactory as is everything that comes from Herschel."

After three years' residence at Datchet, William Herschel moved, in June, 1785, to a house called Clay Hall, near Old Windsor. But his financial aspect at this time was not bright; he had exhausted all the savings from his musical profession and it was only by the most arduous exertions in making telescopes for sale that he was able to meet the expenses of his household and workshops, besides the allowance regularly paid to his mother. Caroline's reminiscences give a moving picture of the troubles and disappointments which fell to their lot before they came to a final anchorage at Slough. She wrote in her Autobiography:

"Sir William Watson, who oftens (in the lifetime of his Father) came to make some stay with us at Datchet, saw my Brother's difficulties and expressed great dissatisfaction. And on his return to

Bath he met among the Visitors there, several belonging to the Court (among the rest Madam Schwellenberg) to whom he gave his opinion concerning his Friend and his situation very freely. In consequence of which, very soon after, my Brother had, through Sir J. Banks, the promise that 2000 pounds would be granted for enabling him to make himself an Instrument.

. . . . . . .

"Now immediately every preparation for beginning the great work commenced. A very ingenious Smith (Campion), who was seeking employment, was secured by my Brother. A workman for the brass work of the optical parts was engaged, and two Smiths were at work throughout the summer on different parts for the 40 feet telescope; and a whole troop of labourers were engaged with the iron tool to the figure for the mirror to be ground upon, (for the polishing and grinding by Machines was not begun till about the ending of 1788). These heavy articles were cast in Town and caused my Brother frequent journeys to London, and were brought by water as far as Windsor.

"But here again was an interruption and no small time was lost by moving and fixing so many heavy articles; for when my Brother was going to lay the foundation for the 40 ft. he found that he had to deal with a litigious woman (Mrs Keppel) who told him he must expect to have the rent raised every year according to the improvements he was making on the premises; on which he thought it advisable to look out for another suitable spot on which a running lease could be obtained, and the house and garden at Slough being to be let, he moved there on the 3rd of April 1786, and here I must remember, that among all this hurrying business each moment after daylight in clear weather was allotted to observing. The last night at Clay Hall was spent in Sweeping till daylight, and the next the Telescope stud ready for observation at Slough. But on such occasions I was oftens put in mind of the old saying 'the more hurry the worse speed' for at Slough no steady out-of-door workman for the Sweeping handle could be met with, and a man-servant was engaged as soon as one could be found fit for the purpose. (Ever since we left Datchet many out, and some of indoor servants had been tried but none had hitherto been found sober and steady enough for the purpose.) Meanwhile Campion assisted, but many memorandums were put down 'Lost a Neb by the blunder of the person at the handle'. If it had not been sometimes for the intervention of a cloudy or moon-light night, I know not when my Brother (or I either) should have got any sleep; for with the morning came also his work-people of which there were no less than between 30 or 40 at work for upwards of 3 months together, some employed with felling and roothing out trees, some digging and preparing the ground for the Bricklayers who were laying the foundation for the Telescope, and the Carpenter in Slough with all his men. The Smith meanwhile was converting a wash-house into a Forge, and manufacturing complete sets of Tools required for the work he was to enter upon; and many expensive tools were furnished by the Ironmongers in Windsor as well for the Forge as for the turner or brassman. In short it was at one time a complete workshop for making optical instruments, and it was a pleasure to go into it to see how attentively the men listened to and executed their master's orders; for doing this I had frequent opportunities when I was obliged to run to him with my papers or Slate when stopped in my work by some doubt or other".

In the memoir of William Herschel written by François Arago, the French astronomer, he referred in the following words to the house at Slough:

"On peut dire hardiment du jardin et de la petite maison de Slough que c'est le lieu du monde où il a été fait le plus de découvertes. Le nom de ce village ne périra pas; les sciences le transmettront religieusement à nos derniers neveux".

This house was now to be William Herschel's home for the rest of his life and the birthplace of his equally famous son.

There are some interesting references to the Herschel family in the Journal of Mrs Papendiek, whose father and husband were both Pages at the Court of George III. Mrs Papendiek's father, Mr Albert, had been for some years the tenant of the house, and she mentions that there was

"a very pretty garden of one acre, at the end of which was a gravel walk, with a row of high elms on each side. On the side next to Windsor, from whence there was an imposing view of the north side of the Castle, was a raised terrace with a few trees, just enough to break the scorching rays of the sun without impeding the view.

"1785. We had all made the acquaintance this summer of Dr and Miss Herschel, who were living on Datchet Common, and it was finally settled that he should take the house at Slough when my father's three years' lease was up: so at next Ladyday (1786) he established himself in this pretty place, much to my satisfaction.

"His first step, to the grief of everyone who knew the sweet spot, was to cut down every tree, so that there should be no impediment to his observations of the heavenly bodies. Then he erected in the centre of his garden his wonderful twenty-feet telescope. Every movement of this ponderous machine was easily accomplished by two persons, even when the mirror was in the tube, the interior of which measured five feet in circumference. The gallery for spectators held six people, and Dr Herschel had a seat which was moved up and down at pleasure by himself. At the bottom of the erection were two small rooms, one for Miss Herschel to write down the observations as they were made by her brother, the other for the man who assisted in the movements required. Company and friends were never denied admittance to view this extraordinary piece of mechanism, nor in the evening to look at the moon and planets, either through the large telescope, or another of ten feet also fixed in the garden.

"The stabling had been converted into a small dwelling-house, where Miss Herschel had her apartments and study contiguous to the smaller telescope. The stable yard was now a pretty garden, on one side of which were buildings for chemistry, rooms for polishing mirrors, &c. These mirrors, both for ten feet and five feet, were sent all over the continent, many of them to Catherine, the renowned Empress of Russia, and by this traffic Dr Herschel established his fame as one of the greatest mechanics of his day, and also set himself at ease in pecuniary matters".

After the lapse of more than a hundred years the garden has suffered much change, yet it retains something of its old character. The terrace walk has disappeared and the view of the Castle is blotted out by modern buildings; flower-beds mark the site of the great 40-foot telescope, but the Cottage, where Caroline spent so many hours at work, still seems to smile through its casement windows at the grass plot where the large 20-foot telescope stood.

## CHAPTER X

## 1786  1789

William Herschel goes to Göttingen—his letters to Caroline. Caroline discovers her first comet. Letters to Blagden and Aubert. William returns to Slough. Making the 40-foot telescope. The King gives another £2000. Discovery of the satellites of the Georgium Sidus, and of two new ones of Saturn. History of the 40-foot telescope.

NO sooner was William Herschel fairly settled at Slough than he was obliged to proceed to Göttingen, by the King's desire, to deliver a telescope of his own making as a present to the University. He left Slough on the 3rd July, accompanied by his brother Alexander, whose wife came with him from Bath and remained at Slough during his absence.

The travellers reached Dover on the 5th July and got to Hanover on the 14th, which seems to have been considered a satisfactorily expeditious journey. William wrote several letters to his sister, all of them short and none of them signed. This peculiarity marks all his letters to his relations, though he sometimes concluded with the words "your... brother". The following ones give all the information which Caroline received about his journey.

WILLIAM HERSCHEL *to* CAROLINE

Dover. Wednesday, July 5th, 1786

"Dear Lina,

"We are safely arrived here at Dover and shall go over the water in the evening with the Mail; the wind is not extraordinary but will do pretty well and the weather is very fine. To-morrow night we shall probably set off on the post waggon and as soon as the post sets out from Holland we will write again, but I believe that cannot be till Saturday as I hear the foreign post does not return till that day, so that it will be next Tuesday or Wednesday before you can hear again from us. When you write I suppose the Griesbachs will take care of the letters for you, and by way of soon having the pleasure of hearing from you, please sit down immediately and write me a tolerable-good-long Letter. Direct to me as follows:

"To Dr Herschel at Mr Herschel's in Hanover. An English direction will do very well. Farewell."

(No signature.)

"Dear Sister,                  Hannover. Friday, 14 July, 1786

"This morning we arrived safely at Hannover. We are a little tired but perfectly well in Health; we travelled extra post all the way through very bad roads. The post is going out in a very little time so that I write in a hurry so that you might hear from us so much sooner. After a night or two of sleep here (by way of recovery) I shall go on to Göttingen; but when I have collected my thoughts better together I will write more. Mama is perfectly well and looks well. Jacob looks a little older but not nearly so much as I expected. In Sophy there is hardly any change but a few white hairs on her head. John[1] is just the same as before; his little boy seems to be a charming creature. Farewell, dear Lina, I hope we shall see you again in a few weeks. I must finish for Alexander to write. Adieu once more."

"Dear Lina,                  (Beginning of August, '86)

"We are still in Hannover and find it a most agreeable place. I have been in Göttingen where Jacob went along with me, and the King's telescope arrived there in perfect order. The Society of Göttingen have elected me a member. We long very much to hear from you as we have never had a letter yet. This is the 4th we have sent you and we hope you received the former ones. This day fortnight we have fixed for our getting out from this place and be assured that we shall be happy to see *old England* again, tho' *old Germany* is no bad place. Yesterday and the day before I have seen the Bishop of Osnabrügge and the Prince Edward.

"If an inquiry should be made about our return you may say (I hope with truth) that we shall be back by about the 24th of August.

Adieu Lina."

"Dear Lina,                  August, 1786

"This is to be the last letter we write from Hannover and therefore you will not receive any more till you see us. We intend to return by

[1] Johann Dietrich Herschel.

way of Calais, which will take us two or three days longer than we were coming hither. But we still hope to be with you the 21, 22 or 23 of this month; the day of our setting out is not quite fixed but will be perhaps—Monday or Tuesday. I hope no accident has happened to you or my sister[1] at Slough. Alexander longs to come home as much as I do that we may go on with the great work; otherwise I think this is a very pleasant place tho' I should not wish to spend much time here. The 40 feet I hope is not hurt by any accident or otherwise; polishing the speculum will be the first work we undertake. Have you found any comet since we left you? There have been some fine nights.

Adieu."

While William Herschel is absent in Hanover it may be a favourable moment to consider Caroline's position as an astronomer. Though she shines chiefly as a satellite to her more resplendent brother, she has claims to recognition as an observer on her own account. In the course of her life she discovered eight comets, as well as many nebulae never before mentioned or catalogued. She kept a careful Journal of her observations, which were made on a methodical plan; one portion of this Journal, together with three of her neatly stitched notebooks, in which she had entered the more noteworthy items in the Journal, she sent to her nephew long afterwards from Hanover. Some extracts from these will show how she worked.

As soon as they were settled at Datchet, her brother put into her hands a small refracting telescope, with the encouraging suggestion that she should hunt for comets. The inducement did not at the moment appeal very brightly to Caroline, whose spirits were sadly depressed by the discomforts and difficulties of their domestic arrangements and her gloomy forecast of what life on £200 a year would mean. However, with her wonted docility, she set to work at once, though at first she could not have much assistance from her brother, who was obliged, during this summer and autumn, to spend many evenings at Windsor, besides having to return to Bath for a time to fetch away the remainder of his instruments.

"I was in his absence from home", she wrote, "left solely to confuse myself with my own thoughts, which were anything else than cheerful;

[1] Mrs Alexander Herschel.

for I found I was to be trained for an assistant Astronomer; and by way of encouragement a Telescope adapted for sweeping, consisting of a Tube with two glasses such as are commonly used in a finder, was given to me. I was to sweep for comets, and by my Journal No. 1, I see that I began Augt. 22, 1782, to write down and describe all remarkable appearances I saw in my sweeps (which were horizontal). But it was not till the last two months of the same year before I felt the least encouragement for spending the starlight nights on a grass-plot covered by dew or hoar frost without a human being near enough to be within call; for I knew too little of the real heavens to be able to point out every object for finding it again without losing too much time by consulting an Atlas. But all these troubles were removed when I knew my Brother to be at no great distance making observations with his various Instruments on D. Stars, planets, &c.; and could have his assistance immediately when I found a Nebula or Cluster of Stars, of which I intended to give a Catalogue, (but at the end of 1783 I had only 14, when my sweeping was interrupted by being employed with writing down my Brother's observations with the large 20-feet)."

Her "Book of Observations" shows that she observed very diligently all through the year 1783. In February she was rewarded with the discovery of a nebula not mentioned in Messier's list (published in the *Connaissance des Temps*), which she kept always beside her for reference. This nebula and two more which she found in August and September of this year are included in William Herschel's "Catalogue of One Thousand New Nebulae", where in a note he gives his sister the credit of their discovery. In July, 1783, he gave her a better instrument, which he describes in the afore-mentioned note as:

"An excellent small Newtonian *Sweeper* of 27 inches focal length and a power of 30".

In her Autobiography she wrote:

"*July* 8, 1781 (a mistake for 1783). I began to use the new Newtonian small sweeper.... But it could hardly be expected to meet with any Comets in that part of the heavens where I swept, for I generally chose my situation by the side of my Brother's instrument that I might be ready to run to the clock or to write down memorandums. But in the beginning of December I became entirely attached to the writing desk and had seldom an opportunity of using my newly acquired Instrument".

When they removed to Slough, better arrangements were made for Caroline's observing. The rooms connected with the stables had been converted into a dwelling-house;[1] the harness room became a writing room and the room or rooms above, Caroline says, were now "Our Observatory". A small staircase led out on to the roof, which was flat, and here the small Newtonian sweeper was installed. To assist her in recording the time of her observations, her brother Alexander had made for her a sort of metronome which ticked seconds loudly; this instrument was known in the family as the "Monkey Clock". Caroline thus describes it in a letter, written long years after to her nephew, from Hanover:

"You mention a Monkey clock or Jack in your paper. I would only notice (if you mean the Jack in the painted deal case) that Alex made it merely to take with me on the roof when I was sweeping for comets, that I might count the seconds by it, going softly downstairs till I was within hearing of the beat of the timepiece on the 1st Floor (at that time our Observatory); all doors being open. Your Father never used it except when refining colcothar or when polishing the 40 feet, seconding the stroke".

Being thus provided with a comfortable observing station, we can imagine that Caroline found some consolation for her brother's absence in the leisure which she might now devote to her "sweeps". Her hope of having something of interest to show him on his return was not disappointed, for before he reached England again she had discovered her first comet.

But though the nights, when fine, were devoted to astronomy, there was no diminution of the assiduity with which she attended to her daily tasks. Her brother had left her in charge to pay the wages of the workmen, and her Journal shows how much she worried herself in her anxiety to see that these were honestly earned. It is worth the space to give a few extracts from her Autobiography to show how this indefatigable woman spent her days.

[1] This building is now known as the Observatory House Cottage.

## CAROLINE HERSCHEL'S AUTOBIOGRAPHY

"1786. I will now return to July 1786, when my Brother was obliged to deliver a 10 feet Telescope as a present from the King to the Observatory of Göttingen. Before he left Slough (which was July 3), the stand of the 40 ft. Telescope stud on two circular walls capt with portland stone,—(which cracking by frost have been since covered with oak)—ready to receive the Tube. The Smith was left to continue to work at the tube which was sufficient employment cut out for him before he should want farther direction. The mirror[1] was also pretty far advanced and ready for the polish; for I remember to have seen 12 or 14 men dayly employed in grinding or polishing, and that a polisher was left ready cast.

"To give a description of the task (or rather tasks) which fell to my share, the readiest way, I think, will be to transcribe a few days' transactions out of a day book which I began to keep at that time, and called '*Book of Work done*'.

"*July* 3. My Brothers Wm. and Alex. left Slough to begin their journey to Germany. Mrs Alex. Herschel was left with me at Slough. By way of not suffering too much by sadness, I began with bustling work. I cleaned all the brass work for 7 and 10 feet telescopes and put curtains before the shelves to hinder the dust from settling upon it again.

"*July* 5. I spent the morning in needlework. In the afternoon went with Mrs Herschel to Windsor. We chose the hours from 2 to 6 for shopping and other business on purpose to be from home at the time most unlikely of any persons calling; but there had been four foreign Gentlemen looking at the instruments in the garden, but did not leave their names.

"In the evening Dr and Mrs Kely (Mr Dollond's daughter) and Mr Gordon came to see me.

"*July* 8. Paid the Smith and the Gardener who worked 5 days this week. N.B. I had a deal of trouble with this fellow; for my orders were not to employ him more than 3 days just to keep the grass-plot in order, for at that time there was but little garden ground, and he had been told so by my Brother; but he would be idling about the premises and gave me the name of stingy in the village, because I objected to his being there when he was not wanted.

---

[1] This mirror, the first one made for the 40-foot, is now at the Science Museum, South Kensington.

"*July* 12. I put paper in a press for a Register and calculated for Flamsteed's Catalogue.

"Mem. When Flamsteed's Cat. was brought into zones in 1783, it was only taken up at 45° from the Pole; the apparatus not being then ready for sweeping in the zenith.

"By July 23 the whole Cat. was completed all but writing it in the clear; which at that time was a very necessary acquisition, as it was not till the year 1789 before Wollaston's Cat. made its appearance, and many sweeps nearer the Pole than the Register of Sweeps, which only began at 45°, being made, it became necessary to provide a Register for marking those sweeps and the nebulae discovered in them.

. . . . . . .

"*Aug.* 1. I have calculated 100 nebulae to-day, and this evening I saw an object which I believe will prove to-morrow night to be a comet.

"*Aug.* 2. To-day I calculated 150 nebulae. I fear it will not be clear to-night, it has been raining throughout the whole day, but seems now to clear up a little.

"1 o'clock; the object of last night *is a Comet.*

"*Aug.* 3. I did not go to rest till I had wrote to Dr Blagden and Mr Aubert to announce the Comet. After a few hours' sleep, I went in the afternoon to Dr Lind, who, with Mr Cavallo, accompanied me to Slough with the intention of seeing the Comet, but it was cloudy and remained so all night.

"*Aug.* 4. I wrote to-day to Hanover, booked my observations and made a fair copy of 3 letters, made accounts.

"The night is cloudy.

"*Aug.* 5. The night was tolerably fine and I saw the Comet.

"*Aug.* 6. I booked my observations of last night. Received a letter from Dr Blagden in the morning and in the evening Sir J. Banks, Lord Palmerston and Dr Blagden came and saw the Comet. The evening was very fine.

. . . . . . .

"*Aug.* 16. My Brothers returned about 3 in the afternoon."

The following are some extracts from the letters which she wrote (and copied to show her brother) before she went to rest, on the night when she ascertained that the object she suspected was indeed a comet.

*To* DR BLAGDEN (*Sec. of the R.S.*)

Aug. 2, 1786

"Sir,

"In consequence of the Friendship which I know to exist between you and my Brother I venture to trouble you in his absence with the following imperfect account of a Comet.

"The employment of writing down the observations when my Brother uses the 20 feet reflector, does not often allow me time to look at the heavens, but as he is now on a visit to Germany I have taken the opportunity of his absence to sweep in the neighbourhood of the Sun, in search of Comets; and last night, the 1st of Augt. about 10 o'clock, I found an object very much resembling in colour and brightness the 27 Nebula of the *Connaissance des Temps*; with the difference however of being round. I suspected it to be a comet, but a haziness coming on, it was not possible intirely to satisfy myself as to its motion till this evening. I made several drawings of the stars in the field of view with it and have enclosed a copy of them, with my observations annexed that you may compare them together....

"These observations were made with a Newtonian sweeper of 27 inch focal length and a power of about 20; the field of view is about $2° 12'$. I cannot find the stars $a$ or $c$ in any catalogue; but suppose they may easily be traced in the heavens whence the situation of the comet as it was last night at $10^h 33'$ may be pretty nearly ascertained.

Your obed$^t$ humble serv$^t$

CAROLINA HERSCHEL."

Dr Blagden, in reply, wrote:

"I believe the comet has not yet been seen by anyone in England but yourself. Yesterday the Visitation of the Royal Observatory at Greenwich was held, where most of the principal astronomers in and near London attended, which afforded an opportunity of spreading the news of your discovery, and I doubt not but many of them will verify it the next clear night. I also mentioned it in a letter to Paris, and in another I had occasion to write to Munich in Germany. If the weather should be favourable Sunday evening, it is not impossible that Sir Joseph Banks, and some friends from his house, may wait upon you to beg the favour of viewing this phenomenon through your telescope".

To Aubert, Caroline wrote, giving the same particulars as to Dr Blagden.

Aubert wrote to her:

London, the 7th August, 1786

"Dear Miss Herschel,

"I am sure you have a better opinion of me than to think I have been ungrateful for your very, very kind letter of ye 2nd August; you will have judged I wished to give you some account of your comet before I answered it. I wish you joy most sincerely for the discovery. I am more pleased than you can well conceive that *you* have made it and I think I see your *wonderfully clever* and *wonderfully amiable* Brother, upon the news of it, shed a tear of joy. You have immortalized your name and you deserved such a reward from the Being who has ordered all these things to move as we find them, for your assiduity in the business of astronomy and for your love for so celebrated and so deserving a brother.

"Notice has been given at home and abroad of the discovery. I shall continue to observe it and will give you bye and bye a further account of it.

"I was glad to hear to-day by my friends at our Club that they had seen you last night in good health. Pray let me know what news you have of your brother and when we may expect to see him.

Yours &c.

ALEX. AUBERT".

Caroline wrote again to Aubert in return for this letter; the warmth of which she felt required a response.

*To* ALEX. AUBERT, ESQ.

Slough. Augt. 9, '86

"Dear Sir,

"I feel myself unable to express all the gratitude due to you, for the many undeserved encomiums which are contained in your kind letter, but as I am convinced they flow intirely from the goodness of your heart, I beg you will accept my sincere thanks for those; and for your careful attention to the comet, in furnishing it with such valuable observations. I could have wished to defer thanking you, Sir, till I might *early* after my Brother's return do it personally at Slough;

but when I recollect how deficient I have been on Sunday night when Sir Jos. Banks with Lord Palmerston and Dr Blagden called here, I find that where I feel the most esteem I have the least power of expressing it. Another inducement I have of troubling you with the above is to answer your inquiry about my Brother's return. I had a letter yesterday whereby I see he has left Hanover the 7th or 8th and he intends coming by Calais and expects to be in England the 21st or 22nd of this month."

William Herschel reached Slough by the end of August, eager to resume the work on his great instrument. His friend, Dr Watson, wrote to him:

Dawlish. 2 Sept. 1786

"My father sent me the agreeable intelligence that you was safely returned to Slough. I hope that the fatigues of travelling did not counteract materially the effects your friends wished for from this excursion. They hoped it would prove a time of rest, in which not only your mind would be unbent but your body likewise recover the same state in which it was previous to your ague. Pray give my congratulations to Miss Herschel upon her discovering a comet in your absence. This must have given her great pleasure as well as to yourself.

"You will now, I presume, proceed in your *magnum opus*. Dr Blagden informed me your apparatus was up but I suppose the tube is not finished and consequently you have not given the metal its last polish and figure, which you defer till you can make use of it the moment it is finished. When that time comes I hope you will favour me with the news".

Dr Watson rightly conjectured that the making of the tube was the work now to be taken in hand. It was not an easy task. An extract from the "Description of the Forty Foot Telescope", already mentioned, will show the difficulties Herschel encountered and overcame.

"The tube," he says, "though simple in its form, which is cylindrical, was attended by great difficulties in its construction. The body of the tube is of rolled iron which has been joined together without rivets by a seaming well known to those who make iron funnels for stoves. The whole outside was thus put together in all its length and breadth so as to make one sheet of nearly 40 feet long and 15 feet 4 inches broad. The tools, forms and machines we were obliged to make for its construction were very numerous.... When the whole sheet was done in a convenient barn not far from my house, the sides were cut perfectly parallel and bent over."

(The tube was finally completed by resting it on a curved wooden support and bending the sheet over circular wooden frames, all of which had to be made on the spot.)

"As the tube was now much lighter than it would be hereafter we transported it into my garden, where a proper apparatus of circular blocks was put under to receive it."

It must have been while the tube was thus lying on the grass plot in front of the house that the incident occurred which Caroline recalled later in a letter to her nephew's wife.

"One anecdote of the old tube I must give you (if you have not already heard it). Before the optical part was finished many visitors had the curiosity to walk through it, among the rest King George III. And the Archbishop of Canterbury following the King and finding it difficult to proceed, the King turned to give him the hand, saying: 'Come, my Lord Bishop, I will show you the way to Heaven'. This was in the year 1787, August 17th, when the King and Queen, the Duke of York and some of the Princesses were of the company. I hope the book where the visitors' names were noted has been preserved."

In the Visitors' Book kept at Slough, under the date 17th August, 1787, appear the names of the King and Queen, the Duke of York, the Princess Royal, Princess Augusta, the Duke of Queensberry and a company of Lords and Ladies; but the Archbishop of Canterbury is not mentioned. This Royal visit may have been occasioned by the petition, which William Herschel had been emboldened to make, through Sir Joseph Banks, for some additional pecuniary assistance to complete the 40-foot telescope. It would seem that the King was satisfied with what he saw, for a week later William Herschel heard from Sir Joseph that the King would grant what he asked. The additional sum given by King George was £2000, for the completion of the telescope, and a yearly sum of £200, for the expenses of its upkeep.[1]

Prof. Pictet of Geneva visited Herschel at Slough in 1786, and afterwards recorded his impressions in a series of short articles in the *Journal*

[1] See Caroline Herschel's Autobiography, *infra*, Chapter XI, p. 172.

*de Genève* for October, 1787. His account of Herschel's process for polishing the mirror of the 40-foot telescope is particularly interesting, as being written by an eyewitness. The following is a translation of portions of the article.

After explaining the optical requirements of a reflecting telescope, he wrote:

"It follows that besides the most perfect polish the mirror of the telescope must have a parabolic figure. The work, as directed by Herschel, to achieve these two conditions, would make a subject for a picture. In the middle of his workshop there rises a sort of altar; a massive structure terminating in a convex surface on which the mirror to be polished is to rest and to be figured by rubbing. To do this the mirror is encased in a sort of twelve-sided frame, out of which protrude as many handles which are held by twelve men. These sides are numbered and the men who are stationed at them carry the same numbers[1] on the strong linen overalls which protect their clothes from the splashes of the liquid, which from time to time is introduced between the mirror and the mould to give the polish.

"The mirror is moved slowly on the mould, for several hours at a time and in certain directions, which by pressing more on certain parts of the surface than on others, tends to produce the parabolic figure.

"It is then removed on a truck and carried to the tube, into which it is lowered by a machine expressly contrived for the purpose. This labour is repeated every day for a considerable time and by the observations he makes at night, Herschel judges how nearly the mirror is approaching the standard he desires".[2]

M. Pictet then described in great detail the mounting of the great telescope; with the mechanical ingenuity displayed in its design he was much impressed.

---

[1] The numbers on the men's overalls were no doubt for the purpose of enabling Herschel to give the word of command when changing the direction of the strokes.

[2] In a letter to Sir John Herschel (February, 1841) Caroline referred to the manner in which her brother tested the figure of his mirrors, "when polishing and the foci (were) to be tried, by three Apertures, they generally wanted to be repaired first; for the 20-ft. they were made of pasteboard, but for the 40-ft. of light deal; and I was directed to hold them before the mirror, and listening to the report of the trial was glad to hear: 'All the three foci perfectly alike'; and then the work was proceeding to a perfect polish".

"Herschel", he wrote, "has shown himself as great a Mechanician as a consummate Astronomer in thus achieving his object from the first attempt; for the mounting was carried out according to his drawings without it having been necessary to make any changes either in the proportions or in a single detail of a construction so vast and complicated."

The general result of the arrangement, he declared, was so perfect "that Miss Herschel could unaided easily raise or lower the mighty telescope by herself".

He then described how Caroline, seated in the little hut at the base of the telescope, was able to register exactly the position of the tube corresponding to the time indicated by the clock in front of her, and so to write down the record of the observations which her brother called down to her through the speaking-tube; the latter was so well placed that the brother and sister could speak to each other without raising the voice.

In conclusion, Prof. Pictet alludes to Herschel's researches into the nature of nebulae of which he had, up to this time, discovered 466 new ones; and alludes, without any expression of dissent, to the hypothesis that the Milky Way may in fact be one of these nebulae, as they, like the Milky Way, have been shown by Herschel's telescopes to be resolvable into vast clusters of stars. Finally, Pictet gives his own impression of Herschel's personality.

"This same wonderful man, whose imagination is so brilliant, whose brains and physical powers are so well balanced, possesses besides these advantages the most agreeable manner and social qualities which would make him sought after if he had no other claims to public esteem. Visitors often take unwarranted advantage of his courtesy and compliance, wasting his time and putting unnecessary and often ridiculous questions, but his patience is inexhaustible and he takes these inconveniences, which are attached to celebrity, in such good part that no one could guess how much they cost him."

The vision which astonished the eyes of such visitors as were permitted to gaze into the great telescope is thus described by one of them.[1]

"When the star Sirius entered the field of the telescope the eye was so dazzled that one could not at once even perceive stars of lesser

[1] See Fourier's *Éloge* of Sir William Herschel, Appendix.

magnitude. More than twenty minutes elapsed before one could observe those lesser lights."

The notes of observations which Caroline made at night, when seated at her table in the little hut at the foot of the telescope, were carried by her into the library of the Cottage, when her brother gave the signal to cease work and they both retired to sleep. The following morning she copied out the observations on to separate sheets of foolscap. These "Register Sheets" of Nebulae are now in the possession of the Royal Society; those of Double Star Observations belong to the Royal Astronomical Society.

Dr Dreyer remarks:

"Caroline Herschel's perseverance in doing the enormous amount of copying required of her was equalled by her carefulness and accuracy in performing her tasks; the books and slips written in her clear and most legible hand seem completely free from slips, either clerical or arithmetical. In fact the only mistakes she ever made in her work seem to have been two or three slips of $1°$ in Polar Distance, in the Zone Catalogue of her Brother's nebulae which she made in her old age".

One peculiarity in the construction of the 40-foot telescope remains to be mentioned. Hitherto all Herschel's telescopes had been constructed on the Newtonian or Gregorian model, having a small mirror within the tube to deflect the image to the side where the observer stood. In the drawings of the 40-foot telescope, it will be noticed that the observer's seat is at the mouth of the tube, not at the side. This is the position which he called front-view and which he now adopted both for the 20-foot and the new 40-foot telescopes. In a note appended to his "First Catalogue of New Nebulae", he explained this method:

"The *Front-view* is a method of using the reflecting telescope different from the Newtonian, Gregorian, and Cassegrain forms. It consists in looking with an eyeglass, placed a little out of the axis, directly in at the front, without the interposition of a small speculum; and has the capital advantage of giving us almost double the light of the former constructions. In the year 1776, I tried it for the first time with a 10-feet reflector, and in 1784, again with a 20-feet one; but the success not immediately answering my expectations, it was too hastily laid aside. By a more careful repetition of the same experiment I find now, that several other considerable advantages added to the brilliant light before mentioned, make it so valuable a construction that a judicious observer may avail himself of it at least in all cases where light is more particularly wanted".

More light was always for Herschel the first desideratum. The slight distortion of the image, caused by the tipping of the mirror, was not such a serious matter to him as it would be now, when accuracy is the chief aim.

His attachment to the front-view method was increased when by its use he discovered two satellites to his own planet. This was in the year 1787. He immediately reported his discovery to the Royal Society, in a paper from which the following extracts are taken:

"The great distance of the Georgian planet and its present situation in a part of the zodiac which is scattered over with a multitude of small stars, has rendered it uncommonly difficult to determine whether, like Jupiter and Saturn, it be attended by satellites.

"In the beginning of last month, however, I was often surprised when I reviewed nebulae that had been seen in former *sweeps*, to find how much brighter they appeared, and with how much facility I saw them. The cause of it could be no other than the quantity of light gained by laying aside the small speculum, and introducing the *Front-view*, an account of which has been inserted, by way of note, to the *Catalogue of Nebulae*.[1]

"It would not have been pardonable to neglect such an advantage, when there was a particular object in view, where an accession of light was of the utmost consequence; and I wondered why it had not struck me sooner. The 11th of January, therefore, in the course of my general review of the heavens, I selected a *sweep* which led to the Georgian planet; and while it passed the meridian, I perceived near its disc, and within a few of its diameters, some very faint stars whose places I noted down with great care".

. . . . . . .

He then describes the series of observations from 14th January to 5th February, and concludes thus:

"though at the end of this time I had no longer any doubt of the existence of at least one satellite, I thought it right to defer this

[1] *Phil. Trans.* vol. LXXII, p. 499.

communication till I had had the opportunity of actually seeing it in motion. Accordingly I began to pursue this satellite on February 7th, about six o'clock in the evening, and kept it in view till three in the morning on February 8th, at which time, on account of the position of my house, I was obliged to give over the chase; and during those nine hours I saw this satellite faithfully attend its primary planet, and at the same time keep on in its own course, by describing a considerable arch of its own proper orbit.

"While I was chiefly attending to the motion of this satellite, I did not forget to follow another small star, which I was pretty well assured was also a satellite.... The interval which the cloudy weather had afforded was however rather too short for seeing its motion sufficiently, so that I deferred a final judgement till the 10th and, in order to put my theory of these two satellites to a trial, I made a sketch on paper to point out beforehand their situation with respect to the planet and its parallel of declination.

"The long-expected evening came on, and, notwithstanding the most unfavourable appearance of dark weather, it cleared up at last. And the heavens displayed the original of my drawing by showing, in the situation I had delineated them, *The Georgian Planet attended by two satellites.*

"I confess that this scene appeared to me with additional beauty, as the little secondary planets seemed to give a dignity to the primary one, which raises it to a more conspicuous situation among the great bodies of our solar system".

Two mirrors were made for the 40-foot telescope; the first had been cast in London in October, 1785, by a founder who persuaded Herschel to have it made of a white metal, the composition of which is not known. Though it weighed nearly half a ton, it was too thin to keep its shape, and the back turned out to be depressed in the middle, yet he proceeded with the shaping and polishing, as he had always intended to have two mirrors and expected to gain experience on the thin one. When the polishing was begun ten to twelve men were employed upon it as described by Prof. Pictet. The operation required constant supervision, as the mirror was not simply turned round on the polisher, but had to be worked backwards and forwards or across, in diagonal strokes, as Herschel's experience in figuring smaller mirrors had taught him. But the men could not be got to work together with the accuracy required,

SIR WILLIAM HERSCHEL
at the age of 56
*From the pastel portrait by J. Russell, R.A.*

and all the experience gained by Herschel himself was wasted. When the second mirror was ready for polishing, Herschel, after a brief attempt with the old method, gave it up, and the mirror was finished on a polisher worked by machinery.[1]

Caroline, in a letter to her nephew many years later, recalled with regret the labour and expense which had been

"wasted on the first mirror, which had come out too light in the casting (Alex more than once would have destroyed it secretly if I had not persuaded him against it), and without two mirrors you know such an instrument cannot be always ready for observing".

The second mirror was also cast in London, but the metal was Herschel's own composition of copper and tin; with the intention of making the metal less brittle, the proportion of tin had been reduced, but the result was unfortunate, as the mirror tarnished quickly and much arduous labour was spent in repolishing.

By the month of February, 1787, the telescope was mounted and, the mirror being placed in the tube, he was able for the first time to test it on a celestial object; he thus describes the experiment which finally decided him to finish it off for front-view only:

"The apparatus for the forty feet telescope was by this time so far completed that I could put the mirror into the tube and direct it to a celestial object; but having no eye-glass fixed, nor being acquainted with the focal length which was to be tried, I went into the tube and laying down near the mouth of it, I held the eye-glass in my hand and soon found the place of the focus. The object I viewed was the nebula in the belt of Orion, and I found the figure of the mirror, though far from perfect, much better than I expected".

Herschel wrote to Sir Joseph Banks, concerning the progress he was making with the mirror for the 40-foot, as follows:

Slough. Augt. 29, 1789

"Dear Sir,

"You will perhaps be glad to hear that my machine performs its work with as much success as I could possibly have expected. After a few days' polishing I tried the speculum upon Windsor Castle and was so much pleased that, being impatient to apply it to

[1] The second mirror and the polishing machine are preserved at Observatory House, Slough.

the Heavens, I would not go on with the polish tho' the speculum is certainly not half come to its lustre. By tryal on the stars I found the figure as much advanced as that of my 20 feet Spec. was when I had observed two years with it. As soon as I could get hold of Saturn it gave me convincing proof of its superior power by shewing me immediately a *sixth satellite*, which has till now escaped the vigilance of astronomers. The discovery was made and compleated on friday Aug$^t$ 28; I must however do my good 20 feet telescope the justice to say that strong suspicions of the sixth satellite's existence were given by it July 17, July 18 and July 27, and as they are now verified the observations of those days will assist me to determine the revolution of the satellite by reckoning from them.

"I have not communicated this discovery to any person but our Friend Mr Watson and to him only with a precaution not to let it go farther till I have heard from you, how to announce it by your means to the astronomical world.

"Would you think it proper, as my paper on Nebulae is now printing, and I believe one of the last things in the volume, to add at the bottom of it?—

P.S. Saturn has six satellites.

40 feet Reflector. Wm. Herschel.

Or will you have even the mention of it reserved for the next meeting of the R. Soc. when I shall present you with a paper containing the particulars? Your advice in this affair will be received with great pleasure by Dear Sir

your most faithful humble serv$^t$

WM. HERSCHEL".

*To* SIR JOSEPH BANKS, BART.

Slough near Windsor. Sept. 4, 1789

"Sir,

"Last week I sent you a letter in which I acquainted you with the discovery of a sixth satellite of Saturn by the 40 feet telescope; and also mentioned that my 20 feet instrument had already given strong suspicions of its existence some time ago. As I recollected that my observations on Saturn in the year 1787 contained also things that might throw a light on the subject, I looked them over carefully and to my surprize now find that the discovery of this satellite was compleatly made the 19 of Aug$^t$ 1787; but being so taken up with the Georgian satellites I had laid by the observation of those of Saturn for want of sufficient time to examine them properly. To this may be added that, having it in view to make a series of capital observations on Saturn with the 40 feet reflector, I reserved all further inquiries till my speculum should be advanced far enough to begin such observations. And indeed the charming view I had of that Planet with all its six satellites in motion may still claim part of the merit of the discovery, tho' I might certainly have announced it above two years ago, if I had not been otherwise taken up. As it is of some consequence that you should be furnished with proofs of the time when the discovery of this satellite was made, in case any other astronomer might also have observed it lately (which is however not probable), I join the time of its greatest western elongation, which Epocha happened

1797, Aug. 19. 21$^h$ 54' 56" sidereal time.

In my last I proposed a P.S. to my paper which is now printing; which might perhaps run thus

"P.S. Saturn has Six satellites. An account of its discovery, its Revolution and Orbit will be given in the next Volume of the *Ph. Tr.*

"But all this I leave intirely to your better judgement and have the honor to remain

Sir

your most obed$^t$ humb$^e$ servant

WM. HERSCHEL."

The postscript, when printed, was slightly different and had the words 'with the forty feet reflector" inserted after "An account of the discovery".

Next month Herschel wrote again to Sir J. Banks:

*To* SIR JOSEPH BANKS, BART.

Slough near Windsor. Oct. 21, 1789

"Dear Sir,

"Perhaps I ought to make an apology for troubling you again with a letter on the same subject as my former one; but if satellites will come in the way of my 40 feet Reflector, it is a little hard to resist discovering them. Now as *a seventh satellite* of Saturn has drawn me into that scrape I must trust to your good nature to forgive me when I acquaint you with it.

"The truth is that I have detected the younger brother of the sixth satellite. It is still nearer the ring than the latter, and I am writing a short paper to give you for the Royal Soc. which will be ready by the time of your return to town.

I am with great Respect, Dear Sir,
Your faithful most humble ser$^t$
WM. HERSCHEL".

Herschel was no doubt very anxious that the great telescope should prove worthy of all the labour and expense that had been lavished upon it, and was therefore disposed to over-emphasize the part which it had played in this discovery. In the paper which was submitted to the Royal Society in November, 1789, he did indeed give the true version, as he had given it to Sir Joseph Banks, but the rather dramatic announcement, printed as a postscript to the paper on Nebulae in August, which attributed the discovery solely to the 40-foot, was read and commented on all over the world. The reputation of the monster telescope was now considered fully established, and great expectations were aroused as to what it might accomplish in the future; expectations unhappily not realized.

Even when he first noticed the new satellite with the 40-foot he was obliged to have recourse to the 20-foot to follow it. In his own record of the observation, he says:

"(The satellite) seems to be bright enough; now it is discovered, I may perceive and follow it in the 20 feet telescope. Having no side motion on, I can follow it no longer than the distance of my ladders will permit".

Dr Dreyer says[1]:

"The seventh satellite (Mimas) was discovered with the 20 foot on the 8th September 1789, and was probably seen again on the 14th.—After that night Herschel continued to the end of the year to follow the satellites of Saturn, and there is no doubt whatever that all these observations... were made with the 20 foot telescope, which instrument must indeed have been in the highest state of efficiency."

[1] "A Short Account of Sir William Herschel's Life and Work", p. liii; Introduction to *The Scientific Papers of Sir William Herschel.*

**THE FORTY-FOOT TELESCOPE**
at Slough
*From an engraving dedicated to King George III*

It must sadly be confessed that the great telescope did not justify the hopes of its maker. Dr Dreyer, who took much pains in searching through the records of Herschel's observations, could find only seven or eight occasions, between the years 1790 and 1815, when it was used on Saturn, and on none of these was there anything new or of importance to note. Herschel gave an explanation of the paucity of observations taken with the 40-foot.

"The forty-feet telescope having more light than the twenty feet, it ought to be explained why I have not always used it in these observations. Of two reasons that may be assigned, the first relates to the apparatus and the nature of the instrument. The preparations for observing with it take up much time, which in fine astronomical nights is too precious to be wasted in mechanical arrangements. The temperature of the air for observations that must not be interrupted, is often too changeable to use an instrument that will not easily adapt itself to the change: and since this telescope, besides the assistant at the clock and writing desk, requires moreover the attendance of two workmen to execute the necessary movements, it cannot be convenient to have everything prepared for occasional lucid intervals between flying clouds that may chance to occur; whereas in less than ten minutes, the twenty feet telescope may be properly adjusted and directed so as to have the planet in the field of view....

"The machinery of my twenty feet telescope is so complete that I have been able to take up the planet at an early hour in the evening and to continue the observations of its own motion, together with that of its satellites, for seven, eight, or nine hours successively."

Dr Dreyer further remarks:

"If we turn to the satellites (real or supposed) of Uranus, we find the same disinclination to use the 40-foot....

"But it is most remarkable that nebulae, a class of objects which played a very important part in Herschel's life-work, and with which his name is more associated than with any other celestial bodies except perhaps double stars, that these striking-looking objects should have been so very little looked at with the great telescope. Herschel was apparently not skilful in drawing, but still the study of the form of the various kinds of nebulae with a large aperture might have given him many hints as to their nature; and in particular he might have discovered the spiral form of some of the nebulae, which ought to have been within the reach of a 48-inch mirror.... No one doubts that Herschel was capable of grinding and polishing even a speculum four feet in diameter, and thereby giving it both an excellent figure and a brilliantly polished surface. But all the same it is likely enough that the instrument did not generally perform well. Besides, whenever it was polished, it became tarnished very rapidly, a fault for which the low composition of the metal was undoubtedly responsible".

For over fifty years the great telescope reared its elaborate scaffolding behind that of the more modest 20-foot. It became a landmark for the country-side, and its position was even noted in the trigonometrical survey of the county. It figured in popular magazines as one of the Wonders of the World, comparable to the Colossus of Rhodes or the Porcelain Tower of Nanking. The American writer, Oliver Wendel Holmes, describes, in his book *The Poet at the Breakfast Table*, his impressions on first catching sight of it:

"I was riding on the outside of a stage coach from London to Windsor in the year—never mind the year, but it must have been in June, I suppose, for I bought some strawberries. But, as I was saying, I was riding down from London to Windsor, when all at once a picture, familiar to me from my New England village childhood, came upon me like a reminiscence rather than a revelation.

"It was a mighty bewilderment of slanted masts, spars and ladders and ropes, from the midst of which a vast tube, looking as if it might be a piece of ordnance such as the revolted angels battered the walls of Heaven with, according to Milton, lifted its mighty muzzle defiantly towards the sky. 'Why! you blessed old rattle-trap,' said I to myself, 'I know you as well as I know my father's spectacles and snuff-box.' And that crazy witch, Memory, so divinely wise and foolish, travels thirty-five thousand miles or so in a single pulse beat, and makes straight for an old house and an old library, an old corner of it, and whisks out a volume of an old cyclopaedia, and there is the picture of which this is the original—Sir William Herschel's great telescope!"

## CHAPTER XI

### 1786  1788

Fanny Burney's *Diary*. Caroline's Autobiography. She receives a salary from the King as assistant to her brother. Mrs Papendiek's Journal. Her account of William Herschel's marriage. Character of Mrs Herschel. Letters—Aubert, Sir W. Watson. Portraits of William Herschel.

THERE are several references to William and Caroline Herschel in Fanny Burney's *Diary*. In an entry for the year 1786 we find:

"1786. In the evening Mr Herschel came to tea. I had once seen that very extraordinary man at Mrs De Luc's but was happy to see him again, for he has not more fame to awaken curiosity than sense and modesty to gratify it. He is perfectly unassuming yet openly happy, and happy in the success of those studies which would render a mind less excellently formed presumptuous and arrogant....

"He seems a man without a wish that has its object in the terrestrial globe. At night Mr Herschel, by the king's command, came to exhibit to His Majesty and the royal family the new comet lately discovered by his sister, Miss Herschel; and while I was playing at piquet with Mrs Schwellenburg, the Princess Augusta came into the room and asked her if she chose to go into the garden and look at it. She declined the offer, and the princess then made it to me. I was glad to accept it for all sorts of reasons. We found him at his telescope. The comet was very small, and had nothing grand or striking in its appearance; but it is the first lady's comet, and I was very desirous to see it. Mr Herschel then showed me some of his new discovered universes, with all the good humor with which he would have taken the same trouble for a brother or a sister astronomer; there is no possibility of admiring his genius more than his gentleness.

"1786, *Dec. 30th*. This morning my dear father carried me to Dr Herschel. That great and very extraordinary man received us with almost open arms. He is very fond of my father, who is one of the Council of the Royal Society this year as well as himself and he has much invited me when we have met at the lodge or at Mr de Luc's.

"At this time of day there was nothing to see but his instruments, those, however, are curiosities sufficient. His immense new telescope, the largest ever constructed, will still, I fear, require a year or two more for finishing, but I hope it will then reward his labour and ingenuity by the new views of the heavenly bodies and their motions, which he flatters himself will be procured by it. Already, with that he has now in use, he has discovered fifteen hundred universes! How many more he may find who can conjecture? The moon, too, which seems his favourite object, has already afforded him two volcanoes; and his own planet, the Georgium Sidus, has now shewn two satellites".

Of Caroline Herschel Miss Burney wrote:

"She is very little,[1] very gentle, very modest, very ingenuous; and her manners are those of a person unhackneyed and unawed by the world, yet desirous to meet and return its smiles. I love not the philosophy that braves it".

We turn back to Caroline's Autobiography for her own experiences after her brother's return from Hanover.

### CAROLINE HERSCHEL'S AUTOBIOGRAPHY

"1786. It would be impossible for me, if it were required, to give a regular account of all what passed around me in the lapse of the following two years, for they were spent in a perfect chaos of business. The garden and workrooms were swarming with labourers and workmen; smiths and carpenters going to and fro between the forge and 40-feet machinery; and I ought not to forget that there is not one screw bolt about the whole apparatus but what was fixed under the immediate eye of my Brother. I have seen him lie stretched many an hour in a burning sun, across the top beam, when the iron work for the various motions was fixed.

"At one time no less than 24 men (12 and 12 relieving one

---

[1] The dress which Caroline wore at her brother's wedding has been preserved. It exactly fits a great-niece whose height is 4 feet 3 inches and who is very slight; such therefore must have been about Caroline's height and figure.

another) kept polishing, day and night, and my Brother of course never leaving them all the while, taking his food without allowing himself time to sit down to table.

"The moonlight nights generally were taken for such like experiments and for the frequent journeys to Town, which he was obliged to make for ordering tools and materials which were constantly wanting, (I may say by wholesale).

"The discovery of the Georgian satellites caused many breaks in the sweeps which were made at the end of 1786 and beginning of 1787, by leaving off abruptly against the meridian passage of the Planet, which occasioned much work, both in shifting of the Instrument and booking of the observations. And much confusion at first prevailed among the loose papers on which their first observations were noted and some of them have perhaps been lost; for I remember configurations of the situations of the satellites having been made by Sir Wm. Watson and Mr Marsden, and only one could be found....

"That the discovery of these satellites must have brought many Nocturnal Visitors to Slough may easily be imagined; and many times have I listened with pain to the conversation my Brother held with his Astronomical Friends, when being quite exhausted with answering their numerous questions. For I well knew that on such occasions instead of renewing his strength by going to rest,—there were too many who could not go on without his direction, among whom I oftens was included. For very seldom could I get a paper out of his hands time enough for finishing the copy against the appointed day for its being taken to Town.

"But considering that no less than seven papers were delivered to the R.S. in 1786 and 1787 it may easily be judged that my Brother's Study had not been intirely deserted. And I had always in hand some kind of work with which I could proceed without troubling him with questions, such as the Temporary Index which I began in June 1787; and some years after, the Index to Flamsteed's observations; calculating the beginning and ending of Sweeps and their breadth for filling up the vacant places in the Registers,—many of which have never been made,—and works of that kind, which filled up the interval when nothing more necessary was in hand.

"My Brother Jacob was with us from April to October (1787), when he returned to Hanover again. Alex. came only for a short time to give his brother the meeting, Mrs H. being too ill to be left long alone. (She died in January of '88.)

"Professor Sniadecky oftens saw some objects through the 20 ft. Telescope, among others the Georgian satellites. He had taken lodgings in Slough for the purpose of seeing and hearing my Brother whenever he could find him at leisure; himself was a very silent man.

"My Brother's bust was taken by Lochee according to Sir Wm. Watson's order. Professor Wilson and my Brother Jacob were present.[1]

"1787. In August an additional manservant was engaged who would be wanted at the handles of the motions of the 40 feet, for which the mirror in the beginning of July was so far finished as to be used for occasional observations or trial, and besides such a person became necessary for shewing the telescopes to the curious strangers, as by their numerous visits my Brother or myself had for some time past been much incommoded.

"In consequence of an application having been made through Sir J. Banks to the King, my Brother had in Augt. a second 2000 pounds granted for completing the 40 ft. and 200 yearly for the expense of repairs, such as ropes, painting &c. &c. and the keep and cloathing of the men to attend at night.

"And a salary of 50 pounds per year was settled on me as an assistant to my Brother, and in October I received £12, 10,—being the first quarterly payment of my salary; and the first money in all my lifetime I ever thought myself at liberty to spend to my own liking, exactly the sum I saved my Brother at Bath in writing music by a clean fireside.

"A great uneasiness was by this means removed from my mind, for though I had generally (and especially during the last busy 6 years) been the keeper, almost, of my Brother's purse, with a charge to provide for my personal wants with only annexing in my accounts the memorandum: '*for Car.*' to the sums so laid out, they did, when cast up, hardly amount to 7 or 8 pounds per year since the time we had left Bath. For nothing but Bankruptcy had all the while been running through my silly head when looking at the sums of my weekly accounts, (and) knowing they could be but trifling to what had been and was yet to be paid in Town. For my Brother had not

[1] The plaster cast of William Herschel's face, taken on this occasion, was given to Sir W. Watson by the artist.

been fortunate enough to meet with a reasonable man for a Caster, who also furnished the Crane &c. as his bills came in greatly overcharged. But more of this in another place. I will only add that, from this time on, the utmost activity prevailed to forward the completion of the 40 feet. An additional optical workman was engaged and preparation made for casting the second mirror. Journeys to Town were made for moulding and at the end of January (1787) a fine cast mirror arrived safely at Slough. And several 7 feet telescopes were finished and sent off.

"The fine nights were not neglected, (and observations oftens interrupted by visitors. Messrs Casini, Mechain, Le Genre and Carochet spent Nov. 26 and 27 with my Brother and saw many objects in the 20 feet).

"The Catalogue of the second 1000 new Nebulae wanted but few numbers in March to being complete; and the observations on the Georgian Satellites furnished a paper which was delivered to the Royal Society in May. And the 8th of that month being fixed on for my Brother's marriage it may easily be supposed that I must have been fully employed (besides minding the heavens) to prepare everything as well as I could, against the time I was to give up the place of a Housekeeper, which was the 8th of May, 1788."

In this last sentence we have the only allusion in her Autobiography to the event which had such a tragic significance for Caroline Herschel —her brother's marriage. Fortunately the Journals of Mrs Papendiek supply the deficiency and give sufficient details to satisfy all legitimate curiosity regarding William Herschel's domestic concerns. In these Journals Mrs Papendiek frequently mentions her friends, Mr and Mrs Pitt, who lived at Upton, a short walk from the house which her father had occupied and which now was the home of the Herschels. This house belonged to Mrs Baldwin, a widow lady, who lived herself in a little house behind the Crown Inn. Her husband, Adee Baldwin, had been a merchant in the City of London and owner of a considerable property at Slough. Mr Pitt, her daughter's husband, owned a house at Upton. He was a well-read and cultivated man, whose acquaintance William Herschel was glad to make, and his wife was at least much better educated than her mother.

## EXTRACTS FROM MRS PAPENDIEK'S JOURNAL

"Dr Herschel showed every kind of attention to Mr and Mrs Pitt. The former being in a declining state of health, the Doctor passed many hours with him in his well-chosen library, and avowed that he derived much instruction from his remarks and great pleasure from his society and conversation. At the end of the summer (1786) this excellent man died, leaving the property to his widow until her death, when it was to revert to their son Paul Adee Pitt, my brother's Eton friend and companion.

. . . . . .

"We were constantly with Mrs Pitt, at Upton, enjoying the homely fare of cake or bread, with wine, in the dear, brick-floored parlour. She, poor woman, complained much of the dullness of her life, and we did our best to cheer her, as did also Dr Herschel, who often walked over to her house with his sister of an evening, and as often induced her to join his snug dinner at Slough.

"Among friends it was soon discovered that an earthly star attracted the attention of Dr Herschel. An offer was made to Widow Pitt, and accepted. They were to live at Upton, and Miss Herschel at Slough, which would remain the house of business. All at once it struck Mrs Pitt that the Doctor would be principally at the latter place, and that Miss Herschel would be mistress of the concern, and considering the matter in all its bearings, she determined upon giving it up. Dr Herschel expressed his disappointment, but said that his pursuit he would not relinquish; that he must have a constant assistant and that he had trained his sister to be a most efficient one. She was indefatigable, and from her affection for him would make any sacrifice to promote his happiness."

Mrs Pitt did not however remain obdurate; in her Journal for the following year, 1787, Mrs Papendiek records:

"A few days after our return home (in the autumn) Mrs Pitt called to tell us that the offer from Dr Herschel had been renewed, and again accepted, under the following arrangement. There were to be two establishments, one at Upton and one at Slough; two maid-servants in each, and one footman to go backwards and forwards, with accommodation at both places, and Miss Herschel to have apartments over the workshops. A tube was arranged for the

Doctor to communicate with her, direct from his post of observation, so she was able to write minutes of his proceedings without being exposed to the open air.

"Sir Joseph Banks and ourselves were the only friends entrusted with the secret of their engagement".

Mrs Papendiek's desire to enhance her own importance probably prompted this statement; Herschel certainly made no secret to his friends of his intended marriage; on the contrary, he seems to have been very anxious to have their approval, and to have asked Dr Watson to tell him frankly how the contemplated step was regarded. Dr Watson wrote in reply as follows:

"I had no opportunity when I was last with you, of speaking a word apart with you, or I should have given you the result of what I have collected relating to the general opinion of your friends upon your future marriage. And I am extreamely happy to inform you, that excepting some little fears with respect to Astronomy, I have not heard anything which you would have disliked; on the contrary, it seemed to meet with the approbation of everyone. For my own part, my dear friend, I need not say you have my warmest wishes for your happiness, and even if it was to occasion you to relax somewhat of the intensity of your present application to Astronomy, I should be the better pleased, not only because your happiness is my *first* concern, but I think it may likewise turn out to the advantage of Science upon the whole, as I fear your endeavours are too fatiguing, both to your mind and body, to be consistent with your precious health".

The Register in Upton Church records the marriage as follows:

"1788, No. 30.

"William Herschel of this Parish, Batchelor and Mary Pitt of this Parish, widow, were married by License in this Church on the eighth day of May, one thousand seven hundred and eighty eight: by me

J. Hand, Vicar.

This marriage was solemnized between us
    William Herschel
    Mary Pitt

In the presence of
    Alexr. Herschel
    Caroline Herschel".

Mrs Papendiek says that Mrs Pitt's brother, Mr Baldwin, gave her away, and that her son was present at the marriage. She adds:

"Sir Joseph Banks acted as friend of the Doctor. Miss Herschel received them at Slough, which was the honoured house for the reception of the newly married pair and where they spent their honeymoon. About six weeks after, cards were sent round dated Upton, so there, of course, the congratulatory visits were paid.

"Not feeling inclined to take a carriage, I walked over with nurse and baby one fine afternoon in May, resting for a few minutes under the yew tree in Upton churchyard, where I put on my white gloves before going on to the house. Dr Herschel and his bride received me warmly, and I was ushered into the well-known tent, where cake and wine were presented. I hoped for some assistance home, but none was offered, so we walked back again, and I was not so fatigued as might have been expected".

Caroline destroyed every page of her Journals which contained any reference to this period of her life, and, in her later correspondence with her nephew, she refers with some bitterness to the rather inquisitive interest which Mrs Papendiek took in the affairs of the Herschel family. Caroline Herschel was never inclined to suffer gossip gladly, and at this crisis in her life, when she was resigning herself silently and heroically to see another woman take the first place in her brother's affection, she must have been more than ever sensitive to any form of prying curiosity; yet we cannot but be grateful to Mrs Papendiek for the particulars she has preserved for us.

Gradually the nobility of Caroline's character and her ardent attachment to her brother, added to the gentleness and amiability of her sister-in-law, succeeded in healing the wound, and when at length a son was born to her brother, the reconciliation was complete; she had now a second object for the passionate devotion of her warm heart.

Mrs Herschel was evidently a very lovable woman, and though without any special intellectual gifts, she could, as her letters and journal of travels prove, write pleasantly and intelligently about what she saw. She liked society, and had a circle of warmly attached friends, and loved to have the company of members of her family. Her mother, Mrs Baldwin, a rather formidable old lady, to judge by her portrait, lived close by, and one or other of her brother's daughters, Sophia or Mary Baldwin, was constantly at Slough. There are also many little indications that she was a good woman of business and ordered her household wisely and kindly. The atmosphere of leisure and peace which she introduced into his home must have been very grateful to her husband, as the natural fatigues of age began to tell on him. We have an appreciation of Lady Herschel's character from one who could speak from intimate personal knowledge—her son's wife. In a letter to the publisher of the *Memoir of Caroline Herschel*, the second Lady Herschel wrote:

> "I would wish to bear personal witness to the graceful and dignified gentleness of Lady Herschel's manner, even in old age, which were the faithful indications of the genuine kindness and amiability of her nature. Miss Herschel's good sense soon got over the startling innovation of an English lady-wife taking possession of her own peculiar fortress, and she who gladdened her husband's home soon won the entire affection of the tough little German sister".

There were not lacking ill-natured gossips to insinuate that Dr Herschel had married the widow Pitt for the sake of her money. Even Fanny Burney could not refrain from a cheap innuendo. Describing a tea-party at Mrs de Luc's, she wrote in her *Diary*:

> "Dr Herschel was there, and accompanied them (the Miss Stowes) very sweetly on the violin; his new-married wife was with him and his sister. His wife seems good-natured; she was rich too! and astronomers are as able as other men to discern that gold can glitter as well as stars".

It is proof sufficient that William Herschel was not contemplating living at ease on his wife's money, that he continued the laborious work of making telescopes for sale, till failing strength obliged him to desist. A list of some of the most important of these is given in the Appendix; for the large 25-foot made for the King of Spain and completed in 1805, he received £3150. One may admit so much, however, that he would not have married if Mrs Pitt's own means had not been sufficient to ensure that she would be able to enjoy the comforts and pleasures to which she had been accustomed. For the rest, she so perfectly fulfilled the ideal of a wife which he had sketched to his brother many years before, that we may confidently attribute the marriage to her own personal qualities and not to such meretricious glitter as Fanny Burney suggests.

The only change which his marriage produced in William Herschel's habits was that he now took regular holidays from his work, travelling about in the summer with his wife and generally accompanied by one of her nieces. On these occasions Caroline was left in charge of the house and instruments; her brother Alexander coming to stay with her.

William Herschel wrote as follows to Alexander Herschel, on hearing of the death of the latter's wife:

Thursday afternoon. Feb. 7, 1788

"Dear Brother,

"This morning I received your sad information of your loss, which occasioned a very sincere grief. Having been up all night Carolina was still in bed when your letter came. Poor Girl, she has hardly had a dry eye to-day; however as our late sister's health has been so very bad we cannot say that she died unexpectedly and therefore we should not grieve too much at her present relief from misery. If the distance were not so great one of us would come to see you, but upon the whole it might only increase your melancholy.

"As her late illness and your present situation may have been more expensive than common, I beg you will freely say if I can be of any service, by lending you a Bill for present use. Carolina is not well enough to write to-day but will either to-morrow or next day endeavour to take up the pen.

"Last week I went to London to cast a 40 feet speculum, much thicker and stronger than my present one. We put 2758 (lbs.) of Metal in the furnace and it just filled our mould; 3 days after when we opened it we found it cracked in cooling. Next week I go to cast again and shall make the Metal less brittle.

"Half an hour ago the rope of my large telescope broke and the tube was a little bulged but not materially injured. Our contrivance for catching it, in case of the ropes breaking, was not put on yet, otherwise this would not have happened. The rope-maker has certainly cheated me and I shall take care my next ropes shall be sufficiently tried before I use them.

"Excuse my endeavouring to amuse you with an account of things that perhaps at this time can be but little interesting to you; and believe me

very affectionately yours<br>WM. HERSCHEL".

As we have none of Herschel's private letters to his friends, we are dependent on theirs to him for light on his social relations. It is clear that he was no hermit, but took pleasure in society. Mrs Papendiek, in her Journal, describes a musical gathering at her house:

"The Griesbachs' quartet-playing", she says, "was exquisite. Among other pretty things, Dr Herschel asked me to take part in a catch of his, which went off excellently well".

Speaking of much later years, his wife's niece, Mary Baldwin, told Herschel's granddaughter that at the frequent musical gatherings at his house at Slough he was the life and soul of the party; sitting playing at the piano and turning round, full of animation, to join in the general talk. She added how dearly he was beloved in his own family for his cheerfulness and geniality. The letters of his special friends, Sir William Watson and Alexander Aubert, testify to the ever-increasing warmth of their affection. Caroline must have been with her brother to visit Aubert at Loampithill, for he wrote to William Herschel, under date "y$^e$ 18th July, 1785":

"My dear Sir,

"I received with joy the news of your Intention to come to Loampithill with your good sister, next sunday; pray come early and stay as long as you can; your sister I hope will put up with my introducing her to dine *as usual* with the family Gibson. I wish I could offer her a Bed but that is not in my power; as to you I have always one at your service in my apartment.

"I long to see how your calculations of my observations of the Georgium Sidus agree with mine, they have been ready some time and if I had not heard from you I should have sent them to you soon.

"I shall ever be ready to go to Bedlam with you for I shall ever glory in resembling you in everything and going wherever you go, both *here* and *hereafter*; more I think cannot be said, but I think I run no danger in being content with sharing the same fate with one I have so sincere a regard for.

I am yours most cordially
ALEX. AUBERT".

The remark in this letter about going to Bedlam together no doubt refers to the reception accorded by the Royal Society to Herschel's first paper about the Milky Way; in the next, Aubert describes the astonishment excited by the second paper, which had been read on the 7th of February.

London. 22nd February, 1785

"My dear Sir,

"I hope you and your sister are in perfect health. I begin by returning you and her my sincere thanks for your kind visit and whenever you are inclined to come again you will make me happy. Your sister should see Loampithill in summer, it is quite another place.... Your last paper has given much satisfaction to ye learned and furnished much amazement to the unlearned; I make no doubt of its being better understood by the last when they see it in print. I had an opportunity of talking much about you with his Royal Highness the Prince of Wales, who in the kindest manner takes much notice of me at our Hanover Square concert. He asked me what I thought of your astronomical abilities and improvement of telescopes, and I leave you to judge what I said. He has promised to come to Loampithill. If I could give him a turn for Astronomy and Philosophy I should reckon myself doing what I think would contribute to his happiness.

I am yours most cordially<br>ALEX. AUBERT."

This allusion to the Prince of Wales is interesting, as it was he, when Prince Regent, and not his father, George III, who bestowed on William Herschel the honour of knighthood.

London, the 19 Oct. 1786

"My dear Friend,

"I have sent this day to Webb & Burt's *Reading* Waggon, which goes from Gerrard's Hall Inn, Basing lane, the Regulator of Shelton and the astronomical Quadrant of Bird, directed to you and I hope they will reach you in good order; they are both carefully packed.

"The Regulator was Shelton's own and which he made for himself; the work is almost as fine as watch-work and it has a gridiron pendulum which upon trial is perfectly compensated for heat and cold.

"The Quadrant is one of Bird's best divided ones; it has only the arc of 90° and was on that account the more carefully divided; he did it when he was in his prime. I lent it some time ago to Count de Bruel and M. de Zach; they found the micrometer screw somewhat bent; I have had it made as good as ever and I hope you will find it a very good thing.

"I am now, my dear Friend, going to try the generosity of your Heart; I believe you will have no objection to give mine one of those sensations which will make me happy; accept of these Two Instruments freely and without scruple as a mark of my sincere affection, proceeding from your merit, your ability, your modesty, and your regard for me. You cannot hesitate in indulging me for I have not bought them for you; I happened fortunately to have them by me; I can spare them very well having of both sorts more, and I can, thank God, do without the value of them. I sincerely wish they may please you and when you use them, and they perform to your mind if you think of me it will be a great pleasure to me to judge something puts you in mind of a sincere friend. Let me know if all got down safe. I hope your brother is not gone and that the clock gets down time enough for him to put it up for you. My best compliments to your sister. I will endeavour soon to pay you a visit.

I am with the greatest regard, &c.

ALEXR. AUBERT."

The Shelton clock given by Aubert to William Herschel is still in the possession of his great-grandson at Slough, and still keeps excellent time. It has ever been regarded with affectionate veneration by William Herschel's descendants for the sake of the generous donor. The quadrant by Bird is also preserved among the Herschel relics at Slough.

Sir William Watson's attachment to William Herschel took the form of inducing him, first to have his bust taken by a sculptor named Lochee, and then to have his portrait painted by Abbott. This is the picture which now hangs in the National Portrait Gallery in London. Sir William Watson wrote, on "Xmas Day, 1784":

"I have had my picture painted and it is so well painted, and is universally said to be so like, that I must beg the favour of you, whenever you are most at leisure, to set to him for me. When you are in town on full moon nights you may perhaps spare an hour early in the morning, and the thing may in this way be done without much inconvenience or loss of time. His name is Abbott and he lives in Great Russell Street (where the Museum is). He will, I doubt not, exert his very best, as he will be sensible that he is painting for posterity as well as for the present time".

In the same letter Sir William Watson alludes to the profile portrait of William Herschel which Wedgewood had just had made from a design by Flaxman. He wrote:

"I have got one (and I could only procure one) medallion of you which I have presented to Sir Joseph, who has put it into a frame and it is hung up over the fireplace in the inner room of his library. The ground is blue. Though I don't think it by any means a bad likeness, I could wish it to be more so".

## CHAPTER XII

## 1783   1785

William Herschel's astronomical work. State of science. Sir Joseph Banks, Priestley, Cavendish. Variability of Algol. Letters—William Herschel to Prof. A. Wilson with summary of "First Paper on the Motion of the Sun", Aubert. Former theories on Construction of the Universe, Thomas Wright. Extracts from Herschel's first and second papers "On the Construction of the Heavens"—reception by the Royal Society. John Herschel's remarks on Prof. Pictet's disapproval. William Herschel as a mathematician.

IN spite of the almost feverish activity over telescope-making and the diligence with which Herschel continued his systematic observations at night, he found time and energy, during the three years which he spent at Datchet, to send to the Royal Society three most important papers, in which for the first time he developed his theories on the subject which was the main object of all his researches—the structure of the heavens.

The paper, "On the proper Motion of the Sun and Solar System", was read at the Royal Society on the 6th of March, 1783; the first paper "On the Construction of the Heavens" on the 17th of June, 1784, and the second paper on the same subject on the 3rd of February, 1785. Besides these he had contributed a long paper on the planet Mars, and a second "Catalogue of 498 new Double Stars".

To appreciate the enormous strides which his mind was taking, one must consider the state of scientific knowledge at the time. It is indeed difficult to realize the condition of science generally a hundred and fifty years ago. It was only towards the end of the eighteenth century that chemistry was being disentangled from the clogging notions of the alchemists and was entering the experimental stage. Linnaeus and Buffon had just inaugurated the systematic classification of plants and animals; but astronomy, based as it is on mathematics, held the place of honour.

It is therefore rather remarkable that the position of President of the Royal Society should have been held at this time by Sir Joseph Banks, who had no claim at all to be ranked as a mathematician. He had devoted all his life to the study of natural history, and had accompanied Captain Cook on his famous voyage to the South Seas. Among his contemporaries he was held in great and, one may say, affectionate regard, both for his munificence in the cause of science and for his open-handed hospitality and kindness. He stood high in the favour of the King, who knighted him on his return from the South Seas, and was influential in securing his election as President of the Royal Society.

To William Herschel, Sir Joseph Banks was ever a loyal friend, while Herschel on his part, though he had chosen the service of astronomy for himself, was far too broad-minded to deny the claims of other branches of research. We have already noted the interest he took in his brother Dietrich's entomological collection.

The study of physics and chemistry was held in better repute than that of natural history, as these sciences admit of mathematical demonstration. Their two most distinguished representatives among the Fellows of the Royal Society at this time were Priestley and Cavendish. The establishment of the new era in chemistry was the work of Lavoisier, Priestley and Cavendish, and it is interesting to notice that the dates associated with their most epoch-making experiments almost synchronize with that of William Herschel's discovery of Uranus. It was in 1781 that Cavendish effected the synthesis of water.

Henry Cavendish was highly esteemed as a mathematician as well as for his physical researches; we have seen how Hornsby referred to his calculations of the orbit of the new planet. In spite of his taciturnity he was on very friendly terms with all the men of science of his day, and earned their gratitude for the generous manner in which he placed his extensive library at their service. Such a permission was much valued, and William Herschel was glad to avail himself of it at a time when he was engaged in a controversy with Michell over the sensitiveness of the human eye.

There was, in the spring of 1783, considerable excitement in astronomical circles over the suggestion that the variability of Algol was due to a dark body, perhaps a planet, revolving round the star. Sir Joseph Banks wrote to Herschel:

Soho Square. May 3, 1783

"Dear Sir,

"I learnt at the R. Society that the periodical occultation of the light of Algol happened last night at about 12 o'clock; the period is said to be 2 days, 21 hours, and the discovery is now said to have been made by a deaf and dumb man, the grandson of Sir John Goodricke, who has for some years amused himself with astronomy. This is all I have yet made out; you may depend on any intelligence which I think likely either to amuse or instruct you.

Your faithfull servant
JOS. BANKS".

In compliance with a request from Sir Joseph, Herschel turned his telescope on Algol and communicated the result of his observations to the members of the Royal Society at an evening meeting. This led to a little unpleasantness, as Mr Goodricke attributed to Herschel the intention of wresting the honour of the discovery from him. Mr E. Pigott, a friend of Goodricke, who had been recently staying at Slough, wrote to Herschel about this and received a long letter in reply, explaining all that had occurred in a manner which seems to have entirely satisfied Mr Goodricke, as he and Herschel carried on a friendly correspondence for some years after. This incident would hardly be worth recording, if it were not one example of occasions when Herschel was accused of not being sufficiently ready to give credit to the work of other men.

Up to Herschel's time, though it was well known the stars were at different distances from the earth, yet for convenience' sake it was still usual to think of them as fixed points on a chart, spread out in imagination on the inside of a globe, within which the observer was supposed to be placed. The perfecting of such charts constituted a great part of the work of professional astronomers. Yet from the time of Galileo onwards philosophical writers had speculated on the true condition and arrangement of the stars in space; that is, on the actual structure of the universe.

Once the fact was generally accepted that the planets revolve round the sun, and Newton had demonstrated that this motion was due to the action of gravitational forces, it was natural that philosophical thinkers should speculate upon the possibility of similar forces acting among other celestial bodies. Some support was given to these speculations when it was found, by Cassini in 1730, and by several later observers, that a good many stars had undoubtedly changed their position since the time of Tycho Brahe.

An English astronomer, Thomas Wright, had propounded the idea that the whole solar system was in motion round a central body, as the planets revolve round the sun. The same idea was held by Lambert, a distinguished German mathematician at Berlin, and it was further elaborated later by the philosopher Kant.

Herschel does not seem, at the time of writing his first paper, "On the Motion of the Sun", to have been acquainted with the works of these writers; but whether the idea of the motion of the solar system in space had occurred to him independently, or had been suggested by others, is not of such importance as the way in which he attacked the problem. It was by close reasoning upon the facts disclosed by actual observations that he was able to prove that the sun was moving towards a particular region of the heavens.

In a letter to Prof. Alexander Wilson, of Glasgow, he gave a synopsis of this paper, which will sufficiently indicate its contents and also show his anxiety to do full justice to other writers.

*To the* REV. DR WILSON, *College, Glasgow*

March 3rd, 1783

"Dear Sir,

"I have the honour of your letter and esteem myself particularly happy to be favoured with the correspondence of a Gentleman so dear to all lovers of science and to Astronomers in particular. Your very interesting tract on the motion of the Sun has given me the greatest pleasure; what is remarkable is that I am just returned from London, where I have been to deliver to Sir Joseph Banks a paper on the same subject. I was apprehensive that what I had wrote on the motion of the Solar System might be thought too much out of the way to deserve the notice of Astronomers; but since I have seen the contents of the valuable tract you have sent me, I am not without hopes that what I have said will be received, by a few at least, with no disapprobation; and if you should be one of that number I shall think myself particularly flattered. You will perhaps like to hear a short report of that paper as it may be a good while before a copy of it can come to you.

"From my last observations I give an account of a great number of changes amongst the fixed stars; such as—stars that are lost—that have changed their magnitude—that have moved from their former place—new ones come to be visible, &c. Hence I conclude that it is highly probable that every star is more or less in motion. By analogy

I apply this to the Sun as one of the fixed stars. I deliver afterwards a method that will easiest enable us to detect the direction and quantity of the solar motion and apply it then to observations that have already been made, and facts that are known; whence at last I draw the conclusion that our Sun is now actually moving with a velocity greater than that which our Earth has in her annual orbit, towards a point of the heavens in the northern hemisphere not far from $\lambda$ or $\rho$ Herculis. I have not seen what Mr De la Lande has said upon the subject, but will endeavour to see it. I ought to mention also that I have used the theory of attraction in support of my Hypothesis of the motion of the Solar System. And as I find the Tract you have favoured me with rests on the same foundation, I shall not fail to add a note, or Postscript to my paper (if it should be honoured with an impression in the *Transactions*) to pay the proper respect of quotation to that Tract as well as to De la Lande's Idea; tho' as to the latter, it does not appear to me at present in what manner the rotatory motion upon an axis is connected with progressive movement.

. . . . . . .

Your most obed$^t$ serv$^t$
WM. HERSCHEL."

The paper was, apparently, favourably received by the Royal Society. Aubert wrote to Herschel:

London. 7th March, 1783

"My dear Sir,

"We finished last night your paper on y$^e$ motion of our whole system; it pleased me much and I think you bring very strong proofs of your Ideas; the paper seemed to give general satisfaction; it was much commended after the meeting and I wish you joy of the rapid increase and accumulation of your fame.

I am yours very sincerely and gratefully
ALEXR. AUBERT".

Other friends, among them Mr Goodricke of York, wrote expressing interest and general approval of Herschel's conclusion. Dr Maskelyne criticized some minor points and advised Herschel to separate his account of the movements observed amongst the fixed stars from his hypothesis of the solar motion, making two papers instead of one; but he did not commit himself, in writing, to agreement with the main argument. Indeed, notwithstanding the fact that Laplace and Prévost, two of the highest authorities, accepted Herschel's conclusion, the general impression, both at the time and for long afterwards, was that it was based on insufficient evidence. In 1833, his son, Sir John Herschel, mentioned it in his *Treatise on Astronomy*, only to say that: "in the general opinion of astronomers it was too soon to arrive at any secure conclusions of this kind". But a few years later two most competent observers, Argelander and Struve, after examining the proper motions of 360 stars, found that the point in the heavens, which Herschel had assigned as that towards which the sun was moving, only differed by less than two degrees from the average of their calculations.

In a second paper on this subject, published in the *Philosophical Transactions* for 1803, Herschel reviewed the further evidence for the motion of the sun and estimated its velocity. The position he then assigned to the apex was not so near the truth as his first one; some element of chance had favoured the choice of the thirteen stars by which he had then determined the point.

The most diligent observer of the heavens before Herschel was Tycho Brahe; it was upon the careful records which he had made of the motions of the planets that Kepler was able to base the calculations by which he explained their true motions. But as a theorist Tycho was not happy; he opposed the Copernican hypothesis, and the system which he proposed as a substitute was absurd. It is rather to Galileo that we must look for a parallel to William Herschel's contribution to scientific progress. Merely as an inventor and optician Herschel would have had a claim to fame, though not so great as Galileo's, but as an observer he far surpassed him. In this department William Herschel has but one rival, his own son. It was a gigantic task which the two Herschels laid upon themselves; the father made a complete systematic survey of the northern visible heavens, examining closely, with vision fortified by new and powerful telescopes of his own construction, every object which in turn presented itself to his view; the son even surpassed his father in this field, for he repeated and confirmed his father's observations in the northern hemisphere, and, taking his telescope to the Cape of Good Hope, he made a similar pioneering survey of the southern sky.

As has been already said, Herschel was not the first to formulate a theory of the construction of the universe. Thomas Wright, in his very remarkable work, *Theory of the Universe*, maintained that the Milky Way was a vast stratum of stars of inconsiderable thickness in which our sun was placed eccentrically. This conception of the Galaxy is practically identical with that advanced by Herschel in his first paper on Construction. Wright ventured even further; though he considered that all the stars visible to us are included in the system of the Milky Way, he hinted at the possibility of other systems existing, resembling it, but far beyond it in the immensity of space.

"That this, in all probability, may be the real case", he wrote, "is in some degree made evident by the many cloudy spots, just perceivable by us, as far without our starry regions, in which, although visibly luminous spaces, no one star or particular constituent body can possibly be distinguished: those, in all likelihood, may be external creations, bordering upon the known one, too remote even for our telescopes to reach."

Herschel, with his large telescopes and high powers, was able to resolve many of these "cloudy spots" into clusters of stars, and thus to give an observational basis to the hypothesis that some at least of them were systems resembling the Milky Way.

For more than six years he had systematically reviewed the heavens before he ventured to publish the deductions he had drawn from his observations. In the introductory paragraphs of his first paper, "On the proper Motion of the Sun and Solar System", he gave an account of his several reviews, as follows:

"...I have now almost finished my third review. The first was made with a Newtonian telescope, something less than 7 feet focal length, a power of 222, and an aperture of $4\frac{1}{2}$ inches. It extended only to the stars of the first, second, third and fourth magnitudes. Of my second review I have already given some account; it was made with an instrument much superior to the former, of 85·2 inches focus, 6·2 inches aperture, and power 227. It extended to all the stars in Harris's maps, and the telescopic ones near them, as far as the eighth magnitude. The catalogue of double stars, which I have had the honour of communicating to the Royal Society, and the discovery of the Georgium Sidus, were the result of that review. My third review was with the same instrument and aperture but with a very distinct power of 460, which I had already experienced to be much superior to 227, in detecting excessively small stars, and such as are very close to large ones. At the same time I had ready at hand smaller powers to be used occasionally after any particularity had been observed with the higher powers, in order to see the different effects of the several degrees of magnifying such objects. I had also 18 higher magnifyers, which gave me a gradual variety of powers from 460 to upwards of 6000, in order to pursue particular objects to the full extent of my telescope, whenever a favourable interval of remarkably fine weather presented me with a proper opportunity for making use of them. This review extended to all the stars in Flamsteed's catalogue, together with every small star about them, as far as the tenth, eleventh, or twelfth magnitudes, and occasionally much farther, to the amount of a great many thousands of stars. To shew the practicability of what I have here advanced, it may be proper to mention, that the convenient apparatus of my telescope is such, that I have many a night, in the course of eleven or twelve hours of observation, carefully and singly examined not less than 400 celestial objects, besides taking measures of angles and positions of some of them with proper micrometers, and sometimes viewing a particular star for half an hour together, with all the various powers of my telescope".

Herschel's first paper, "On the Construction of the Heavens", was read before the Royal Society in June, 1784. In it, he describes the method of observation which he had adopted, and which he called "gaging the heavens". It consisted in simply counting the stars of all magnitudes in single fields of view, of 15 inches diameter, taken indiscriminately in every part of the sphere visible in our latitude, and carefully recording the number of stars in each. He hoped by this process to arrive at an approximate measure of the depth of space occupied by stars in each region. He believed that as his telescope would be able to reveal stars too faint, by reason of their distance, to be seen with ordinary powers, he would by this method be able to drop, as it were, a sounding line into space and to gain knowledge of the form and extent of the universe of stars in which we exist.

It will be seen that this method involved the assumption that all stars are fairly equally distributed in space and that it would be justifiable to regard fainter and smaller ones as more distant. Such guesses at truth are like scaffoldings on which men may raise themselves to gain higher levels of knowledge. Herschel expected that it would be criticized and defended it in his second paper in these words:

"It may seem inaccurate that we should found an argument on the stars being equally scattered, when in all probability there may not be two of them in the heavens, whose mutual distance shall be equal to that of any two other given stars; but it should be considered that when we take all the stars collectively, there will be a mean distance (apart) which may be assumed as the general one; and an argument founded on such a supposition will have in its favour the greatest probability of not being far short of the truth".

In the same manner he defended the assumption that, on the whole, the stars might be considered as having an average size, and therefore that the smaller they appear the more distant they are.

The paper began with a short description of his new 20-foot telescope, after which he continued:

"It would perhaps have been more eligible to have waited longer in order to complete the discoveries that seem to lie within the reach of this instrument, and are already in some respects pointed out by it. By taking more time I should undoubtedly be able to speak more confidently of the *interior construction* of the heavens and its various *nebulous and sidereal strata* (to borrow a term from the natural historian) of which this paper can give as yet only a few outlines, or rather hints. As an apology however, for this prematurity, it may be said that the end of all discoveries being communication, we can never be too ready in giving facts and observations, whatever we may be in reasoning upon them.

"On applying the telescope to a part of the *via lactea* I found that it completely resolved the whole whitish appearance into small stars, which my former telescope had not light enough to effect. The portion of this extensive tract, which it has hitherto been convenient for me to observe, is that immediately about the hand and club of Orion. The glorious multitude of stars of all possible sizes that presented themselves here to my view was truly astonishing; but as the dazzling brightness of glittering stars may easily mislead us so far as to estimate their number greater than it really is, I endeavoured to ascertain this point by counting many fields and computing, from a mean of them, what a certain portion of the milky way might contain. Hence—we gather—the quantity which I have often seen pass through the field of my telescope in one hour's time, could not well contain less than fifty thousand stars, that were large enough to be distinctly numbered.

"The excellent collection of nebulae and clusters of stars which has lately been given in the *Connaissance des Temps* for 1783 and 1784 leads me next to a subject which, indeed, must open a new view of the heavens. As soon as the first of these volumes came to my hands, I applied my former 20-feet reflector of 12-inch aperture to them; and saw with the greatest pleasure that most of the nebulae, which I had the opportunity of examining in proper situations, yielded to the force of my light and power, and were resolved into stars.... When I began my present series of observations I surmised that several nebulae might yet remain undiscovered, for want of sufficient light to detect them and was, therefore, in hopes of making a valuable addition to the clusters of stars and nebulae already collected and given us in the work already referred to, which amount to 103. The event has plainly proved that my expectations were well founded, for I have already found 466 new nebulae and clusters of stars, none of which, to my present knowledge, have been seen before by any person.

"It is very probable that the great stratum, called the Milky Way, is that in which the sun is placed, though perhaps not in the very centre of its thickness. We gather this from the appearance of the Galaxy, which seems to encompass the whole heavens, as it certainly must do if the sun is within the same. For suppose a number of stars arranged between two parallel planes, indefinitely extended every way, but at a given distance from each other; and calling this a sidereal stratum, an eye placed somewhere within it will see all the stars in the direction of the planes of the stratum projected into a great circle which will appear lucid on account of the accumulation of the stars; while the rest of the heavens, at the sides, will only seem to be scattered over with constellations, more or less crowded according to the distance of the planes or number of stars contained in the thickness or sides of the stratum".

In this paper Herschel goes no further than to consider the Milky Way as a dense stratum of stars, narrow in width and depth but of uncertain radial extension. He now worked on in the hope of plumbing this stratum and of discovering whether it stretched beyond his field of vision or could be shown to have a limit and a form. Early in the following year he sent a second paper on the same subject to the Royal Society, which was read on the 3rd of February, 1785.

During the intervening months he had not only continued to add to his "Catalogue of Double Stars", of which he sent a second instalment to the Royal Society in December, 1784, containing 434, mostly new, but had also examined attentively an immense number of those mysterious bodies, the nebulae, and had classified them into groups. The idea now forced itself upon his mind that our sidereal system, which, in his first paper, he had described simply as a stratum of stars, was in fact an example of a nebula, comparable to one of his third class. In order to explain this view, and to indicate how he imagined such an arrangement of stars could have arisen, he began his second paper with a sketch of what would have been the action of gravitation upon an original chaos of evenly distributed stars of various sizes. He described the evolution of nebulae on this supposition, and then claimed that the appearance which one of them would present, to an inhabitant within it, would be just what we now see in the constellations and luminous tracts of the Milky Way.

A few extracts from this second paper "On the Construction of the Heavens" will show how he developed these ideas:

"That the Milky Way is a most extensive stratum of stars of various sizes admits no longer of the least doubt; and that our sun is actually one of the heavenly bodies belonging to it is as evident. I have now viewed and gaged this shining zone in almost every direction and find it composed of stars whose number, by the account of these gages, constantly increases and decreases in proportion to its apparent brightness to the naked eye. But in order to develop the ideas of the universe, that have been suggested by my late observations, it will be best to take the subject from a point of view at a considerable distance both of space and of time.

"THEORETICAL VIEW

"Let us then suppose numberless stars of various sizes, scattered over an indefinite portion of space in such a manner as to be almost equally distributed throughout the whole. The laws of attraction, which no doubt extend to the remotest regions of the fixed stars, will operate in such a manner as most probably to produce the following most remarkable effects."

He then proceeds to show that condensation about one or more stars larger than the rest will take place and produce a variety of forms which he classifies as follows:

Form I. A globular cluster about a single centre of attraction.

Form II. An irregular cluster composed of several condensations drawn together around a common centre of gravity.

Forms III and IV. Still more irregular clusters formed by combinations of the foregoing.

And lastly he considers the formation "as a natural consequence of the former cases" of "great cavities or vacancies, by the retreat of the stars towards the various centers which attract them".

"At first sight", he continues, "it will seem as if a system, such as it has been displayed in the fore-going paragraphs, would evidently tend to a general destruction, by the shock of one star's falling upon another. It would here be a sufficient answer to say, that if observation should prove this really to be the system of the universe, there is no doubt but that the great Author of it has amply provided for the preservation of the whole, though it should not appear to us in what manner this is effected. But I shall moreover point out several circumstances that do manifestly tend to a general preservation; as, in the first place, the indefinite extent of the sidereal heavens, which must produce a balance that will effectually secure all the great parts of the whole from approaching to each other. And here I must observe, that though I have before, by way of rendering the case more simple, considered the stars as being originally at rest, I intended not to exclude projectile forces; and the admission of them will prove such a barrier against the seeming destructive power of attraction as to secure from it all the stars belonging to a cluster, if not for ever, at least for millions of ages. Besides, we ought perhaps to look upon such clusters, and the destruction of now and then a star, in some thousands of ages, as perhaps the very means by which the whole is preserved and renewed. These clusters may be the laboratories of the universe, if I may so express myself, wherein the most salutary remedies for the decay of the whole are prepared."

Herschel believed that his telescope, which he claimed could show distinctly stars 497 times the distance of Sirius, had enabled him to reach to the utmost confines of the Milky Way, and he stated the conclusion at which he had arrived with considerable confidence.

"*We inhabit*", he wrote, "*the planet of a star belonging to a Compound Nebula of the third form*...."

"It is true that it would not be consistent confidently to affirm that we were on an island unless we had actually found ourselves everywhere bounded by the ocean, and therefore I shall go no farther than the gages[1] will authorise; but considering the little depth of the stratum in all those parts which have been actually gaged, to which must be added all the intermediate parts that have been viewed and found to be much like the rest, there is but little room to expect a connection between our nebula and any of the neighbouring ones."

He then gives a diagram of what he conceived would be the form of a section drawn through the Milky Way at right angles to its plane. The form which he assigned to it is that generally known as the disc or grindstone theory of the Galaxy, as distinct from that of a ring encircling the sun, the improbability of which he had demonstrated. The rest of the paper is taken up with the description of the more remarkable nebulae which he had observed, concluding with those which he called "Planetary Nebulae"; "these," he said, "from their singular appearance, leave me in doubt where to class them".

From the point of view of a biographer, the most remarkable thing about these papers is the simple and ingenuous manner in which he sets forth his astounding theory: that the Milky Way with all the visible sidereal system is just one nebula among many. His mind had soared to a height from which he could survey the whole universe of sun and planets, stars and nebulae, in a manner which, to many of those who listened to the reading of his papers, must have seemed fanciful or even presumptuous. In the letter already quoted, dated 22nd February, 1785, Aubert wrote to him: "Your last paper has given much satisfaction to ye learned and furnished much amazement to the unlearned; I make no doubt of its being better understood when they see it in print".

There is little allusion to the subject from other English correspondents. Goodricke wrote in a non-committal manner, after reading the first paper:

"I have read your curious paper on the construction of the

---

[1] William Herschel consistently spells "gauges" as in the text, and this spelling has been followed in all quotations from his writings.

Heavens with great pleasure. It seems as if all the riches of the heavens are now opened to us by means of your large telescopes and I heartily wish you success in the farther pursuit of the subject".

In condonation of this indifference, it should be remembered that the wonderful and varied appearances of the nebulae, which so excited Herschel's imagination, had as yet been seen by him alone; no other telescope than his own new 20-foot could reveal these marvels. For instance, Messier, the only person who had given much attention to these objects, had described the 98th nebula in his Catalogue as "nébuleuse sans étoiles, ronde et apparente".

To Herschel it appeared:

"A cluster of very close stars; one of the most beautiful objects I remember to have seen in the heavens. The cluster appears under the form of a solid ball, consisting of small stars, quite compressed into one blaze of light, with a great number of loose ones surrounding it, and distinctly visible in the general mass".

Besides this, the astronomers were still wholly obsessed by planetary notions; they had as yet hardly digested the discovery of Uranus, and were eagerly discussing the suggestion that the star Algol was accompanied by a satellite. A few years later the discovery of the little planets, filling the gap between Mars and Jupiter, and even more the publication of Laplace's monumental work, the *Mécanique Céleste,* all contributed to concentrate attention on our own little world; the time was hardly ripe for Herschel's grandiose visions, but his pioneering conception of the Galaxy as a gigantic nebula has been fully established in our own century.

But perhaps the attention which has been concentrated on Herschel's description of the Milky Way has made commentators overlook the interest and importance, for its influence on the philosophy of science, if one may so speak, of his attempt to trace the *process of evolution* by which the nebulae have assumed their present form and of the vast ages which such processes demand. In this attempt he was breaking new ground and opening up a new field for scientific inquiry. Hitherto, all scientific research had been directed to the investigation of *things as they are*; accumulating facts, systematizing their arrangement into classes and orders, discovering the forces existing in nature and the laws of their action; but to the question, if any one had been so bold as to ask it: "How did these things come to be?" the answer would have been simply: "God made them so when He created the world".

Herschel was the first to introduce a disturbing factor into this view of Creation by his suggestion that it was a long process, not a sudden and completed act. Perhaps one reason for the coldness with which these papers were received by the Royal Society may be the fact that thought was not yet so free in England as in France and Germany. At this time no one could hold any high position in either of the English Universities or in the teaching profession unless he took orders in the Church. The Astronomer Royal was a clergyman as well as Dr Hornsby at Oxford and the Professor of Astronomy at Edinburgh. Under these circumstances it was natural that there was a certain reluctance to show approval of Herschel's theories, which seemed to run counter to the accepted interpretation of the Biblical account of Creation. It would seem that in Geneva, also, Herschel's theories were considered dangerous. In 1824, John Herschel wrote to Piazzi:

"I understand from M. Arago that an article has appeared in the *Bibliothèque Universelle* from the pen of Prof. Pictet of Geneva, in which my Father's doctrines respecting the condensation of the nebulous matter are characterized as tending to irreligion. The charge would be contemptible were it not associated with the name of Pictet; but I hope you will take care to let it be distinctly understood that my Father, so far from contemplating such consequences, was a sincere believer in, and worshipper of, a benevolent, intelligent and superintending Deity, whose glory he conceived himself to be legitimately forwarding by investigating the magnificent structure of the Universe".

William Herschel made no attempt to defend his theory by argument; having once stated it he turned back to his telescopes to gather fresh evidence either to confirm or refute his hypothesis. At first he found nothing antagonistic to it. Four years later he gave to the Royal Society, in 1789, a fresh catalogue of 1000 new nebulae and clusters, and reiterated his opinion that these all showed evidence of a clustering tendency amongst the stars, which he still considered the only denizens of the heavens. But in the next year, while revising his former observations, he made a very important discovery, which threw an entirely new light upon the constitution of stars and nebulae and caused him to envisage a different sequence of events in the evolution of these bodies. Extracts from his later papers will be found in a subsequent chapter.

It looks as if Herschel's friends were a little anxious lest his reputation as a sober thinker might suffer from the publication of these theories about the universe. A letter from Sir William Watson, written about this time, would seem to have been prompted by some such feeling. It is dated:

"Xmas Day, 1784"

"I happen'd in conversation with Sir Joseph and Dr Blagden to mention that in your publication you had not had any opportunity of showing the world how well versed you are in the deeper parts of Mathematicks. They both agreed with me that, if an occasion should offer and you was to embrace it, your reputation would immediately rise with great rapidity. I cannot but be extremely solicitous that the world should know you to be the man you are in that branch of science; this would besides add great weight to your other publications".

William Herschel paid no attention to his friend's suggestion; he was indifferent to mere fame and was modest enough to think that his mathematical attainments, acquired entirely by private study, though adequate for his practical work, would be the better for professional testing. He was in the habit of referring his calculations to his friend, Prof. Vince, at Cambridge, before publishing them; though he did not always discard his own method for Vince's. He was also grateful to the French mathematicians for help and corrections. Writing to M. Lalande at Paris in 1788 he said:

"My dear Friend,

"I have received your obliging letter and give you many thanks for the honour you have done me to look over my paper on the new planet and its satellites. Do me the favor also to give my best compliments to Mr De Lambre, to whom I am no less obliged for the trouble he has taken to go over the tedious calculations contained in that paper. I have so little leisure for practice that it would be no wonder, on account of the multiplicity of things that take up my time and continually disturb my thoughts, when I am calculating, if I had made many more blunders than I have made.

. . . . . .

"Your guess that I had written 80 instead of 88 is perfectly right.

"Upon the whole I once more return you and Mr De Lambre thanks for your friendly notice of my blunders, which I hope you will ascribe to my busy situation.

&c."

This confession recalls one which he made in a letter to Dr Blagden concerning one of his sister's comets. He admitted that he had made a mistake in a logarithm, and added:

"When I consider that I was obliged to make these calculations while I was polishing, with 12 men about me who were talking, and sometimes singing, on all sorts of subjects, it is a wonder that no more mistakes found their way to my pen.—Pray tell Sir J. B. that I hope when he hears of this blundering addition, it will cure him of saying: 'I never knew Herschel in the wrong'".

"Slough. Nov. 27, 1788"

In the Postscript to his paper, "On the proper Motion of the Sun and Solar System", Herschel had drawn attention to a curve which he described as a "non-descript, as far as I can find at present, and may be called a Spherical conchoid from the manner of its generation".

In a letter to M. Jeaurat, of the Royal Observatory at Paris, dated 10th March, 1784, he wrote (in French, of which the following is a translation):

"Do me the favour to present my thanks to M. De la Place for having communicated to me the elements of the Georgium Sidus. I wish very much that that celebrated mathematician would examine the nature of the curve which I have called 'spherical conchoid' and of which he will find the description on the 229th page of the memoir which I take the liberty of sending to you for M. De la Place....

"Concerning the curve of which I have just spoken, it would give me much pleasure to see an algebraic analysis of it, in which however the abscissa and ordinate would be curved lines. I find that the orthographic projection gives a curve of four dimensions".

(The full description of this curve will be found in *Phil. Trans.* vol. LXXIII, 1783, p. 279, or *C.S.P.* vol. I, p. 128.)

## CHAPTER XIII

## 1784    1794

Foreign correspondence—Lalande, Schroeter, Bode, etc. Abstract of Herschel's first paper "On the Construction of the Heavens" in letter to Bode. Lichtenberg's appreciation. Abstract of second paper "On the Construction etc." in letter to Lalande. Correspondence with Schroeter. Observations on Venus. Volcanoes in the moon.

WILLIAM HERSCHEL corresponded with most of the eminent astronomers in France and Germany, and sent copies of his papers, or abstracts of them, to Lalande, Laplace and Messier in France, and to Bode, Lichtenberg and Schroeter in Germany.

Bode was at this time pre-eminent among European astronomers, both for his original work and as editor of the *Astronomisches Jahrbuch* of Berlin. He had in his early years helped to draw attention to the importance of getting observations of the coming transits of Venus (in order to have data for estimating the distance of the sun from the earth) from different stations on the globe. But it is to Lalande that the credit is due for having organized this first international effort in the cause of science. It will be remembered that Capt. Cook's expedition to the South Seas, in 1769, was sent out by the British Government for this object. Lalande was Professor of Astronomy at the Collège de France. He twice visited Herschel at Slough and took a warm interest in Caroline and her work.

Herschel tried in vain to induce his German correspondents to write to him in French or English. His aversion to German may have been only the difficulty of reading the writing; an aversion with which anyone will sympathize who has spent weary hours in deciphering Bode's handwriting. Its illegibility was recognized by his friends, one of whom was kind enough to enclose a transcript into English characters of a letter which Bode had entrusted to him for Herschel.

The extracts which follow will show how William Herschel's discoveries and theories were regarded by these astronomers.

WILLIAM HERSCHEL *to* M. DE LA LANDE

Datchet près de Windsor. March 19, 1784

"Monsieur,

"En vous envoyant par cette occasion un mémoire sur le mouvement du soleil j'ai pensé qu'une idée qui s'accorde si bien avec ce que vous en avés dit autre fois, en parlant du mouvement de rotation, pourroit bien vous engager à trouver quelque plaisir à le lire. En même tems j'aurai l'honneur, Mons. de vous donner un petit extrait de mes observations sur la Planète de Mars tiré d'un mémoire que j'ai communiqué à la Société Royale sur ce sujet il y a quelque mois.

Monsieur,
Votre très humble ser.
Wм. Herschel."

M. DE LA LANDE *to* WILLIAM HERSCHEL

Paris. 7 avril, 1784

"Monsieur et illustre confrère,

"J'ai reçu avec bien du plaisir et de la reconnaissance votre lettre du 19 mars avec votre mémoire sur le système solaire. Ce n'est plus qu'à vous qu'il appartient de faire des découvertes, mais je suis bien flatté que mon idée, publiée en 1776, vous ait donné l'occasion de faire un si beau travail, et que vous ayés daigné me citer. M. de Luc m'a écrit que vous aviés estimé 44000 étoiles dans un espace de 8° sur 3°, je voudrois bien savoir dans quelle partie du ciel.

"Croyés vous que la blancheur de la voye lactée vienne de les petites étoiles? ou croyés vous que ce soit la matière céleste, celle qui produit les nébuleuses?

"On a calculé de nouveaux elemens pour votre planète, et l'on en a donné des tables détaillées soit dans la *Connoissance des Temps* de 1787, soit dans les Éphémérides de Milan pour 1785....

"Nous avons fait graver un nouveau caractère pour votre planète; ♅ C'est un globe surmonté de la première lettre de votre nom; vous voilà associé malgré vous à toutes les divinités de l'antiquité.

"je vous prie de faire mille complimens à mon cher ami M. de Luc, et de lui dire que quand il voudra me donner des nouvelles de vos observations, il ne les mette pas dans une lettre à M. de la

Place, que je n'ai vu que très longtems après la date; je voudrais bien savoir aussi des nouvelles de ses travaux en physique. Je le trouve heureux d'être à portée de vous voir si souvent, ainsi que M<sup>lle</sup>. votre soeur, de qui nous devons espérer bientôt quelque nouvelle comète.

"je vous fais bien mon compliment sur votre nouvelle dignité de Esq<sup>r</sup>; c'est une justice de votre digne monarque qui doit être bien glorieux de vous avoir dans ses états.

"je suis avec la plus haute considération et le plus sincère attachement,

    Monsieur et illustre confrère,

      Votre très humble et très obéissant serviteur,

            De la Lande."

### OBERAMTMANN SCHROETER *to* WILLIAM HERSCHEL

(Translated from the German)

Lilienthal. 14 Jan. 1784

"...I cannot properly describe to you how much interest is felt in Germany about your excellent instruments and how many more important discoveries we hope to hear of from you. May God preserve your health for many many years for the sake of Astronomy. Your great discovery of the movement of the Solar System towards λ Herculis seems to me very remarkable. How would the great Lambert rejoice to see that, which his far-seeing mind only guessed, brought in the same century to a nearly mathematical certainty.

"Pardon me, best patron and friend, that I have written this time in German. I thought of it too late. If you will now and then favour me with a few lines you will make the French language doubly agreeable to me.

      Your most obedient servant,

         Schroeter."

(Schroeter wrote one letter to Herschel in French, but no more. His German writing is large and clear.)

Herschel sent copies of his paper "On the proper Motion of the Sun and Solar System" to Bode and to Lichtenberg and in his letters to both he added an abstract of his first paper "On the Construction of the Heavens". That to Bode is the fullest. Some portion is given here, translated from the French (in which language he always wrote to Bode). It is of interest, as being Herschel's own summing up of the results of his observations; as far as they had as yet led him.

WILLIAM HERSCHEL "*to* MR BODE *at Berlin*"
(Translated from the French)

Datchet, near Windsor. May 18, 1784

"Sir,

"The honour you have done me in sending me your Ephemerides for 1784, and 1785, deserved an earlier acknowledgment, but the hope of being able at the same time to send you in return the Memoir on the Motion of the Sun caused me to delay. I now have the honour to send it.

. . . . . .

"A short while ago I also contributed a Memoir on the Construction of the Universe. It contains the details of observations made with a Newtonian telescope of 20 feet, having an aperture of $18\frac{3}{4}$ inches English measure. The subjects with which it deals are: the Milky Way, the Nebulae, and the layers or *Strata* of the heavens.

"I point out that the brilliance and whiteness of the Milky Way arise from small stars; that there are at least 50 thousand stars in a space 15 degrees by 2 degrees; counting only those which are large enough to be distinctly numbered. I then give the resolution of nearly all the nebulae of the *Connaissance des Temps*, which are nothing more than clusters of small stars. After this I point out that there are in the heavens several *strata* of nebulae and clusters of stars, and that I have already seen 466 nebulae and clusters of stars not hitherto known, of which I give a Catalogue.

"I give drawings of several kinds of nebulae.

"I then point out that the sun is probably situated in a *Stratum*; I mark its place, from which it follows that one can with facility explain all the phenomena of the Milky Way.

"One sees also that the situation of the sun in this *stratum* provides the means of explaining the cause of the movement of the system.

"Finally, I give the method I have employed to arrive at the knowledge of what I have just said, and which I shall in future employ to verify what, up to this moment, I can only propose as *an attempt to arrive at the knowledge* of the construction of the heavens.

"Pardon, Sir, this scrawling writing, and believe me &c.

your much obliged and obed.<sup>t</sup> serv.<sup>t</sup>

WM. HERSCHEL."

Bode, in his reply, thanked Herschel for his papers, but made no comment on their contents; he was chiefly interested in the new planet and in trying to induce Herschel to give him some original matter for his *Jahrbuch*.

Lichtenberg, to whom Herschel had sent a similar résumé of his papers, and who seems to have read the whole paper "On the Construction of the Heavens" in the *Philosophical Transactions*, was much more effusive.

### LICHTENBERG *to* WILLIAM HERSCHEL

(After receiving W. Herschel's paper on Mars, and a résumé of the first paper "On the Construction of the Heavens".)

(Translated from the German)

Göttingen. 2 August, 1784

"I have read your letter with intense satisfaction and consider it to be one of the greatest happinesses of my life to correspond with a man of so much worth. Your discoveries on Mars pleased me much and I will shortly make them public. As you now observe stars down to the 12th magnitude I have been hoping that you would present us with an 8th planet.

"Your theories concerning the structure of the Universe are excellent; the celebrated Lambert guessed at something similar and published this in his *Cosmological Letters*.... You seldom mention in your writings Mayer's *Opera Inedita*. It is the fault of the English booksellers, who do not trouble themselves about foreign publications. If you do not possess it yourself, I beg you will take the first opportunity to let me know, when I will send it by the first available courier.

. . . . . . . . .

"How much those in England recognize your merit is clear from one example. When you were staying in London, last year or whenever it was, a learned Englishman, whose name I will not mention, wrote to me: 'Your countryman, Mr Herschel, is now in Town to teach *Our* astronomers how to see'.

"If you have among your papers anything suitable for my monthly publication, will you be so good as to communicate it to me? I know you could find much, and the slightest remark or suggestion would be welcome to me. Someone told me recently that you had once continued polishing a mirror for 24 hours without desisting; if this is so, a particular account of it would be here uncommonly welcome. Also I have heard of a portrait of yourself, could I not get one? You see I am rather importunate, but in this matter I am so only for the honour of my Fatherland, which is proud of you and has good cause to be so, and longs to hear and read more, but whose curiosity cannot be satisfied with the tardy *Transactions*.

G. LICHTENBERG."

In the following year William Herschel sent copies of his second paper "On the Construction of the Heavens" to Bode, Schroeter and Lichtenberg. To Lalande he sent his first paper on that subject and an abstract of the second in a letter which is given here:

WILLIAM HERSCHEL *to* M. DE LA LANDE

(Translated from the French)

Datchet. March 17, 1785

"Sir,

"Having, by the favour of M. de Luc, the opportunity of sending you the Memoir on the Construction of the Heavens, I write at the same time in order to give you some slight details of a second Memoir on the same subject which I have sent to the Royal Society of London and which will be printed, together with a second Catalogue of about four hundred double stars, in the *Transactions* of the Society.

"In this second Memoir I treat first of the formation of the nebulae, which I classify under four forms. Then I give a table of *Star gages*; and I shew that by this manner of counting the stars present at one time in a field of view of a given diameter, I have succeeded in determining approximately our situation in the universe. I find that, by all appearance, our system of stars constitutes a nebula of the third form. I give a section of this great Nebula, which shows that it is limited rather narrowly at the sides, but much extended in length and height.

"The principles upon which I base this idea are derived from observation assisted by calculation. I have already seen 1015 nebulae, among which there are also some of the third form and which perhaps not only equal but even surpass in size the one of which our sun is one of the stars which compose it.

"Among the curiosities of the sky worth noticing is one which consists of a sort of ring of stars; it is the 57th of the *Connaissance des Temps*.

"At the end of this Memoir I give the places of half a dozen Planetary nebulae (as I have called them). They are celestial bodies of which we have not at present any very clear idea. They may be of a kind quite different from anything with which we are acquainted in the heavens. I have already found four that have a visible diameter of 15 up to 30 seconds. These bodies present a disc much like a planet, that is, equally illuminated throughout; round or somewhat flattened, and nearly as clearly defined as the disc of the planets; a light strong enough to be seen with an ordinary telescope of only one foot, but then they only appear as a star of the 9th magnitude. I hope in a few weeks to have a copy of this Memoir to send you.

"You probably know, Sir, that we have two more variable stars, namely β Lyrae and η Aquilae. I have myself seen them change their apparent magnitude, though I have not had occasion to verify the periodic times assigned to them by Messrs Goodricke and Pigott.

Believe me &c.

your humble and obed$^t$ serv$^t$

WM. HERSCHEL".

The following extract from a letter of Bode's of this date may be given as a specimen of the repeated applications for items of astronomical gossip which filled Herschel's correspondence:

BODE *to* WILLIAM HERSCHEL

(Translated from the German)

Berlin. 13 Aug. 1785

"Highly honoured Sir and Friend,

"For a long time I have hoped to be honoured with a letter from you and to hear of many astronomical novelties as the fruit of your tireless researches. Prof. von Zach led me lately to expect something, but I have waited in vain. Oberamtmann Schroeter in Lilienthal wrote to me that last autumn a ship was wrecked in which was a packet from you; this I regret deeply, but thank you for your kind trouble. Would it be possible to have another copy of this paper or memoir?

"As a slight token of my esteem and gratitude I am sending a copy of my *Jahrbuch* for 1786 and 1787, through his Excellency, Count von Bruhl.

"My most earnest request is that you would give me some information about your newest discoveries and astronomical observations, especially about a *star* which, according to Von Zach, you have *in petto*, if this is possible; or at least to tell me if it is to be kept secret as yet. You would not believe how I am pestered by lovers of astronomy about this, and I should like to give these too-curious ones something to quiet them in my next *Jahrbuch*. I hope you will not refuse me this small petition.

"I should like to be informed of the prices of your excellent telescopes, in order to be able to answer inquiries; what Mr Magellan tells me is too vague.

Your obed$^t$ &c.

BODE".

In a later letter, Bode acknowledged the arrival of the packet containing Herschel's "Catalogue of Double Stars" and the paper "On the Construction of the Heavens", and said that he would make their contents known through his *Jahrbuch*, but he expresses no interest in Herschel's theories. He was puzzled by Herschel's request for another copy of his "catalogue", but sends one of his *Celestial Atlas*; the price of which, he is careful to remark, is 4 thalers, 12 groschen. "If it is not what you want, Prof. Von Zach will doubtless be able to dispose of it."

Bode's *Celestial Atlas* was a very fine work, issued in twenty sheets and giving the position of 17,240 objects.

Lichtenberg alone, of Herschel's German correspondents, showed any real interest in Herschel's theory of the Milky Way. He wrote:

### LICHTENBERG *to* WILLIAM HERSCHEL

(Translated from the German)

Göttingen. 20 October, 1785

"Highly honoured Sir,

"Our mutual friend, Herr Planta, has brought me your priceless letter and present,[1] for which I render you hearty thanks; it was as if I received a new pair of eyes to look at the heavens. When I first looked at the engraving of the star-cluster, I took it for a nebula which you had perhaps seen outside our system, when however I read on and found that this represented a section of the nebula within which our sun is a minute point—I cannot lie—I was so enraptured at the grand conception and so filled with wonder that I felt again all the delight that I had tasted when I first understood clearly the structure of our own world-system. In one word, the reading of your paper has given me the pleasantest hour of my life for the last twenty years, for which I have to thank you. If I ever visit England again it will be solely with the intention of seeing you and your instruments before I die.

Your obedient servant,

G. LICHTENBERG".

It is rather difficult to get a clear idea of Herschel's relations with Schroeter, as the correspondence is so very one-sided; all Herschel's original letters being lost, we have only the five copies in his Letter Book to put with forty-three letters from Schroeter. Schroeter was an enthusiastic amateur astronomer who had set up a very well-appointed observatory at Lilienthal, about 12 miles from Hanover. As he kept up a lively acquaintance with all the astronomers in Germany, his residence became a central resort; it was here that Von Zach, some years later, collected the little band of observers who undertook the systematic

---

[1] W. Herschel's second paper, in *Phil. Trans.*, "On the Construction of the Heavens".

search for the missing planet, which, if Bode's law were true, must exist between Mars and Jupiter.

Herschel and Schroeter never met; Schroeter had, to his great disappointment, missed seeing Herschel when the latter visited Göttingen in 1786. He had introduced himself as being acquainted with Herschel's brothers in Hanover, and asked Herschel to make a mirror for a 4-foot reflector in his possession. Afterwards he ordered a 7-foot telescope complete, to be made by Alexander Herschel. It was probably entirely constructed under William's eye, if not by his hand. Schroeter was greatly excited about this telescope; there are five consecutive letters during the autumn of 1785, expressing his impatience. One may be quoted as an example:

OBERAMTMANN SCHROETER *to* WILLIAM HERSCHEL
(Translated)

Lilienthal. 22 Nov. 1785

"The information which you have been so good as to send me through our excellent mutual friend, Herr Best, that the fittings for my 7-foot telescope were not ready for the Courier, relieved me from a great anxiety caused by recent storms at sea; for I feared that the mirror and other apparatus might already have started.

"I hope you have received the three letters which I have written to you since August. I am daily looking for the arrival of my excellent telescope. If there is yet time I beg that you will send it by Post (and will also insure it), both for the sake of safety and of speed; addressed

Schroeter, Grand Bailli de S.M. Brit.
à Lilienthal près de Bremen, en Allemagne.

"My longing for it really oversteps the limits of philosophy.

. . . . . .

"I have shown your excellent memoirs to several astronomers and I am often besought to ask you whether these can be bought from any London Bookseller.

"Forgive me, highly esteemed patron and friend, my numerous questions; they are not all selfish. Again I beg that you will let me know if there are any publications in Germany which you wish to have.

"I am having a Lamp-Micrometer made, according to your invention, by Herr Drechsler; but a reliable Time-Regulator with a compensating Pendulum can only be got in England. Probably one such without a case would cost not more than 14 guineas. The favour of some indication to what maker I should apply, would oblige me greatly.

"May God long preserve you, for the good of the noblest science, in health and strength. I have the honour to be, with the sincerest regards,

Your obed$^t$ serv$^t$
J. H. SCHROETER".

To this letter is attached a note, in W. Herschel's hand, of the articles "sent to Mr Schroeter in a box".

A special interest attached to this telescope, as it is the only one of all those which Herschel made for sale which we know to have been used for systematic observations. Schroeter has been called the Herschel of Germany, and he deserves the name not only for his great interest in the improvements of telescopes, but more so in that he was the first purposefully to devote himself to the study of the appearances which the telescope revealed, emulating Herschel in his perseverance and enthusiasm, though he had not Herschel's discrimination in interpreting what he saw. The treatises in which he recorded his results are as prolix as his letters. He had paid special attention to the planet Venus, and his observations were noticed in the *Philosophical Transactions* of the Royal Society. Herschel also contributed a paper on the same subject, dissenting from Schroeter's views. Fortunately, copies of the letters which he wrote to Schroeter himself about this time are in the Letter Book and are given here with Schroeter's reply.

WILLIAM HERSCHEL *to* OBERAMTMANN SCHROETER

Slough. May 2nd, 1792

"Dear Sir,

"About a fortnight ago I received your letter and waited till I could have an opportunity of seeing $a$ Lyrae in order to determine whether the small star which is marked in my Catalogue had changed its situation.

"There is no mistake in the Catalogue as I have 4 observations upon $a$ Lyrae made in 1782, where the measure of the position is accompanied by a figure that leaves no room for doubt as to its accuracy. However, to satisfy you more completely I can assure you that the small star is still in the same position in which it was in 1782, with only some little variation in distance, of not many seconds. The other star you saw to the north, is also mentioned in my observations of 1782, and was not put into the Catalogue because it would be endless to mention all the small stars that are near large ones. In my 20-feet reflector I can see no less than 130 stars in the field with α Lyrae, and all which therefore must be less than 15′ from that star.

"You mention your new *Projection-Micrometer*. As I suppose that you have undoubtedly taken notice of my Camera-eye-piece &c.[1] whereby I project objects on a sheet of paper, upon a wall, upon a measuring scale, upon a set of discs, peripheries, draw images of objects, set the point of a pair of compasses so that they will exactly fit into any two holes that a person makes upon a card fixed at a distance &c. &c.—As I suppose you are acquainted with all these things I should be glad to know in what respect *your new* differs from *my old* Projection-Micrometer.

"In the years 1776, 7, 8, and 9, I measured several hundreds of the mountains of the moon. I used three different methods for that purpose. (1) The illuminating ray, (2) the length of the shadow, (3) the actual projection of the mountain. The result of my measures was that none of the mountains which I measured was quite two English miles high. A very short abstract of this you have probably seen in the *Phil. Transact.*

"I have not had time to peruse your large work on the moon, but promise myself much pleasure from the reading of it, and admire beforehand the great assiduity and labour you have shewn in such a work.

"Many thanks for the copies of your portrait; but I had much rather see the original and hope for the pleasure.

Believe me to be with great esteem,

Dear Sir, your most faithful &c. servant

WM. HERSCHEL."

[1] This mention of a *Camera-eyepiece* is interesting. It seems to have been the same as the *Camera-lucida*, by means of which Sir John Herschel made so many interesting drawings—among them the sketch of the house at Slough.

WILLIAM HERSCHEL *to* OBERAMTMANN SCHROETER

Slough. Aug. 20, 1793

"Dear Sir,

"In answer to the favour of your letter I can say that to be admitted as a member of the Imperial Academy of Sciences will be esteemed the highest honour that can be conferred on me, and I beg of you therefore to communicate these sentiments to the President of that Society.

"I shall be very happy to receive your observations on the rotation of Venus on her axis, as I confess that at present I look upon that motion as undetermined.

"I have lately given to our Society a paper on the Planet Venus, which I shall have the honour of sending to you as soon as it is printed. You will find that I have never been able to see any mountains in Venus; that to me the horns were always equally sharp; that I rather suppose you cannot have seen *flat spherical* forms on Saturn, as with my instruments, which I ought not to believe inferior to yours, I never could see any such.

"You will also find that I take notice of your not allowing for the sun's diameter, and of some other little inaccuracies, which in the course of my paper I could not help remarking. All these things I mention and beg of you not to ascribe them to any inclination of criticising, which is far from my mind, but as astronomy is an experimental science, it is incumbent upon us to give our observations as they are, without being misled by partiality to our friends.

"It will give me great pleasure to hear of your success in the great undertaking you mention, as I suppose it to be an astronomical or optical work of consequence. Pray favour me with a line as soon as you see the two satellites of the Georgian Planet, and the sixth satellite of Saturn. As to the seventh you will probably not get a sight of it till the ring becomes again to stand edgeways towards us.

Believe me to be with the greatest esteem
your faithful and most obed[t]
humble servant
WM. HERSCHEL."

OBERAMTMANN SCHROETER *to* WILLIAM HERSCHEL

(Translated from the German)

Lilienthal. 29 November, 1793

"Worthy Sir,

"The Imperial Leopold Academy of Science has entrusted to me the honourable task of conveying to you the enclosed Diploma with their best compliments, and I do this with the greatest pleasure and am ready to forward your acknowledgement.

"I greatly desire to see your essay on Venus, which will certainly interest me very much; I doubt not that Herr Best has given you mine.

"It is to be expected that observers, who have only truth faithfully and eagerly at heart, should publish their observations, even if they give different results, without regard and without reference to persons. Thus will truth prevail. So in my *Selenfragmente* I have put forward my calculations, which differ greatly from yours, without mentioning yours, though well known to me, or even suggesting the conflict between them. Those who are well acquainted with the subject can then judge for themselves; and truth will not be obscured by partisanship.

"In every respect then your Venus observations will be valuable and very welcome. Meanwhile the truth in this case depends on this; whether *both* observers had, not only at the same time but had *both specially* and with equal care, directed their attention *to one* and *the same object*, which indeed is not easily seen. Though I have observed Venus for 10 or 11 years, I have never, because I did not particularly attend to it, noticed such changes; though the varied appearance of the horns, that one is much narrower than the other—a much less noticeable phenomenon than this—has often been seen by me in the last few years.

. . . . . .

"I therefore have good ground for hoping as well as wishing that my observations on Venus will in due course receive confirmation from you as well as from other authorities.

"Official duties and bodily infirmities have prevented me lately from doing much in the way of observation. But picture to yourself, my dear Sir, a very bold undertaking! I have during the past year begun the construction of a 24-foot telescope, the polished surface of the mirror to be about 20 inches and its weight 170 pounds. The enterprise has been successful from first to last and with all its apparatus it is now finished.

. . . . . . .

"I intend to use it with your beautiful invention of 'front view' which I have already tried, and am having the fittings made for this.... I should be most grateful if you would inform me what inclination of the object glass you have found after many trials to be the best.

"May God preserve you as the glory of the noblest of sciences, in health and prosperity, is the earnest wish of

your faithful admirer

J. H. SCHROETER."

WILLIAM HERSCHEL *to* OBERAMTMANN SCHROETER

Slough near Windsor. Jan. 4, 1794

"Dear Sir,

"I have the favour of your letter and the enclosed Diploma for being admitted a member of the Imperial Academy; and beg of you to 'return my most grateful thanks to the President and members of that Academy for the high honor they have conferred upon me, and to assure them that it will be my study to render myself worthy of so great a distinction'.

"It has given me great pleasure to hear of your success in the 24 feet telescope, and I hope you will make some valuable discoveries with it. The eye glass, in the *front-view*, must be two inches more than the semi-diameter of the speculum from the center of the tube, and inclined so as to be directed to the center of the speculum; which latter, of course, must be inclined in such a manner as to throw a whole pencil of rays into the eye glass. As my tube is an octagon I have placed the slider which carries the eye glass into one of the corners; which makes it very convenient for the right eye. I have another on the opposite side, in case I should want the left eye for any long time; but, of course, the inclination of the speculum will require an alteration.

"You mention two very high mountains that were visible in the last eclipse of the sun; I have given a paper on the subject of this eclipse to the Royal Society some time ago, in which these mountains are mentioned; as soon as I can obtain a copy of the paper I will send one for you.

"With this letter you will also receive my observations on the planet Venus. I shall be very glad to hear what you will say on the subject of my Planetary Nebulae when you come to view them with your 24-feet telescope. They are very curious objects.

Believe me with the greatest regard,
Dear Sir, your faithful and most obed$^t$ servant
WM. HERSCHEL."

There are passages in the paper on Venus, which Herschel sent to Schroeter, alluding to the latter's observations on that planet, which he intended to be taken, to use an expression of his own, in a "bantering" sense. That he saw nothing in them likely to offend is obvious from his sending the paper himself accompanied by such a friendly letter. No letters from Schroeter are to be found in the collection for the rest of the year 1794, nor for the next, though there are manuscript copies of observations, showing that the correspondence was maintained. In the following year, 1796, Herschel sent a copy of his "Description of a Forty-feet Telescope" and kept Schroeter's reply, which ran as follows:

OBERAMTMANN SCHROETER *to* WILLIAM HERSCHEL

(Translated from the German)

Lilienthal. 30th Sept. 1796

"Highly honoured Doctor,

"By a chance absence from home your precious, splendid gift has only lately come into my hands; and I hasten to offer my liveliest, heartfelt, unbounded thanks for it. Indeed my pleasure in receiving it is for many reasons indescribable. Your forty-foot Reflector is in truth a monument of astronomical and mechanical ingenuity. It shows how far human perseverance and zeal for the sublimest science can attain; though I am quite of your opinion that wide apertures do interfere with distinctness and that size renders the use of very large instruments more difficult and limited, yet in special cases, where increase of light is called for, they are of the greatest service, as your many important discoveries abundantly prove".

The letter concludes in Schroeter's usual manner with a prayer to Providence for Herschel's continued well-being, and is signed "With boundless heartfelt regard, your obed$^t$ serv$^t$ H. Schroeter".

In the following year, 1797, Schroeter secured the services of a very clever young observer, Harding, and together they made many observations on the newly discovered little planets, which they refused to call by the name "asteroids" given them by Herschel. The last letter is written in 1803. The outbreak of war is probably accountable for the absence of any more letters. The correspondence broken off was not renewed and we do not know what Herschel's feelings were if he heard later of the terrible misfortune which befell his friend, when the French soldiery, in 1813, sacked the town of Bremen and broke up and destroyed the observatory, as well as all the records of Schroeter's observations. He never recovered from the shock of this catastrophe and died three years later.

Lalande was almost as constant a correspondent of Herschel as Schroeter, but his letters, though written in an almost microscopic hand, are more legible and much shorter. He frequently sent a message to Caroline, for whom he had so great an admiration that he named his niece, who happened to be born on the day when Caroline discovered a comet, by her name. In November, 1786, before he was personally acquainted with the Herschel family, he wrote (in French):

> "Recevés un hommage que je me suis empressé de rendre à votre mérite et à votre célébrité, en faisant graver votre portrait; la gravure n'est pas digne de votre nom, mais je n'ai pas eu mieux à ma disposition; Mad. du Piery, qui a fait le dessein, est une savante qui tâche de marcher sur les traces de miss Caroline Herschel, mais qui n'a pas découvert de comète; elle prie son heureuse rivale de recevoir ses hommages, et je lui demande la même grâce; j'espère l'été prochain aller lui rendre mes hommages, et voir votre superbe télescope de 40 pieds".

William Herschel, in his reply, desired his sister's compliments to Mme de Piery, adding that Caroline would think herself happy indeed if she were able to calculate fluxions, as she heard that her *happy rival* could do; but that following such a glorious example, she would not cease to beg her brother to teach her that sublime science.

Shortly after William Herschel's marriage, Lalande visited Slough, and on his return to London wrote the following rapturous letter:

M. DE LA LANDE *to* WILLIAM HERSCHEL

À Londres, le 8 août, 1788

"Monsieur et cher confrère,

"je suis arrivé à Londres tout occupé, tout pénétré de vos bontés, de vos leçons, de votre aménité, de votre génie, de vos prodiges. Mon premier soin est de vous remercier, ainsi que Madame H. et l'aimable miss, que j'ai eu tant de plaisir de voir occupée avec vous d'une manière qui me servira pour modèle auprès de nos frivoles françoises.

"je vais bientôt regagner mon pays, y célébrer ce que j'ai vu dans le vôtre et en faire usage pour la perfection d'un livre que vous m'avés rendu plus cher par l'usage que vous avés daigné en faire.

"j'espère avant mon départ recevoir de vous une position du second satellite de Saturn, qui puisse me servir à vérifier celles de M. Bernard. je vais aller à Portsmouth et à Cambridge, je partirai vers le 20 pour Paris. je suis avec autant de respect que d'attachement et de reconnaissance,

Monsieur et cher docteur,

votre très humble et très obéissant serviteur

DE LA LANDE".

William Herschel must have sent him a letter, of which we have no copy, beginning, as does his next, with "My dear Friend". Lalande hastened to write again:

M. DE LA LANDE *to* WILLIAM HERSCHEL

Great Castle Street, Cavendish Sq$^r$. Le 16 août, 1788

"je vous remercie de tout mon coeur, cher et aimable docteur, de la lettre aimable dont vous m'avés honoré. le titre d'ami que vous voulés bien m'accorder est bien flatteur pour moi, je vous en remercie, et je l'accepte avec autant de plaisir que de reconnaissance".

Herschel had about this time been examining some appearances in the moon which he believed to be active volcanoes. He wrote on 26th April, 1787, to Sir Joseph Banks:

"Sir,

"Inclosed I have sent you an account of three volcanos in the moon, one of which is now actually burning with great violence, and probably disgorging an immense lava. I may perhaps be able to see traces of its course when the mountain comes to be illuminated by the sun".

He was so convinced of the existence of this volcano that he sent a message to the King by Mr Ernest, one of the Pages:

"Sir,
"Last month I discovered three volcanos in the moon, and saw the actual eruption, or fire, of one of them; yesterday I examined the same place again and found that one of these volcanos is not yet quite extinct. Will you do me the favour to acquaint the King with these circumstances; and if his Majesty would wish to see the moon, the best time for viewing the Crater, which continues still to be considerably luminous, will be this evening between 9 and 10 o'clock.

"I will be at Windsor in good time to see the King's ten-feet telescope brought out and prepared, if it should please his Majesty to have it done.
. I remain, Sir, your most humble servant
WM. HERSCHEL ".

May 20, 1787

Herschel's accuracy as an observer was so highly regarded that the existence of these volcanoes was not immediately disputed, though Lichtenberg and others pointed out that, as it seems fairly proved that there is no air on the moon, it was difficult to see how there could be any fire. Lalande, in a letter to William Herschel dated 21st May, 1788, suggested what is probably the true explanation of the phenomenon.

M. DE LALANDE *to* WILLIAM HERSCHEL

(Translated from the French)

21 May, 1788

"...The volcano in the moon has been visible these last few days; but there are astronomers who are inclined to believe that Mount Aristarchus, which is naturally very brilliant, might very well reflect the light of the earth in such a manner as to produce this bright appearance across the pale light of the moon".

We do not know how Herschel took this suggestion; the two astronomers must have discussed it when Lalande visited Herschel at Slough in August of this same year.

One more letter may be given out of Herschel's correspondence. Mechain, who was Director of the Royal Observatory of Paris, and editor of the scientific journal, *Connaissance des Temps*, wrote very frequently. He had visited Slough, and took a very friendly interest in Caroline's comets.

M. MECHAIN *to* WILLIAM HERSCHEL

À Paris, le 25 8$^{me}$, 1789

"Monsieur,

"Me permettrez vous de me rappeler à l'honneur de votre souvenir et de vous présenter un exemplaire de la *Connoissance des Temps* pour 1791, comme un faible hommage de ma haute estime pour vous? Je me rappelle toujours avec la plus sensible reconnoissance l'accueil favorable que vous avez daigné me faire, et les marques d'amitié que vous m'avez données, Monsieur, lorsque j'ai eu l'avantage de vous voir il y a deux ans. Ce seroit une bien grande satisfaction pour moi que d'être à portée de vous revoir; vous nous aviez fait espérer que vous feriez un petit voyage à Paris; ce seroit une bien grande fête pour nous tous que de vous y posséder. Les circonstances actuelles ne sont pas engageantes pour les étrangers; mais il faut espérer que le Calme renaîtra parmi nous, et peut-être alors vous déciderez vous à venir voir vos amis. En attendant cette satisfaction je n'ai point de moments plus agréables que ceux où je me rappelle le tems que j'ai passé auprès de vous, toutes vos bontés et vos complaisances. Je suis souvent avec vous de coeur et d'esprit, il me semble être témoin de vos étonnantes découvertes et je partage bien sincèrement tous vos succès.

"J'ai appris avec bien de l'intérêt que vous avez découvert un sixième satellite à Saturne avec votre fameux télescope de 40-pieds qui fera toujours l'étonnement et le désespoir des opticiens....

"Avant de finir cette longue lettre, permettez moi de faire agréer l'hommage de mon respect à Miss Caroline votre digne soeur; sa célébrité sera en honneur dans tous les siècles, mais ceux qui ont l'honneur de la connoître savent combien ses respectables qualités du coeur ajoutent à son mérite. Je n'ai pu observer ici que deux fois la comète qu'elle a découverte en Décembre dernier, mais beaucoup d'affaires, de troubles et de chagrins m'ont empêché jusqu'à présent de perfectionner la détermination de l'orbite de cette comète; j'espère pouvoir reprendre ce travail incessamment, et dès que je l'aurai terminé je me ferai un devoir de faire hommage des résultats à Mlle votre soeur.

"J'ai l'honneur d'être avec le plus inviolable attachement.
&c. &c.
MECHAIN.

"P.S. Nous n'avons encore aucun indice de la réapparition de la comète de 1661."

## CHAPTER XIV

### 1786  1802

William Herschel's discovery of nebulous matter. "First and Second Catalogues of New Nebulae and Clusters." Laplace's remarks on Herschel's nebular theory. Present view.

THE object which Herschel had in view when he first began his systematic review of the heavens was the search for double stars, and he continued to note and catalogue these throughout his life; but when the much improved 20-foot telescope with an 18-inch mirror, which he made at Datchet, was completed, he used it to sweep the sky in a search for nebulae and star clusters. The observation of each object in turn was noted down by Caroline on a separate sheet, and the record copied next day into a special "Sweep-book". The entries in these books were continued up to the year 1802. Some idea of the vast contribution to astronomical knowledge made by William Herschel may be gathered from the fact that whereas when he began his search for nebulae, only about 150 of these bodies had been systematically catalogued, the number of these objects recorded in his sweeps amounts to 2908.

The regular nightly sweeps were occasionally interrupted when new discoveries, such as the satellites of the Georgium Sidus or of Saturn, diverted his attention for a time; but the systematic review was afterwards resumed where it had been dropped. The results of these observations were given to the Royal Society in the "First and Second Catalogues of New Nebulae and Clusters" which have been already referred to, and in the paper, "On Nebulous Stars", which he sent to the Royal Society in 1791.

The aim which he had set before him in all this work was much more than merely to catalogue the position of a number of objects; it was rather that of a naturalist, who, in making a collection of individual specimens, seeks to group them, by similarities of character, into species, genera and orders; and, carrying his reasoning yet further, sees in these group systems evidence of the evolution of the more highly organized forms from the simpler ones.

Herschel prefaced the "First Catalogue of One Thousand New Nebulae" with a short description of his improved 20-foot telescope and an account of his method of taking his "Sweeps". He wrote:

"When I had seen most of the objects I wished to examine I proceeded to the work of a general review of the heavens. The first method that occurred was, to suffer the telescope to hang freely in the center; then, walking backwards and forwards on the movable gallery, I drew the instrument from that position by a handle fastened to a place near the eye-glass, so as to make it follow me and perform a kind of very slow oscillations of 12 or 14 degrees in breadth, each taking up generally from 4 to 5 minutes of time. At the end of each oscillation I made a short memorandum of the objects I chanced to see.... This being done the instrument was, by means of a fine motion under my hand, either lowered or raised about 8 or 10 minutes, and another oscillation was then performed like the first. Thus I continued generally for about 10, 20 or 30 oscillations, and the whole was then called a *Sweep*, and as such numbered and registered in my journal.

"When I had completed 41 Sweeps, the disadvantages of this method were too evident to proceed any longer. By going into the light so often as was necessary to write down my observations the eye could never return soon enough to that full dilation of the iris which is absolutely required for delicate observations;...(besides) the fatigue attending the motion, upon a not very convenient gallery, with a telescope in my hands of no little weight, were sufficient motives to induce me to look out for another method of sweeping.... I therefore began now to sweep with a vertical motion; and as this increased the labour of continually elevating and depressing the telescope by hand, I called in the assistance of a workman to do this part of the business, by which means I could observe very commodiously, and for a much longer time than before.

"Soon after I removed also the only then remaining obstacle to seeing well, by having recourse to an assistant whose care it was to write down, and at the same time loudly repeat after me, everything I required to be written down. In this manner all the descriptions of nebulae and other observations were recorded; by which I obtained the singular advantage that the descriptions were actually being written and repeated to me while I had the object before my eye and could at pleasure correct them whenever they disagreed with the picture before me without looking from it ".

He then explained the various devices he employed to get both the Polar distance and Right Ascension correct, in spite of such practical difficulties as "the unequal tension of the great ropes, and their expansion or contraction by moisture or dryness".

"To those who are accustomed to the accuracy of transit instruments in regular observatories, this telescope, notwithstanding the above-mentioned improvements, may perhaps appear far from being brought to perfection; but they should recollect the size of the instrument as well as its extensive use, since I can not only follow any object for near a quarter of an hour, without disturbing the situation of the apparatus, but can at pleasure, in a few minutes, turn it to any part of the heavens, and view a celestial object wheresoever it may chance to be situated, the zenith not excepted.

"If I had been willing to delay giving this catalogue[1] till, by a repeated review of the heavens, the places had been more accurately determined, the work would undoubtedly have been more perfect; but whoever considers that it requires years to go through such observations will perhaps think with me, that it is the best way to give them in their present state, if it were but to announce the existence of such objects by way of inducing other astronomers also to look out for them. Another motive for not delaying this communication is to shew that my late endeavours to delineate the construction of the heavens have been guided by a careful inspection of them; and probably a catalogue which points out no less than one thousand of such systems as those are, into which I have shewn the heavens to be divided, will considerably support what has been said on the subject in my two last Papers."

He then points out how the description he is able to give of nebulae supports the hypothesis he had advanced that the Milky Way (our sidereal system) is a nebula.

"...the view of so many sidereal systems, some of which we may

[1] "Catalogue of One Thousand New Nebulae and Clusters of Stars", read 27th April, 1786.

discern to be of most surprising extent and grandeur, will...add credit to what I have proposed with regard to the condition of our situation within a system of stars."

Three years later he sent a second catalogue of a thousand new nebulae from which the following extracts are taken:

*Catalogue of a second Thousand New Nebulae and Clusters of Stars; with a few introductory remarks on the Construction of the Heavens*

Read 11th June, 1789

"By the continuation of a review of the heavens with my twenty-feet reflector, I am now furnished with a second thousand of new Nebulae.

. . . . . . .

"The method I have taken of *analyzing* the heavens, if I may so express myself, is perhaps the only one by which we can arrive at a knowledge of their construction.... To begin our investigation in some order, let us depart from the objects immediately around us to the most remote that our telescopes, of the greatest *power to penetrate into space*, can reach".

He then shows that with regard to planets, satellites and comets, there is evidence enough to convince us

"that bodies shining only with borrowed light can never be seen at any great distance. This consideration brings us back to the sun, as a refulgent fountain of light, whilst it establishes at the same time beyond a doubt that every star must likewise be a sun, shining by its own native brightness. Here then we come to the more capital parts of the great construction.

"These suns, every one of which is probably of as much consequence to a system of planets, satellites and comets, as our own sun, are now to be considered in their turn as the minute parts of a proportionally greater whole. I need not repeat that by my analysis it appears that the heavens consist of regions where suns are gathered into separate systems, and that the catalogues I have given comprehend a list of such systems; but may we not hope that our knowledge will not stop short at the bare enumeration of phaenomena capable of giving us so much instruction? Why should we be less inquisitive than the natural philosopher, who sometimes, even from an inconsiderable number of specimens of a plant or an animal, is enabled to present us with the history of its rise, progress and decay? Let us then compare together and class some of these numerous sidereal groups, that we may trace the operations of natural causes as far as we can perceive their agency. The most simple form in which we can view a sidereal system is that of being globular. This also, very favourably to our design, is that which has presented itself most frequently, and of which I have given the greatest collection.

. . . . . . .

"Having established that the clusters of stars of the 1st Form and round nebulae are of a spherical figure, I think myself plainly authorized to conclude that they are thus formed by the action of central powers. To manifest the validity of this inference the figure of the earth may be given as an instance, whose rotundity, setting aside small deviations the causes of which are well known, is without hesitation allowed to be a phaenomenon decisively establishing a centripetal force.... Since then almost all the nebulae and clusters of stars I have seen, the number of which is not less than three and twenty hundred, are more condensed and brighter in the middle; and since from every form it is now equally apparent that the central accumulation of brightness must be the result of central powers, we may venture to affirm that this theory is no longer an unfounded hypothesis, but is fully established on grounds which cannot be overturned.

. . . . . . .

"Let us then continue to turn our view to the power which is moulding the different assortments of stars into spherical clusters. Any force that acts uninterruptedly must produce effects proportional to the time of its action. Now as it has been shown that the spherical figure of a cluster of stars is owing to central powers, it follows that those clusters which, *ceteris paribus*, are the most compleat in this figure, must have been the longest exposed to the action of these causes.... An obvious consequence that may be drawn from this consideration is, that we are enabled to judge of the relative age, maturity, or climax of a sidereal system from the disposition of its component parts.

"This method of viewing the heavens seems to throw them into a new kind of light. They are now seen to resemble a luxuriant garden, which contains the greatest variety of productions, in different flourishing beds; and one advantage we may at least reap from it is, that we can, as it were, extend the range of our experience to an immense duration. For, to continue the simile I have borrowed from the vegetable kingdom, is it not almost the same thing, whether we live successively to witness the germination, blooming, foliage, fecundity, fading, withering, and corruption of a plant, or whether a vast number of specimens, selected from every stage through which the plant passes in the course of its existence, be brought at once to our view?
WILLIAM HERSCHEL".
Slough, near Windsor. May 1, 1789

Up to the time of writing the last paper, Herschel had surmised "that every visible object in the heavens was of the starry kind". Having proved to his own satisfaction that the Milky Way was composed entirely of stars, it seemed not inconsistent to conclude that any other extensive nebulosities, such as that in Orion, were similar galaxies— "telescopic Milky Ways", he called them—consisting of multitudes of stars. On this supposition he had based his hypothesis that nebulae were the result of the action of gravitation on scattered stars, and that the gradations in the process of evolution would have been—first, scattered stars; next, clusters; and, finally, nebulae.

But after the discovery of nebulous matter, *not stellar*, as described in the next paper, he adopted a view of the relation between nebulae and stars the exact reverse of his former one. Nebulae now appear as the parents of stars; and the evolutionary process is—first, nebulous matter; then nebulae (star dust); lastly, stars. In the paper read before the Royal Society on the 10th February, 1791, he described the discovery of the star, of such unusual appearance, which led him to this conclusion. The following extracts from this paper indicate the line of his argument:

1791.

*On Nebulous Stars, properly so called*

"[1] Cloudy or nebulous stars have been mentioned by several astronomers; but this name ought not to be applied to the objects

[1] *C.S.P.* vol. I, p. 415.

which they have pointed out as such; for, on examination, they proved to be either mere clusters of stars, plainly to be distinguished by my large instruments, or such nebulous appearances as might reasonably be supposed to be occasioned by a multitude of stars at a vast distance. The Milky Way itself, as I have shewn in former papers, consists entirely of stars, and by imperceptible degrees I have been led on from the most evident congeries of stars to other groups in which the lucid spots were smaller but still very plainly to be seen; and from them to such wherein they could but barely be suspected, till I arrived at last to spots in which no trace of a star was to be discerned. But then the gradations to these latter were by such well-connected steps as left no room for doubt but that all these phaenomena were equally occasioned by stars, variously dispersed in the immense expanse of the universe".

He then describes how, in 1784, he observed a small star which appeared to be surrounded "by a milky nebulosity or chevelure".

"It must appear singular, that such an object should not have immediately suggested all the remarks contained in this Paper; but about things that appear new we ought not to form opinions too hastily; and my observations on the construction of the heavens were then but entered upon. In this case, therefore, it was the safest way to lay down a rule not to reason upon the phaenomena that might offer themselves, till I should be in possession of a sufficient stock of materials to guide my researches."

He then cites from his journals observations of stars "with a milky chevelure" recorded between the years 1784 to 1790, till he comes to the crucial example seen in November of that year.

"[1]*November 13, 1790. A most singular Phaenomenon! A star of about the 8th magnitude, with a faint luminous atmosphere, of a circular form, and about 3′ in diameter.*[2] *The star is perfectly in the center and the atmosphere is so diluted, faint and equal throughout, that there can be no surmise of its consisting of stars; nor can there be a doubt of the evident connection between the atmosphere and the star. Another star not much less in brightness, and in the same field with the above, was perfectly free from any such appearance.*

. . . . . . .

[1] *C.S.P.* vol. I, pp. 421, 422.  [2] NGC. 1514.

"Supposing the connection between the star and its surrounding nebulosity to be allowed, we argue, that one of the two following cases must be admitted. In the first place, if the nebulosity consist of stars that are very remote,... then what must be the enormous size of the central point, which outshines all the rest in so superlative a degree? In the next place, if the star be no bigger than common, how very small and compressed must be those other luminous points that are the occasion of the nebulosity which surrounds the central one? As, by the former supposition, the luminous central point must far exceed the standard of what we call a star, so, in the latter, the shining matter about the center will be much too small to come under that denomination; we therefore have either a central body which is not a star, or have a star which is involved in a shining fluid of a nature totally unknown to us.

"I can adopt no other sentiment than the latter, since the probability is certainly not for the existence of so enormous a body as would be required to shine like a star of the 8th magnitude, at a distance sufficiently great as to cause a vast system of stars to put on the appearance of a very diluted, milky nebulosity.

"But what a field of novelty is here opened to our conceptions! A shining fluid, of a brightness sufficient to reach us from the remote regions of a star of the 8th, 9th, 10th, 11th or 12th magnitude, and of an extent so considerable as to take up 3 to 6 minutes in diameter! Can we compare it to the coruscations of the electrical fluid in the aurora borealis?

. . . . . . .

"More extensive views may be derived from this proof of the existence of a shining matter. Perhaps it has been too hastily surmised that all milky nebulosity, of which there is so much in the heavens, is owing to starlight only. These nebulous stars may serve as a clue to unravel other mysterious phenomena. If the shining fluid which surrounds them is not so essentially connected with these nebulous stars but that it can also exist without them, which seems to be sufficiently probable, and will be examined hereafter, we may with great facility explain that very extensive, telescopic nebulosity, which, as I mentioned before, is expanded over more than sixty degrees of the heavens, about the constellation of Orion; a luminous matter accounting much better for it than clustering stars at a distance.

. . . . . . .

"The nature of planetary nebulae, which has hitherto been involved in much darkness, may now be explained with some degree of satisfaction, since the uniform and very considerable brightness of their apparent disc accords remarkably well with a much condensed luminous fluid; whereas to suppose them to consist of clustering stars will not so completely account for the milkiness or soft tint of their light.... The surmise of the regeneration of stars, by means of planetary nebulae, expressed in a former Paper, will become more probable, as all the luminous matter contained in one of them, when gathered together into a body the size of a star, would have nearly such a quantity of light as we find the planetary nebulae to give.

"...If the point be a generating star, the further condensation of the already much condensed luminous matter may complete it in time.

. . . . . .

"I hope it may be found that in what has been said I have not launched out into hypothetical reasonings; and that facts have all along been kept sufficiently in view. But, in order to give everyone a fair opportunity to follow me in the reflections I have been led into, the place of every object from which I have argued has been purposely added, that the validity of what I have advanced might be put to the proof by those who are inclined, and furnished with the necessary instruments to undertake an attentive and repeated inspection of the same phenomena.

W. HERSCHEL."

Slough. Jan. 1, 1791

The further Herschel pushed his researches the more complicated did the appearances of the several objects which he encountered seem to be; he found himself compelled to distinguish many different categories instead of the simple division into single stars, clusters, and nebulae. In the year 1802, he submitted to the Royal Society a fresh "Catalogue of 500 New Nebulae, nebulous stars, planetary Nebulae and Clusters of Stars". Of those called "Planetary Nebulae" he wrote:

"This seems to be a species of bodies that demands particular attention.... In my second Catalogue of Nebulae a single instance of a planetary nebula with a bright central point was mentioned; and in the annexed one, No. 73 of the 4th Class, is another of very nearly the same diameter, which also has a lucid, though not quite so regular, a center. From several particulars observed in their construction, it would seem as if they were related to nebulous stars. If we might suppose that a gradual condensation of the nebulosity about a nebulous star could take place, this would be one of them, in a very advanced state of compression. A further discussion of this point, however, must be reserved for a future opportunity. The nature of these remarkable objects is enveloped in much obscurity. It will probably require ages of observation before we can be enabled to form a proper estimate of their condition".

The importance of Herschel's discovery of nebulous matter and of the inferences which he drew from it did not attract much attention at the time. Laplace indeed gave Herschel's theory the sanction of his approval. In the short abstract of the history of astronomy with which he concluded his *Exposition du Système du Monde*, published in 1796, Book V, Chapter VI, is the following passage:

"Herschel, while observing the nebulae by means of his powerful telescopes, has followed the progress of their condensation, not in a single instance, as such changes would not be sensible to us till after the lapse of centuries, but among the whole collection, as in a vast forest one could follow the development of a tree from the condition of the several individuals found there. He first observed the nebulous matter dispersed about in various masses, in different parts of the sky, where it occupies a considerable space. He also noticed in some of these masses that the matter was faintly condensed round one or more nuclei. In other nebulae these nuclei are brighter in comparison with the nebulosity which surrounds them. As, by further condensation, the atmospheres surrounding each nucleus draw apart, multiple nebulae are formed, consisting of closely associated nuclei, each surrounded by an atmosphere of its own; sometimes the nebulous matter, by condensing uniformly, produces those nebulae which have been named planetary. Finally, by further condensation, all these nebulae are converted into stars.

"The nebulae, classified according to this philosophical view, indicate with great probability, both their future transformation into stars, and the anterior nebulous state of existing stars. So one descends by the contemplation of the condensation of nebulous matter, to considering our sun as originally surrounded by a vast atmosphere; a view to which I have arrived through my examination

Lick Observatory

THE NEBULOUS STAR
(N.G.C. 1514) described by W. Herschel. See p. 227

R.A.S.

A GLOBULAR CLUSTER
(M. 3 Can. Ven.)

of the phenomena of the solar system, as will be seen in the final note of this book. Such a remarkable convergence arrived at by opposite lines of reasoning gives great probability to the existence of this anterior condition of the sun".

It seems plain from this passage that Laplace had read with approval Herschel's papers in the *Philosophical Transactions*.

The term "nebular hypothesis" is nowadays generally applied, in the limited sense, to Laplace's theory of the origin of the solar system only; but Herschel's conception of the evolution of the sidereal heavens more strictly deserves the name, which was so used during his lifetime. John Herschel, who was always very cautious in seeming to agree too readily with his father's theories, wrote, in 1845, to Prof. Sedgwick:

"As the 'nebular hypothesis', I believe, originated with my Father and was from him taken up by Laplace, I am bound to treat it with due respect and I readily grant that there *is* a certain happiness about it, '*grantatis grantandis*',.... By the way, in the 'nebular hypothesis' the *primum mobile* is gravitation, thence conglomeration, (accompanied with rotation), thence pressure, thence heat, resisting further conglomeration and so retarding the process. The physical fact of the development of heat by pressure of elastic materials was unknown to my Father when he framed his nebulous hypothesis, as much as the melting of limestone by heat, under pressure, was to Hutton. Whatever support therefore the reality of central heat gives to the 'hypothesis' is in the nature of a 'consilience of inductions' and must be taken '*quantum valeat*' in that light as tending to establish it. Laplace said to me one day at Arceuil, 'Mons. Herschel, ces idées de M. votre Père sur la condensation des nébuleuses, m'ont toujours paru très philosophiques et très vraies'".

Later research has strengthened the arguments for Herschel's theory. There has been no lack of response among astronomers to the challenge expressed at the end of the paper on Nebulous Stars; they have followed along the path he traced with no less ardour and perseverance, and with the aid of instruments and methods of investigation of which he could never have dreamt. The question whether any nebulous matter is self-luminous still remains *sub judice*, but Herschel's assumption of this is a minor consideration compared to the revolution in astronomical thought effected by his discovery of the existence of such matter, and his consequent theory of the evolution of stars.

One quotation from the writings of Sir James Jeans will be enough to show how nearly Herschel's theory approached the modern view:

> "Our string or sequence of nebulae starts with something that looks like a fluffy, featureless ball of gas and ends up as a city of stars. It would in any case be hard to resist the conjecture that such a sequence is one of advancing development, so that, as we pass along it, what was originally a cloud of formless gas has been condensing into stars".[1]

---

[1] *The Stars in their Courses*, by Sir James Jeans, Cambridge University Press, 1931, p. 128.

## CHAPTER XV

### 1792   1802

Birth of John F. W. Herschel. William goes on a tour to Glasgow with Count Komarzewski. Death of Paul Adee Pitt. Letters—Lalande, Méchain. William Herschel made a member of the French Institute. Letters—B. Greatheed, De Luc.

WE are not able to follow very closely the daily life of the Herschel family immediately after William Herschel's marriage, as Caroline destroyed all her Journals for nine years after that event. The family circle received an important addition when, on the 9th of March, 1792, a son was born to William and Mary Herschel. The child was baptized in the old church at Upton, and received the names of John Frederick William. His godfathers were his father's old friend, Sir William Watson, and General Komarzewski, a Polish nobleman whose acquaintance William Herschel had made some time before and for whom he had a high regard.

Early in this year, 1792, William Herschel accepted an invitation from the University of Glasgow to visit their city and to receive in person the honour of the degree of Doctor of Laws. (The University of Edinburgh had already, in 1786, conferred the same degree upon him.)

In August, 1792, William Herschel and General Komarzewski set out on a leisurely tour of the north, their ultimate goal being Glasgow. Herschel kept a very detailed Journal, in pencil, of this tour; some extracts may be of interest. He invariably noted the distances which they covered each day in their carriage; one example will be sufficient:

Tour with General Komarzewski.

"*May* 29, 1792.
Set off from Slough.

| | |
|---|---:|
| Henley | 15 |
| Benson | 11 |
| Oxford | 12 |
| Woodstock | 8 |
| | 46 |

"*May* 30.        Chapel House      10
                  Halford Bridge    14
                  Warwick           13
                                    ──
                                    37

"Called upon Mr Greatheed (at Guy's Cliff) about 1½ miles from Warwick. It is beautifully situated".

Mr Greatheed was one of William Herschel's intimate friends; it was at his request that Herschel sat to Artaud for the portrait which still hangs at Guy's Cliff. A replica, by the artist, is in the possession of the present Sir John Herschel.

From Warwick the two friends proceeded to Coventry and Birmingham. William Herschel was much interested in all the factories which they visited and made many drawings of machinery. He particularly described all the devices for using water power, as well as the few steam engines he met with.

At Birmingham he met his brother-in-law, Mr Baldwin, with his stepson, Paul Adee Pitt; and visited James Watt at his house and factory at Soho. It is tantalizing that there is nothing more about this visit beyond the meagre entry:

"*June* 3. Dined with Mr Watt, Mr Kerr, Gatton, &c.
"*June* 4. Saw the works at Soho.
"*June* 8. We saw the making of a steam boiler fifty feet long.[1]
"Dudley.

"Mr Finch's house is very pleasantly situated. In the evening, Mrs Finch, &c., &c., saw Jupiter and its satellites through my aerial 7-feet telescope".

After visiting many factories and seeing the processes of iron and lead smelting, which William Herschel noted down in great detail with drawings of the machinery, the friends turned aside to see the beauties of Wales. Of the road from Corwen to Kernioge Mawr he writes:

"In this last stage is a very beautiful passage of rocks on one side

---

[1] The acquaintance thus begun with James Watt was not allowed to drop. Caroline several times mentions in her Journal that Mr Watt, with his wife and daughter, came to stay at Slough. Under the date "1819, June", she wrote: "I went with my brother to town. Mr and Mrs Watt dined one day with us at Mr Beckwith's. It was the last time the two old friends did meet, for Mr Watt died soon after".

and a precipice on the other close to the road. A fall of water down the rocks leading to the precipice adds to the beauty of the landscape.

"The Inn resembles in its situation those of the country about Hannover. They burn turf or peat, a farm yard with trees and cows in front and garden by the side or back.

"LLanorust—11 miles.

"This being market day I found that a great number of women had butter in baskets, others eggs, &c., in the manner of the German country villages. The air of these women was however totally different from that of the Germans; most of them are lean but of agreeable features and good transparent complexions, but rather pale and emaciated.

"A Welsh Harp is in every Inn; generally a blind man plays on it and sometimes not amiss.

"*June* 17. We went to Belgelart. 14 miles.

"There is a most beautiful walk between the rocks, which are extremely high on both sides and only leave a stream, running upon a rocky foundation, meandering between them. At the foot of Snowdon we left our carriage and having taken three guides we set off to go as far as we could, for as it was very cloudy we knew it would be of no use to go to the top. After we had ascended about three quarters of a mile on our way, we came to the clouds that hang on the mountain; we proceeded in the mist, which soon became so thick that we could not see above a hundred yards before us, and having ascended nearly two miles, found that it would be very imprudent to go on, as we were now already wet through on our feet, the ground being quite boggy.

"Our men carried the 7-feet telescope, but there was no hope of its turning out a fine night; we descended, and getting into our carriage returned to Carnarvon.

"*July* 3, 1792. Glasgow.

"This morning the Lord Provost, Mayor of this place, presented Genl. Komarzewski and me with the freedom of the City. He gave an elegant collation of fruits and wines and introduced the intended gift by drinking a health to us as Burgesses of Glasgow. In a short speech we returned our thanks.

"The Principal and Professors gave a dinner to-day in compliment to Genl. Komarzewski and myself.

"After dinner the Principal addressing himself to me desired to drink my health as Doctor of Laws of the University of Glasgow and read a diploma[1] to that purpose, which he afterwards presented to me. I drank to the Principal and Professors in return with a proper acknowledgement of the honour they had conferred upon me.

"*July* 4. Kelsyth. 13 miles.
            Mr Bruce. 14 miles.

"We were received by Mr Bruce[2] with the greatest hospitality, and saw many valuable things in his collection. Among others two horn cups that were given by a young damsel at the source of the Nile.

"Edinburgh. We saw the Castle, the Council chamber, &c. The Observatory contains many of Mr Short's Gregorian reflectors; a 2-feet, two 6-feet and a 10 or 12-feet one. I viewed several land objects thro' them. The large one which has the reputation of being a very bad instrument appeared to me not to be very defective, but the presumption seems to be in favour of its being good. I could only see the top of a steeple and even there great part of the light was lost by being confined in the opening of the roof, which is like our equatorial ones at Greenwich and Cambridge.

"I saw Dr Black, Professor of Chemistry, Dr Robinson, Prof. of Nat. Philos., Mr Playfair, Prof. of Mathematics, Mr Stewart, Prof. of Ethicks, Dr Hutton, Sir Wm. Miller, Lord Daer, Mr Clark on Naval Tactics, Mr Russel, Mr McGowan, Dr Robertson, Principal.

"I saw again the large reflector at the Observatory, and measured its power, which was said to be 800. I found it 130; the aperture is 12 inches. The 6-feet reflector had a power of 90 to 95. An achromatic had a power of $23\frac{1}{2}$. Its aperture was but small.

"We dined with Dr Black. Dr Hutton, who was there, shewed me some specimens of granite invading schistus; they were taken from a very large Quarry at Cairnmuir, Galloway. In another part of the country, (Lord Bute's estate) he had found limestone, schistus and granite in strata.

"I saw the library, the Castle where Queen Mary resided; the furniture which was there in her time; the bedroom and little dressing room where David Rizzio was found with the Queen; the place where

---

[1] The diploma, Patrick Wilson told Herschel, was written "with an Eagle's quill; the Box containing our Seal is of standard Silver, &c."

[2] James Bruce, the celebrated traveller, author of *Travels to discover the Source of the Nile*, died in 1794.

he was killed. They show the stains on the floor said to be made by his blood.

"*July* 8. Haddington. 16 miles.
   Dunbar. 11 miles.

"In our first stage I saw a common steam engine. Upon the piston was kept a coat of water to prevent its losing steam. The condensation was made in the cylinder itself, by a pipe entering the bottom to inject cold water. The pressure of the atmosphere brought down the piston and the steam lifted it up".

From Dunbar, the travellers proceeded by Berwick, Alnwick and Newcastle to Sunderland, which had been William Herschel's headquarters in his early days.

"*July* 10. Sunderland.

"On the road we saw a steam engine to draw water out of a pit. I could find no trace of my former habitation in this place.

"Durham. 14 miles.

"We saw the Cathedral. I enquired after Mr Garth, an old acquaintance, and found that he still lives at Bishop Auckland.

. . . . . .

"Richmond. 14 miles.

"I saw Mr Jackson and several other persons I had seen 32 years before.

. . . . . .

"*July* 13. Thornhill.

"We saw Mr Michell's telescope, it is on an equatorial stand; being without cover behind. I put my hand into the opening and felt the face of the object speculum so wet as to moisten my fingers.

"Mr Michell was very indifferent in health.

"*July* 15. Chesterfield. 12 miles.

"The steeple in this place is constructed with waving triangular ribs which always represent it hanging over to the beholder's right hand.

"Derby. We saw Mr Darwin. He shewed me a curious specimen of lead confined like a nucleus in a shell of iron ore."

They stopped at Birmingham to pay another visit to James Watt, and from there went on to Stratford, where, William Herschel says, "We saw the House where Shakespear lived and his chair".

They hastened their homeward journey, and reached Slough on the 19th July, after a long drive of 64 miles, passing through Oxford.

Mrs Herschel's son, Paul Adee Pitt, died in February, 1793, probably from consumption.

Mrs Papendiek mentions "Young Pitt" several times in her Memoirs. She calls him "my brother's Eton friend and companion, an openhearted friendly acquaintance, who endeared himself to us in every social tie". She says that, about the time when his mother remarried, he had started business with his uncle, Thomas Baldwin, a wholesale chemist in London. In the year 1792, she wrote: "Young Pitt was often at Slough for change of air, as he was getting into delicate health, his mother being made to see it with great difficulty".

There is (or was) a letter written by him to his stepfather on the occasion of John Herschel's birth, in which he expressed in a very frank and natural manner his satisfaction "on the birth of a son to you and a brother to me".

It was always a matter of regret to John Herschel that he had lost the advantage of the companionship and affection of an elder brother.

During these years Herschel continued his correspondence with foreign men of science.

### M. LALANDE *to* WILLIAM HERSCHEL

Au collège royal. 1 avril, 1792

"Mon cher confrère et ami,

"Je m'empresse de vous annoncer une nouvelle justice que l'académie vient de rendre à votre mérite et à votre célébrité. L'Assemblée nationale ayant fondé un prix de 1200 livres pour l'ouvrage le plus utile ou la découverte la plus importante pour les sciences ou les arts, le premier vous a été déféré, et le second à M. Mascagni qui a fait un beau traité des vaisseaux lymphatiques. On avait proposé plusieurs autres personnes, et j'avois indiqué entre autres les cinq comètes découvertes par ma chère commère; mais on n'a pas cru devoir mettre les deux prix dans la même famille.

"Les 1200 livres en assignats perdront beaucoup en vous parvenant puisque le change n'est qu'à 18 au lieu de 30 qu'il devroit être, mais la gloire ne se calcule pas en argent.

"Vous verrés, mon cher docteur, dans la 3me édition de mon Astronomie, un grand nombre de passages où je rens justice à votre supériorité, et où je parle de la savante miss....
LALANDE."

The war which broke out between France and the rest of Europe, after the execution of Louis XVI and of Marie Antoinette in 1793, interrupted all correspondence between England and Paris. In 1796 Herschel received one brief letter from Lalande, of which the following is a copy:

À paris, le 24 mars, 1796

"Il y a si longtemps, mon cher et illustre confrère, que je n'ai reçu de vos nouvelles, et de celles de votre aimable Caroline, que je profite d'une occasion que m'offre l'échange des prisonniers pour vous adresser mes voeux et mes félicitations sur la continuation de vos travaux. j'ai toujours travaillé de mon côté, avec mon neveu. Les troubles de la france n'ont rien changé à ma situation ni à mon courage. je vous envoie mon dernier chapitre de l'histoire de l'astronomie; mettés moi à portée d'enrichir le suivant par la notice de vos travaux, et recevés mes plus tendres embrassemens. LALANDE".

In November, 1801, he wrote to thank Herschel for sending him his papers on sunspots, on heat and light, and the description of the 40-foot telescope. He added:

"Vous avés été choisi comme de raison, pour un des premiers dans les 24 associés étrangers de l'Institut national de france, avec les hommes les plus célèbres de l'europe et de l'amérique.

"...Bonaparte nous promet 2000 livres de platine pour que Caroche entreprenne de faire un télescope à votre imitation, si vous y consentés.

"je suis avec autant d'attachement que de considération, &c.

DE LA LANDE".

William Herschel wrote,[1] in reply to Lalande's letter, to acknowledge the honour done him by the members of the National Institute of France, adding:

"I can also assure you that this satisfaction is considerably enhanced by the authority which it now gives me to call you again my highly esteemed confrère.

"I am very happy to hear that Mr Caroche is to make a large telescope. When I have the pleasure of seeing you at Paris, which I hope will not be many months hence, I will communicate to you many experiments I have made on Platina, with a view to its optical use, and

[1] In English.

if my opinion can any way contribute to the success of an undertaking that must be favourable to astronomy, you may be assured it will be readily given.

"Believe me ever to remain with the highest consideration, &c.

W. HERSCHEL".

### M. MECHAIN, DIRECTOR OF THE OBSERVATORY OF PARIS *to* WILLIAM HERSCHEL

À paris, le 17 juillet, 1790

"Monsieur et cher Confrère,

"Je vous ai beaucoup d'obligation de m'avoir procuré la connaissance d'une personne aussi intéressante que M. le Général de Komarzewski. Il a visité notre observatoire et je l'ai conduit à Passy pour lui faire voir le télescope de 22-pieds de M. l'abbé Vochon; je regrette bien qu'il ne séjourne pas plus longtems à paris, mais il nous fait espérer qu'il y reviendra dans le courant de l'automne.

"J'ai communiqué à l'académie le résultat de votre belle découverte de la rotation de l'anneau de Saturne, qui est une preuve de la perfection de votre étonnant télescope et de l'effet extraordinaire qu'il produit. M. de la Place, qui s'est occupé de la théorie de cet anneau, désireroit bien savoir si c'est la rotation de l'anneau intérieur que vous avez déterminée; vous verrez dans le mémoire qu'il a publié dans le volume de l'académie pour l'année 1787 qu'il avoit trouvé la durée de la rotation de la partie intérieure d'environ dix heures, et il m'a dit qu'ayant fait depuis les calculs avec plus de rigueur, il approchoit de $10\frac{1}{2}^h$; mais vos observations ne laissent plus de doute sur la véritable quantité. Vous nous ferez grand plaisir si vous voulez bien avoir la complaisance de me mander si votre détermination se rapporte à l'anneau intérieur ou extérieur. Il y a tout lieu d'espérer que vous nous ferez connoître aussi la durée de la rotation de Saturne. Vous allez enrichir l'astronomie des découvertes les plus intéressantes et les plus inattendues.

"J'ai observé la comète que Mlle. votre soeur a découverte le 17 avril,.... Je finis en vous priant de me conserver votre amitié et d'agréer l'assurance des sentimens de la plus haute estime et de l'inviolable attachement avec lesquels j'ai l'honneur d'être,

Monsieur et cher confrère

MECHAIN."

From the large number of friendly letters in the Herschel collection, two examples have been chosen, which afford some interesting glimpses of the personal character of two of Herschel's scientific friends and help us to picture the society among which he and Caroline moved. The first is from his great friend, Mr Greatheed, of Guy's Cliff, Stratford-on-Avon.

BERTIE GREATHEED, ESQ. *to* WILLIAM HERSCHEL

Göttingen. June 3d, 1798

"My dear Sir,

"I can no longer withhold my congratulations on the rich discoveries with which you are enriching our world. So Uranus is no longer a feeble power, inferior to his neighbours in splendour and decoration, but magnificently attended and encircled with a double crown! How very interesting! What would I give for one such night as I have frequently passed with you! Has the fresh polish of the great speculum detected these wonders and enabled you to see farther and better into space than ever? And if so has Venus, in the favourable situation where she stood a great part of this winter, afforded nothing new? Excuse all these questions, and, if you would make me very happy, find some idle moment for a few lines in reply.

"We have been settled in this quiet place nine months, and employ the whole of our time in literary pursuits. It is now twenty years since I studied here and twenty such years of revolution! Within that time empires have risen and fallen, but at Göttingen it has not wrought the least alteration. I find scarcely any of my former acquaintance dead, and, more wonderful still, they do not seem to me to be grown older. One would think here, that the same number of seconds had not beat at Göttingen as in London or Paris. Lichtenberg goes on just in his old way, and the utmost of his locomotion is a garden house a few hundred yards without the Wehnder gate, where he delights to pass his Saturdays and Sundays. I expected to enjoy much of his society and to find in him a great resource; but such is his seclusion and, I must say, paltry unphilosophical shame of his figure, that he never will go out, so that, notwithstanding my intimacy, Mrs Greatheed has never yet seen him. I frequently sit with him between eleven and twelve, and now attend his excellent lectures with my son. What a pity that a man of such intellectual powers, so formed for conversation, should, by giving way to a foolish notion, be thus lost to society.

"Blumenbach I see a good deal of and from him it is I have learnt your late discoveries; but he likewise leads too sedentary a life.

"One professor more I must speak of before I have done, who puzzles us all very much; I mean the new De Luc. That he, so great a favourite at Windsor, at his age, should leave his family to come here, is quite inexplicable; yet it was much talked of, and he has been wandering about the north of Germany all this winter. It was said he was to write against some opinions, what I know not, and meant to superintend the translation of his book. This is too absurd to be possible.

"Mrs Greatheed and Bertie unite with me in kindest wishes to yourself and Mrs and Miss Herschel, not forgetting our nice little friend Johnny. Let me repeat that we shall all have the greatest pleasure to hear that you are all going on well, and believe me when I assure you that the instructive and happy hours I have passed in your society will never be forgotten by me.

I am, ever sincerely yours,

BERTIE GREATHEED."

### WILLIAM HERSCHEL *to* BERTIE GREATHEED

"My dear Sir,

"It will hardly be necessary to say how much I have been gratified by your kind remembrance, and as a very convenient opportunity now falls in my way of acknowledging the favour of your letter, I give this to the son[1] of a friend of mine (Dr Parry) who is coming to Göttingen, and who I find has already some other letters of introduction to you.

"I do not despair of once more showing you some objects in my 40-feet telescope, and I wish I could as easily partake of your literary enjoyments. At present the shortness of the nights has allowed me to make a little tour to the seaside, and I am now on my way to Dawlish near Exeter.

"We are no less surprised than you seem to be at the excursion of Mr de Luc and as little able to guess at its intention. There is a theory

---

[1] Edward Parry, the Arctic explorer.

of the earth now lately proposed which bids fair to overset Mr de Luc's system; perhaps he may be engaged by endeavours to keep up the ingenious fabric laid down in his *Lettres sur l'homme et sur la terre*. I have a very high opinion of his knowledge concerning the atmosphere of this globe and of his general acuteness in other philosophical pursuits.

"Mrs Herschel joins with me in respects to Mrs Greatheed and yourself and on my asking my little John, who is at my elbow, whether he will unite with us in the same to your son, he gives his hearty consent.

Believe me ever with the highest esteem, my dear Sir,

Yours faithfully,

(No signature)."[1]

De Luc, mentioned in this letter, was a Swiss by birth, who had settled in England and for some time held a post in the Royal household as reader to Queen Caroline. It was at Windsor that William Herschel had made his acquaintance. De Luc was a Fellow of the Royal Society, and in his younger days had done some good research work in physical science; he had contrived a portable barometer, which he carried to the summit of some Alpine peaks to register atmospheric pressure. He had also made experiments at Paris on heat with an improved thermometer of his own invention. But later he turned his attention to geology, and wrote a treatise propounding some very hypothetical theories. His friends, who were many, for he seems to have been a very lovable old gentleman, looked indulgently on his speculations without taking them seriously.

The new theory of the earth to which Herschel alludes in his letter to Mr Greatheed was that of Hutton, the geologist, whose complete work, *The Theory of the Earth*, had been published only three years before. Hutton entirely rejected all hypothetical explanations of the condition of the earth's crust, and was the first to attribute all former changes to the action of natural agents only. He may be said to have done for geology what Herschel did for the study of the sidereal heavens, by directing the attention to the study of facts, and deducing from these the necessity of attributing immense duration to past ages, in which these natural agents

[1] William Herschel's letters are all taken from the copies in his Letter Book and are unsigned, or the signature copied by Caroline.

have been acting to produce the present conditions. Hutton's views were considered by many to be opposed to revealed truth, and the bitter controversies which they excited were carried on to a later generation; but with this we are not here concerned.

Herschel had always taken an interest in mineralogy and geology, and this was further stimulated by his perusal of Hutton's book. There are many notes referring to geological formations in his diaries of travel.

## CHAPTER XVI

## 1788   1798

Caroline Herschel's astronomical work. Extracts from her *Book of Observations*. She discovers seven comets. Letters—to and from Maskelyne, William Herschel to the Earl of Salisbury, Aubert, Sir J. Banks, Lalande. Caroline's Index to Flamsteed's Catalogue. Letter to Dr Maskelyne.

THOUGH Caroline Herschel has left no record in her Autobiography of the years succeeding her brother's marriage, we are fortunately able to supply the deficiency from the notes which she made of her astronomical work. This period was, from the astronomical point of view, the most important of her life, being that in which she established a reputation as an independent observer and discoverer of comets.

After her first success in finding a comet during her brother's absence, she concentrated her attention on this branch of observation. In December, 1788, she wrote to Dr Maskelyne to announce her discovery of a second comet, and received a very cordial reply.

CAROLINE HERSCHEL *to the* REV. DR MASKELYNE

"Dear Sir,

"Last night, December 21st, at $7^h\ 45'$, I discovered a comet, a little more than one degree south, preceding $\beta$ Lyrae. This morning, between five and six, I saw it again, when it appeared to have moved about a quarter of a degree towards $\delta$ of the same constellation. I beg the favour of you to take it under your protection.

"Mrs Herschel and my brothers join with me in compliments to Mrs Maskelyne and yourself, and I have the honour to remain,

Dear Sir,
Your most obliged, humble servant
CAROLINA HERSCHEL."

Slough. Dec. 22, 1788

DR MASKELYNE *to* MISS HERSCHEL

"Dear Miss Caroline,     Greenwich. December 27, 1788

"I thank you for your favor of the 22d, containing an account of a *second* Comet on the 21st, and recommending it to my attention. I received it only on the 24th at 10 in the morning, owing to the slowness of our penny post. I delayed acknowledging it till I could inform you at the same time I had seen it. The frost, unfortunately for us astronomers, broke up the very same morning that your letter arrived, in consequence of which the weather has been so bad that I could not get a sight of your Comet till last night the 26th, when at $6^h\ 24'$ it followed α Lyrae in the ℞ $3'\ 7''$ of time and was $2°\ 30'$ S. of it.... Its motion is fortunately very favorable for our keeping sight of it for some time, which may be very useful, especially if it should be moving from us, which there is an equal chance for as the contrary. It appeared to me very faint and rather small but the air was hazy. By its faintness and slow motion it is probably at a considerable distance from the earth. Time will explain these things. Let us hope the best and that it is approaching the earth to please and instruct us and not to destroy us, for true Astronomers have no fears of that kind, witness Sir Harry Englefield's valuable tables of the apparent places of the Comet of 1661, expected to return at this time, with a delineation of its orbit, who, in page 7, speaks of the possibility of seeing a curious and beautiful transit of it over the sun's disk, should the earth and comet be in the line of the nodes at the same time, without *horror* at the thought of our being involved in its immense tail. I would not affirm that there may not be some astronomers so enthusiastic that they would not dislike to be whisked away from this low terrestrial spot into the higher regions of the heavens by the tail of a comet. But I hope you, Dear Miss Caroline, for the benefit of terrestrial astronomy, will not think of taking such a flight, at least till your friends are ready to accompany you.

"Mrs Maskelyne joins me in best compliments to yourself and Dr and Mrs Herschel.

I remain, Dear Miss Caroline, your obedient and obliged
humble servant,

N. MASKELYNE."

William Herschel mentioned Caroline's comet in a letter to Sir Harry Englefield, who had pointed out to him a comet observed by Messier. In his reply, Sir Harry Englefield wrote:

Dec. 25, 1788

"I am much obliged to you for your account of the Comet and beg you to make my comp[ts] to Miss Herschel on her discovery. She will soon be the great Comet finder and bear away the Prize from Messieurs Messier and Méchain".

William Herschel also wrote, probably in reply to a question, to the Earl of Salisbury, Hatfield, Hertfordshire:

"My Lord,

"The great Comet[1] which is expected has not yet made its appearance, but we are in hopes every day to meet with it.

"Your Lordship will find the small comet which we are now observing, (which was discovered by my sister the 21st Dec. when she was on the look out for the expected one) in the Constellation of the Lyre, and the enclosed figure will point it out. An achromatic 3-feet telescope of Dollond's will show it as a very faint, small nebulous patch, but in a 7-feet reflector of mine it is pretty large and considerably bright. The best time of seeing it, while the moon is in the way, is in the morning about 6 or 7 o'clock, towards the north-east, but by next Tuesday or Wednesday, it may conveniently be observed in the north-west as soon as it is dark in the evening.

"The papers have ascribed to me a foreknowledge of the weather, My Lord, which I am not so happy as to be in possession of; and I should be glad if I could foretell a speedy end to the present frost; which I can assure your Lordship, does not much agree with my feelings; as the number of additional Cloaths and Pellices, when I am observing, is pretty burthensome to the shoulders of

My Lord, your most obed[t]
and most humble servant

WM. HERSCHEL".

Slough. Jan. 8, 1789

[1] This was the comet of 1661, which was expected about this time but did not appear. See above, Dr Maskelyne's letter to Caroline Herschel of 27th December, 1788.

There are not many records of observations made in the year 1789. Caroline was interested in the satellites of Saturn and made drawings of their positions. She noted in her book:

"*Nov.* 2, 1789. $23^h$ 50'. My Brother shewed me ♄ (Saturn) when the 4th satellite was just detached from the body, and I saw the shadow upon the belts, where I have marked it (in a rough drawing).

"*Jan.* 19, 1790. My Brother shewed me the Georgium Sidus in the 20-feet telescope and I saw both its satellites very plainly".

On 7th January, 1790, Caroline discovered her third comet; it was a very small one and soon vanished. William Herschel notified Sir Harry Englefield, who wrote in reply:

"I have informed M. Méchain of the comet by to-day's Post. Pray congratulate Miss Herschel from me on her new career, which I hope she will follow and share the glory with you".

William Herschel was absent when Caroline encountered her fourth comet; she wrote at once to her friend, Alexander Aubert, and to Dr Maskelyne, and afterwards to Sir Joseph Banks.

CAROLINE HERSCHEL *to* ALEXANDER AUBERT

Slough. April 18, 1790

"Dear Sir,

"I am almost ashamed to write to you, because I never think of doing so but when I am in distress. I found last night, at $16^h$ 24' sidereal time, a comet and do not know what to do with it; for my new sweeper is not half finished, and besides I broke the handle of the perpendicular motion in my brother's absence (who is on a little tour in Yorkshire). He has furnished me to that instrument with a Rumboides, but the wires are too thin and I have no contrivance for illuminating them. All my hopes were that I should not find anything which would make me feel the want of these things in his absence; but as it happens here is an object in a place where there is no nebula or anything which could look like a comet, and I would be much obliged to you, Sir, if you would look at the place where the annexed eye-draft will direct you to. My brother has swept that part of the heavens and has many nebulae there but none which I must expect to see with my instrument.

. . . . . . . .

"I will not write to Sir Joseph Banks or anybody till you, Sir, or Dr Maskelyne (to whom I give a line of information by this post) have seen it; but if you could, without much trouble, give my best respects and that part of this letter which points out the place of the comet, to Mr Wollaston, you would make me happy.

I am, dear Sir, &c.

C. H."

Aubert, in reply, said he had so far failed to find the comet. Dr Maskelyne wrote:

"Dear Miss Herschel, Greenwich. April 22, 1790

"I return you many thanks for your two favours and the present of a comet, which I hope will be an improving one....

"The weather has not yet permitted me to see anything of the comet, but it seems now mending and I hope to be able to make something of it to-morrow morning. Your second communication at the same time that it gave me fresh spirits as to the certainty of its being a comet, will certainly assist me in more readily finding it. I feared that your using your new telescope might make that a bright comet to you which might prove but a very faint one, if at all visible, in a common night-glass, which is what we first use to discover a comet with. As soon as I shall have seen it, I will send you a line. I sent intelligence of your discovery to M. Méchain at Paris last Tuesday and will send to him your further communication next Friday. Mrs Maskelyne joins me in best compliments to yourself and Mrs Herschel, and Dr Herschel on his return. Dr Shepherd sent advice of it from me last Thursday to the Master of Trinity at Cambridge, which perhaps may convey the agreeable intelligence to your brother.

"I remain, Dear Miss Herschel, and my worthy sister in Astronomy,

Your faithful and obliged, humble servant

N. MASKELYNE".

CAROLINE HERSCHEL *to* SIR JOSEPH BANKS

"Sir, April 19, 1790

"I am very unwilling to trouble you with incompleat observations and for that reason did not acquaint you yesterday with the discovery of a comet. I wrote an account of it to Dr Maskelyne and Mr Aubert, in hopes that either of those gentlemen, or my Brother whom I expect every day to return, would have furnished me with the means of pointing it out in a proper manner.

"But as perhaps several days might pass before I could have any answer to my letters, or my brother return, I would not wish to be thought neglectful; and therefore if you think, Sir, the following description is sufficient, and that more of my Brother's astronomical Friends should be made acquainted with it, I should be very happy if you would be so kind as to do it, for the sake of astronomy.

"The comet is a little more than $3\frac{1}{2}°$ following $\alpha$ Andromedae, and about $1\frac{1}{2}°$ above the parallel of that star. I saw it first April 17, at $16^h\ 24'$ Sidereal time. And the first view I could have of it last night was at $16^h\ 5'$; as far as I am able to judge it has decreased in P.D. nearly $1°$ and increased in R.A. something above $1'$.

"These are only estimations from the field of view, and I only mention it to shew that its motion is not very rapid.

I am &c. &c.

C. H."

(This and her former letter are from her own copies.)

### SIR JOSEPH BANKS *to* MISS CAROLINE HERSCHEL

Soho Square. April 20, 1790

"Madam,

"I return you many thanks for the communication you were so good as to make to me this day of your discovery of a Comet. I shall take care to make our astronomical Friends acquainted with the obligations they are under to your diligence.

"I am always happy to hear from you but never more so than when you give me an opportunity of expressing my obligations to you for advancing the science you cultivate with so much success.

I am, Madam,

Your faithful serv$^t$

Jos. Banks."

Caroline was able to keep the comet in view during the whole month of May. On the 28th May she wrote in her *Book of Observations*:

"I saw the comet thro' the clouds, when I could not see a single star in the heavens, with the naked eye.

"*June* 4*th*, 1790. 16ʰ 23′ S.T. I saw the comet in the same field with a star, but the weather turned so cloudy that I could not be sure of the star. The comet was brighter than ever".

But after the 10th of June she lost sight of it entirely.

The following letter from Lalande was written in response to one from her concerning this comet:

À MADEMOISELLE CAROLINA HERSCHEL, *astronome célèbre, à Slough*, 12 *juillet*, 1790

"Ma chère et savante commère,

"J'ai reçu avec la plus délicieuse satisfaction la première lettre dont vous m'avés honoré, je ne pouvois attribuer votre silence à une timidité que votre réputation condamne mais je l'aurois attribué à mon peu de mérite si vous aviés continué de me refuser une réponse; vous écrivés si bien que vous ne pouvés pas avoir à cet égard une excuse légitime.

"Vous verrés bientôt M. Ungeschick qui a baptisé votre filleule Carolina; dites lui qu'elle se porte beaucoup mieux ainsi que le petit Isaac (je l'ai ainsi nommé en mémoire d'Isaac Newton); pour sa soeur je ne pouvois lui donner un nom plus illustre que le vôtre, c'est ce que j'ai fait remarquer en annonçant sa naissance dans notre moniteur ou gazette nationale du 31 janvier.

"Je ne pouvois vous donner un compère d'un plus grand mérite que M. de Lambre; il fait actuellement des tables des satellites de jupiter qui surpassent de beaucoup celles de M. Argentin.

"Votre commère, ma nièce, calcule des tables pour trouver l'heure en mer par la hauteur du soleil.

"Mad. du Piery calcule des observations d'éclipses.

"Pour moi je suis occupé des étoiles boréales, j'en ai déjà 6000, votre compère Le François y met beaucoup de soin, nous tâchons tous de seconder vos heureux travaux et ceux de votre illustre frère; nous vous prions tous de recevoir vous-même et de lui présenter nos respects.

"Remerciés-le bien de la complaisance qu'il a eu de m'envoyer la rotation de l'anneau dont j'étois bien curieux.

je suis avec autant d'attachement que de respect
Savante Miss,
votre humble et très obéissant serviteur
DE LA LANDE.

"Plusieurs de mes étoiles ont servi à comparer votre comète qui a disparu le 30 juin, mais que M. Messier et Méchain ont suivi sans interruption jusques dans le crépuscule".

Au Collège royal, le 12 juillet, 1790

### CAROLINE HERSCHEL *to* M. DE LA LANDE

Slough. Sept. 12, 1790

"Dear Sir,

"Our good friend, General Komarzewski, will persuade me to believe that I am capable of giving you pleasure, by writing a few lines; but I am under an apprehension that he is overrating my abilities. You, my dear Sir, certainly overrated them, when you thought me deserving of expressing your esteem for me in so publick a manner, as the General and Mr Ungeschick have informed me of.

"I do not only owe you my sincerest thanks for your good opinion of me, but my utmost endeavours to make myself, if possible, worthy of it. My good brother has not been omissive in furnishing me with the means of becoming so, in some respects. An excellent Newtonian sweeper, of 5 feet focal length, is nearly compleated; which being mounted at the top of the house, will always be in readiness for observing, whenever my attendance on the 40 or 20 feet telescopes is not required.

"I hope the little goddaughter is in good health, and wish she may grow and give happiness and pleasure to her parents and Oncle.

"I beg to present many respectful comp$^{ts}$ to the ingenious ladies you mentioned in your letter.

"Mrs Herschel desires to be remembered to you, Sir; we do not give up the hopes of seeing you again at Slough and are wishing it may not be long before you visit England again.

I remain, dear Sir, with the greatest esteem &c.

C. HERSCHEL."

Caroline's fifth comet was discovered on the 15th December, 1791; the sixth on the 7th October, 1793, and the seventh on the 7th November, 1795. A special interest attaches to this comet of 1795. It is the one commonly known as Encke's, as this astronomer demonstrated its periodicity. The period is very short, the comet returning at regular intervals of three and three-quarter years. It had been observed by Méchain in 1786, and can therefore not be counted, strictly speaking, as one of Caroline's discoveries. It is very faint and can only be seen with a telescope. On the 9th November, William Herschel observed it when passing over a small double star of the 11th or 12th magnitude. He reported:

"With a power of 287 I can see the smallest of the two stars perfectly well; this shows how little density there is in the comet, which is evidently nothing but what may be called a collection of vapours".

All Caroline's discoveries were duly notified at the time to Sir Joseph Banks and Dr Maskelyne; and doubtless drew forth congratulatory letters, but these have not been preserved. On the copy of her own letter to Sir Joseph Banks, November, 1795, she has written: "N.B. The letter of thanks of this and most of the comets, I have given as keepsakes to my Friends".

There are fewer entries for the later years, but this does not necessarily imply any relaxation in her perseverance in observing, for in one place she wrote:

"*May* 1, 1795. In future when any great chasms appear in my journals it may be understood that sweeping for comets has not been neglected at every opportunity which did offer itself. But as I always do sweep according to the precept my Brother has given me, and as I oftens am in want of time, I think it is very immaterial if the places where I have seen nothing are noted down".

She did not confine her notes entirely to comets; her brother had evidently desired her to keep a look out for any signs of volcanoes in the moon, and she frequently noted that she had examined the dark part of the moon and "found nothing luminous". Once only was she able to record:

"*Jan.* 4, 1794. A very bright spot in the dark part of the moon; circular".

She had two telescopes: the "small sweeper", and the 5-foot Newtonian, which she called her "large sweeper"; and both these she kept in use. On one occasion she wrote:

"Finding myself unable to bear the fatigue of standing, I took the small sweeper, which was in good order and shewed objects very well".

Caroline's own account of the discovery of her eighth and last comet may fitly close this chapter of her life. After she left the Cottage and took to living in lodgings, she seems to have found it impossible to keep up the practice of regular sweeping for comets, though her notebooks show that she continued to observe. The last entries made at her brother's house are as follows:

"*August* 14, 1797. At $9^h$ 30′ common time, being dark enough for Sweeping, I began in the usual manner with looking over the heavens with the naked eye, and immediately saw a comet nearly as bright as that which was discovered by Mr Gregory, Jan. 8, 1793.

"I went down from the observatory to call my Brother Alexander, that he might assist me at the clock. In my way in the garden I was met and detained by Lord Storker (?) and another Gentleman who came to see my brother and his telescopes. By way of preventing too long an interruption I told the gentlemen that I had just found a comet and wanted to settle its place. I pointed it out to them and after having seen it they took their leave. $19^h$ 30′ correct Sid. time = $9^h$ 59′ 13″ Mean time.

"I made the following eye draft.

"*August* 16, 1797. Memorandum. Tuesday morning, the 15th, I went to Greenwich and carried my memorandums to Dr Maskelyne. When I arrived there, he had had no intelligence yet of the comet. In the evening Dr M. received a letter from Stephen Lee, Esq., who had also seen it on the 14th. I had the pleasure to find that my observations agreed perfectly with Mr Lee's".

Though she discovered no more comets herself after she left the Cottage, Caroline's notebooks show how eagerly she watched for reports of any discovered by other observers. In the year 1807, when she was living at Upton, she wrote:

"*Jan.* 27, 1807. I brought this morning a memorandum with me from Windsor, which I copied from the Hamburgher papers at the Castle. The place of a comet for the following dates:

"*Jan.* 15. R.A. 25 14′. Decl. 39 18. Discovered by Mr Ponse,
"  ,,  25.      19 40.       29 34.    assistant to Mr Schroe-
"*Feb.* 2.     17 33.       23 58.    ter zu Lilienthal.

"This being a part of the heavens which I could get sight of at the house at Upton, I took my sweeper down in the garden and began to sweep as soon as daylight was gone and at

"$6^h$ 52′. I saw it in the field of view;...."

"It is one of the smallest comets I have seen, but not the smallest and would undoubtedly have appeared to be very considerable if it had been observed at an higher altitude; for it was actually setting at the time I was looking. It appeared like an oblong nebula, rounded at the corners, shaped like an egg.

"*Jan.* 28. I had my sweeper carried to Slough, and went there with an intention of showing the comet to my Brother, but the heavens were totally overcast.

"*Jan.* 31. I was at Slough time enough to prepare my sweeper. The weather coming on very fine I kept on sweeping near the place and found the comet again. The wind was very high and the weather very cold but my brother (though very ill) came out to make proper observations on the comet and its situation, &c., for which see his Review".

She observed another comet in October, 1807, with her brother's 7-foot telescope, in his absence. Again in September, 1811, she made some lengthy observations of a comet of which her friend, Mr Pigot, had given her notice. It was a very fine one; she described it as having

"a nucleus of white light, which appeared to me equal in size to 47 Ursae (which is of a reddish light); then to the nucleus adhered a circular white nebulosity, which I estimate about 3 times the breadth of the nucleus".

The last entry in her *Book of Observations* has a pathetic significance, as showing how her interest in the branch of astronomy which she had made specially her own survived the shock of her brother's death and her removal to Hanover. It runs thus:

"*Hannover, Jan.* 31, 1824. Eyedraft of the situation of the Comet which has been for some time past (since Jan. 6) observed here in Hannover. It is at present surrounded with a diffused light of no very great extent and with difficulty seen by the naked eye".

About the time when she discovered her last comet, Caroline was engaged during the daylight hours on a task for which her talent for laborious and accurate application peculiarly fitted her. This was a revision of Flamsteed's Catalogue of fixed stars.

Flamsteed, the first Astronomer Royal, at the Royal Observatory, Greenwich, had left, in the third volume of his great work, the *Historia Coelestis*, a Catalogue of 2935 stars observed by himself. His own observations, made during the years 1689 to 1719, on which the positions of the stars are based, were given in another volume of the work. Flamsteed's Catalogue, generally known as the *British Catalogue*, was regarded at this time as the final authority for reference as to the position of any star; but Herschel's observations had shown that it needed revision. His own words, introducing "An Index to Mr Flamsteed's Observations of the Fixed Stars", read before the Royal Society in May, 1797, show how Caroline was induced to undertake and to accomplish this task.

William Herschel wrote:

"In my earliest reviews of the heavens, I was much surprised to find many of the stars of the British catalogue missing. Taking it for granted that this catalogue was faultless, I supposed them to be lost. The deviation of many stars from the magnitude assigned to them in that catalogue, for the same reason, I looked upon as changes in the lustre of the stars. Soon after, however, I perceived that these conclusions had been premature, and wished it were possible to find some method that might serve to direct us from the stars in the British catalogue to the original observations which have served as a foundation to it. The labour and time required for making a proper index withheld me continually from undertaking the construction of it: but when I began to put the method of comparative brightness in practice, with a view to form a general catalogue, I found the indispensable necessity of having this index recur so forcibly, that I recommended it to my Sister to undertake the arduous task. At my request, and according to a plan which I laid down, she began the work about twenty months ago, and has lately finished it".

Caroline kept for her own satisfaction a copy of the note which Mr Planta, the Secretary of the Royal Society, wrote to her brother regarding the printing of the Catalogue of omitted stars, and of her Index. This was the note:

Decr. 21, 1797

"At a Council of the Royal Society

"The President laid before the Council a Catalogue of Stars omitted in the *Atlas Coelestis*, collected by Miss Herschel out of Flamsteed's observations, with an Introduction by Dr Herschel; and proposed that should the Dr approve of it, it be printed at the expense of the Society, in the same form as, and by way of Appendix to, the said *Atlas Coelestis*. The Council unanimously agreed to this proposal; and being moreover informed that Miss Herschel had, in order to form this catalogue, drawn up an Index to Flamsteed's Observations, they ordered a letter to be written to Dr Herschel to request the communication of the same, and his and his Sister's permission to add it to the above intended publication.

J. PLANTA".

In the following letter to Dr Maskelyne, Caroline refers to the printing of her Index, and also to a manuscript list of omitted stars which she had made for her brother, and of which she asks Dr Maskelyne to accept a copy. Wollaston's Catalogue, which she also mentions, was not published till 1800, but Herschel and other astronomers had received copies of the proof-sheets; it was a small but very useful catalogue of circumpolar stars only.

CAROLINE HERSCHEL *to* DR MASKELYNE

Slough. Sept. 1798

"Dear Sir,

"I have for a long while past felt a desire of expressing my thanks to you, for having interested yourself so kindly for the little production of my industry, by being the promoter of the printing of the Index to Flamsteed's observations. I thought the pains it had cost me were and would be sufficiently rewarded in the use it had already been, and might be of in future, to my brother. But your having thought it worthy of the press has flattered my vanity not a little. You see, Sir, I do own myself to be vain because I would not wish to be singular, and was there ever a woman without vanity?—or a man either? only with this difference, that among gentlemen the commodity is generally stiled ambition.

"I wish it were possible to offer something which could be of more use to our Royal Astronomer than merely thanks. Perhaps the enclosed Catalogue may be of some little service on some occasion or other. I was obliged to bring it into that form by way of scrutinizing the real number of omitted stars; and find it now very useful when my brother in sweeping, &c., observes stars which are not contained in Wollaston's Catalogue, to know immediately by this order of R.A. if they are any of Flamsteed's omitted stars, and if they are, what number they bear in the Catalogue of omitted stars; which number we find in the first column. The rest of the columns will want no explanation except the last; which would not be complete, or not even intelligible, without the assistance of the Catalogue of omitted stars, and the Notes to that Catalogue; for they are short memorandums collected from the descriptions in the Catalogue and from the notes to some of the stars.

"As our Index contains all the corrections and informations which I possibly could collect, those corrections and memorandums of which I had the pleasure about 18 months ago to write a copy for Dr Maskelyne, will consequently be laid aside; else I ought to take notice that there are one or two errors, and several omissions which should have been corrected in that copy; but with which it will now be needless of troubling you with, Sir.

"What has laid me under particular obligation to you, my dear Sir, was your timely information, the August before last, of your having proposed the printing of the Index to the *P.R.S.* The papers were then in so incomplete a state, that it needed each moment which could possibly be spared from other business to deliver them with some confidence of their being pretty correct.

"Many times do I think with pleasure and comfort on the friendly infitations Mrs Maskelyne and yourself have given me, to spend a few days at Greenwich. I hope yet to have the pleasure next spring or summer. This last has passed away and I never thought myself well, or in spirits enough to venture from home. If the heavens had befriended me and afforded us a comet, I might have undertaken its convoy; perhaps have ventured on an emigration. However I cannot help thinking but I shall meet with some little reward for the denial it has been to me not coming this summer, in seeing the improvements Miss Maskelyne has made (more perceptibly) in those accomplishments she seemed to be in so fair a way of attaining when I was there last.

"With my best respects and compliments to Mrs M.

I remain with the greatest esteem dear Sir, &c.

C. HERSCHEL."

# CHAPTER XVII

## 1793    1804

William Herschel's scientific work. Paper "On the Nature of the Sun and fixed Stars". Letter to Prof. Wilson. Discovery of heat rays. Letters—Sir J. Banks. Asteroids. Letters—Piazzi, Bode, Olbers, Laplace. William Herschel's experiments for determining the diameters of small spheres.

WILLIAM HERSCHEL was much occupied during the latter years of the century with observations on the planets Venus, Jupiter and Saturn. After the striking discovery in 1789 of the two new satellites of Saturn, he made many very valuable observations on the ring, on the belts, on the discs both of Saturn and of Jupiter, and of the rate of rotation of these planets and of their satellites. But though he had attributed his discovery of Saturn's satellites to the new 40-foot telescope, he does not seem to have used it in these observations. In a note to one of his papers he explained his preference for a smaller telescope.

### Observations on the Belts of Saturn[1]

"*Nov.* 11, 1793. In the course of these observations, I made 10 new object specula, and 14 small plain ones, for my 7-feet reflector; having already found that with this instrument I had light sufficient to see the belts of Saturn completely well; and that, here, the maximum of distinctness might be much easier obtained, than where large apertures are concerned."

His attention was now also directed to the sun. In December, 1794, he contributed to the Royal Society a paper "On the Nature of the Sun and fixed Stars".[2]

The subject of the physical constitution of the sun had not hitherto received much attention from astronomers. Dr Alexander Wilson, Professor of Astronomy at Glasgow and father of Herschel's friend, Patrick Wilson, had made some very valuable observations of a huge sunspot which had appeared in the year 1769, and deduced from these

[1] *C.S.P.* vol. I, p. 459.    [2] *Phil. Trans.* 1795, pp. 46–72.

that sunspots are depressions in the sun's surface, through which the non-luminous body of the sun could be discerned. Lalande controverted this view and maintained that on the contrary the dark appearances were prominences projecting through the luminous atmosphere. William Herschel took the same view of the dark spots as Dr Wilson, but in the introduction to his paper he did not mention the latter by name. Patrick Wilson, who had succeeded his father in the Professorship at Glasgow, was pained at this omission, which he seems to have looked upon as a slight to his father's memory. The letter which he wrote to Herschel is not among those preserved; it would seem that Herschel destroyed it, as was his custom with any that might recall unhappy memories or contain expressions which the writer might later regret to have used. His answer, given here, was copied by Caroline into the Letter Book, and evidently quite satisfied Patrick Wilson.

WILLIAM HERSCHEL *to* PROF. PATRICK WILSON

Slough near Windsor. April 1, 1796

"My dear Friend,

"It has partly not been in my power to answer your kind letter relating to the solar spots, not being in possession of the Volume which contains the observations of the late excellent Professor Wilson upon them. I could not enter into the subject as I ought to do without a proper examination of that paper. Very lately I have had an opportunity of reading it, and find that it contains indeed very valuable observations. They amount to an evident proof of the depression of the dark spots below the shining part of the sun. Some judicious and well authorised consequences are also drawn from those observations.

"In my paper on the nature and construction of the sun and fixed stars, I wished not to enter into a particular examination of what had already been said upon the subject by former astronomers, and therefore only mentioned in a general way that we ought now to profit by the many valuable observations of which we are in possession. A list of successive eminent astronomers might be named from Galileo down to the present time (in which certainly the name of the excellent author of the observations contained in the 74th Vol. of the *Phil. Trans.* could not have been omitted, had it been proper to mention names or to enter into particulars) who had furnished us with materials for examination. To this I added that *I had availed myself of the labours of all these astronomers*: leaving it in this manner open for all the present astronomers to ascribe as much of the merit (if there should be any in my paper) to the consequences that might be drawn from their own observations, or from the observations that have been given us by such other astronomers as have entered into the subject. In confessing that I had availed myself of the labours of all other astronomers, I avowedly disclaimed every merit as a first discoverer, and this, my dear Friend, if you consider the case properly, I hope will be a satisfactory answer to your letter. Perhaps you may wish to know why I did not judge it proper to enter into particulars. Had I mentioned the observations in the 74th Vol. I must have also mentioned the answer made by Mr De la Lande; I must in the next place have turned to a tedious treatise on the solar spots written but lately by Mr Schroeter, which must infallibly have brought on a controversy, as that Gentleman has sufficiently shewn in his last paper on Venus a disposition to take hold of every opportunity to defend his erroneous as well as his good communications. Another person, I find, has shewn that he was the first who noticed the depression of the dark solar spots, in the year 1771.

. . . . . . .

"I have just begun to make a 30 feet telescope for the King of Spain which is to be finished in a short time. This will, I am sorry to say it, put off our intended visit to Scotland till another summer, as there will be no possibility of leaving the workmen this twelvemonth. I shall have a fine opportunity to complete my experiments upon the construction of large mirrors, and hope soon to have my own instrument in a very superior state to what it now is.

"It will give me the most particular happiness to hear often from you, and including Mrs Herschel's affectionate compliments to Miss Wilson and yourself, I beg you will ever believe me to be with the greatest esteem, my dear Friend,
                          your faithful humble servt
                                    WM. HERSCHEL."

In the paper "On the Nature of the Sun and fixed Stars", after giving many records of his observations, Herschel summed up his conclusions thus:

"It will now be easy to bring the result of these observations into a very narrow compass. That the sun has a very extensive atmosphere cannot be doubted; and that this atmosphere consists of various elastic fluids, that are more or less lucid or transparent, and of which the lucid one is that which furnishes us with light, seems also to be fully established by all the phaenomena of the spots, of the faculae, and of the lucid surface itself. There is no kind of variety in these appearances but what may be accounted for with the greatest facility, from the continual agitation which we may easily conceive must take place in the regions of such extensive elastic fluids".

The hypothesis which he put forward to explain the origin of the spots is stated to be that there are clouds in the atmosphere of the sun originating, as he supposed, in a manner somewhat similar to the formation of those in the earth's atmosphere "from the decomposition of some of the elastic fluids of the atmosphere itself".

He did not, however, lay much stress on this hypothesis, but said of the lucid clouds: "They plainly exist, because we see them; the manner of their being generated may remain an hypothesis; and mine, till a better can be proposed, may stand good".

Altogether, it should be said that Herschel's ideas regarding the nature of the sun are not now regarded as well founded.

In pursuing his observations on the sun, Herschel was obliged to try various methods for obviating the danger to the eye of exposure to the heat of the sun's rays. Admiral Smyth, in his book *The Cycle of Celestial Objects*, says that Sir William Herschel lost an eye in this service. Admiral Smyth must have got the knowledge of this fact, if fact it was, from Sir John Herschel, with whom he was very intimate; yet it was never mentioned in the Herschel family, nor is there any record of such an occurrence in any of the Herschel papers. Probably the loss of sight, if it did occur, was only temporary; but the danger of such injury was real.

It had hitherto never been doubted that the heating and illuminating powers of the sun's rays exist concurrently; Herschel, however, when he experimented with differently coloured darkening glasses, inferred that this could not be the case. In a paper[1] which he contributed to the Royal Society in March, 1800, he wrote:

"In a variety of experiments I have occasionally made, relating to the method of viewing the sun with large telescopes, to the best ad-

[1] "Investigation of the powers of the prismatic colours to heat and illuminate objects", *Phil. Trans.* 1800, p. 255; *C.S.P.* vol. II, p. 53.

vantage, I used various combinations of differently coloured darkening glasses. What appeared remarkable was that, when I used some of them, I felt a sensation of heat, though I had but little light, while others gave me much light, with scarce any sensation of heat. Now, as in these different combinations the sun's image was also differently coloured, it occurred to me, that the prismatic rays may have the power of heating bodies very unequally distributed among them; and as I judged it right in this respect to entertain a doubt, it appeared equally proper to admit the same with regard to light.—At all events, it would be proper to recur to experiments for a decision".

The arrangement which he made for testing the heating power of the several prismatic colours was very simple. A row of thermometers was arranged behind a board, in which was a slit which allowed the rays of one colour only of the spectrum to pass through at a time. One of the thermometers was so placed as to receive the coloured ray, while the others remained in shade. Herschel then noted the rise in temperature indicated by the thermometer under the light, during a given time; and by exposing it to different parts of the spectrum he was able to record the different heating effect of the different rays. He found, by this method, that in an interval of eight minutes, the red rays gave a rise of temperature of $6\frac{1}{4}$ degrees, the green rays gave a rise of $3\frac{1}{4}$ degrees and the violet of only 2 degrees. He concluded from this experiment "that the heating power of the prismatic colours is very far from being equally divided; and that the red rays are chiefly eminent in that respect".

He then considered the illuminating power of the several rays, and for these experiments he used a microscope and threw the different coloured rays on to a great variety of objects, such as red paper, a piece of brass, black paper, a black nail, &c. The remarks which he made about his choice of a nail show in so frank a manner the lively pleasure which he derived from his experiments that they seem worth quoting:

"The appearance of the nail in the microscope is so beautiful, that it deserves to be noticed; and the more so as it is accompanied by circumstances that are very favourable for an investigation, such as that which is under our present consideration.

"I had chosen it on account of its solidity and blackness, as being most likely to give an impartial result of the modifications arising from an illumination by differently coloured rays; but on viewing it, I was struck with the sight of a bright constellation of thousands of luminous points, scattered over its whole extent, as far as the field of the microscope would take in. Their light was that of the illuminating colour, but differed considerably in brightness; some of the points being dim and faint, while others were luminous and brilliant.... An object so well devised by nature into very minute and differently arranged points on which the attention might be fixed, in order to ascertain whether they would be equally distinct in all colours, and whether their number would be increased or diminished by different degrees of illumination, was exactly what I wanted".

The general results of these experiments he summed up as follows:

"From these examinations, which agree uncommonly well, with respect to the illuminating power assigned to each colour, we may conclude that the red-making rays are very far from having it in any eminent degree. The orange possess more of it than the red; and the yellow rays illuminate objects still more. The maximum of illumination lies in the brightest yellow or palest green; but from the full deep green the illuminating power decreases very sensibly...the violet is very deficient".

In a second paper on the same subject, which Herschel sent to the Royal Society a month later, he gave details of further experiments which he had made with a view to confirming a suspicion already forming in his mind that "the range of the refrangibility of radiant heat is probably more extensive than that of the prismatic colours".

From these investigations he arrived at the conclusion that no rays beyond the violet had any perceptible power, either of illuminating or heating, but that beyond the red end of the spectrum:

"There are rays coming from the sun, which are less refrangible than any of those that affect the sight. They are invested with a high power of heating bodies, but with none of illuminating objects; and this explains the reason why they have hitherto escaped unnoticed".

Herschel's demonstration of the existence of the non-luminous heat-producing rays beyond the red end of the spectrum was a real discovery, and deserves to be considered a remarkable contribution to scientific knowledge. But he was misled in the views which he adopted concerning the nature of radiant heat, which he conceived to be intrinsically different in kind from visible radiation. Indeed, in his fondness for classification he even attempted, in a later paper, to draw distinctions between six different kinds of "heat". Herschel shared the opinion, generally held since Newton's time, that light consisted of material particles; hence his conception that radiant heat also consisted of particles, different in kind from light, was intelligible. Herschel was above all things a pioneer. His imagination was very active, and his eagerness sometimes, as in this case, led him to believe that he had solved problems which required much more delicate scrutiny than he could possibly give them. His inferences concerning the nature of radiant heat were generally accepted for many years after their publication; it required nearly a century of scientific advance before the true significance of the apparent separation of the heating, illuminating and chemical activities of radiation was fully recognized.

In the course of the first paper on this subject he threw out a suggestion very prophetic of later developments.

"It may be pardonable", he wrote, "if I digress for a moment and remark that the foregoing researches ought to lead us on to others. May not the chemical properties of the prismatic colours be as different as those which relate to light and heat? Adequate methods for an investigation of them may easily be found; and we cannot too minutely enter into an analysis of light, which is the most subtle of all the active principles that are concerned in the mechanism of the operations of nature."

Herschel's primary object in this investigation was to find a method of observing the bright image of the sun's surface without danger to the observer's eye. In this he was successful. In a later paper "On the Nature of the Sun and fixed Stars", which was read before the Royal Society in May, 1801, he described a water eyepiece which he found very efficacious. The following entries explain its construction:

"1801, *March* 4. I viewed the sun with a skeleton eye-piece, into the vacancy of which may be placed a movable trough, shut up at the ends with well polished plain glasses, so that the sun's rays may be made to pass through any liquid contained in the trough, before they come to the eye-glass.

"*April* 4. I viewed the sun through a mixture of ink diluted with water and filtered through paper. It gave an image of the sun as white as snow; and I saw objects very distinctly without darkening glasses.

"*May* 3. Ink mixture. Through this mixture I can observe the sun in the meridian, for any length of time, without danger to the eye or to the glasses".

The following correspondence passed between Sir Joseph Banks and Herschel on the subject of these papers:

SIR JOSEPH BANKS *to* WILLIAM HERSCHEL

Soho Square. Mar. 24, 1800

"My dear Sir,

"I have shown your second paper to Mr Cavendish and to some other of my friends, previous to its being read, which will take place immediately, and all are struck with the discovery, of Radiant Heat being separable from Radiant Light; experiments, it seems, have been tried, with a view to ascertain this fact, on the heat which radiates from our ordinary fires, but it is very remarkable that the heat of ignited bodies could not be made to pass through glass, and therefore could not be subjected to experiments of refraction.

"I think all my friends are of opinion that the French system of Chemistry, on which the names lately adopted by their Chemists is founded, already totters on its base and is likely soon to be subverted. I venture therefore to suggest to you whether it will not be better for you in both Papers to use the term Radiant Heat instead of Caloric; by the use of which latter word it should seem as if you had adopted a system of Chemistry which you have probably never examined.

Believe me, dear Sir, with sincere esteem and regard,
very faithfully yours
JOS. BANKS".

WILLIAM HERSCHEL *to* SIR JOSEPH BANKS

Slough near Windsor. March 26, 1800

"Sir,

"I have the honour of your letter and shall be very ready to change the word caloric for radiant heat, which expresses my meaning extremely well. I shall very soon have completed a series of experiments that will I hope throw a good deal of light upon our subject; and perhaps you will be surprised to hear that the very thing you mention as not to be done, namely making culinary heat pass through glass, I have done with great ease, and made experiments upon it which will be among those I mention. I am analising radiant heat, as well solar as culinary, every way, and hope soon to complete my account of them.

&c.,

WM. HERSCHEL."

### SIR JOSEPH BANKS *to* WILLIAM HERSCHEL

Soho Square. April 9, '00

"My dear Sir,

"I have no thoughts of settling at Spring Grove till Whitsun holidays, but I will readily come and meet you there if a day can be fixed upon that will be suitable. I have no engagement in Easter week and will try, if you will name the day that suits you best, to prevail upon Mr Cavendish, who knows what has been done with the Rays of the sun far better than I do, to come with me and perhaps another or two friends who you approve.

"The public will appreciate as they see fitting the value of your discovery; for my own part I hope you will not be affronted when I tell you that, highly as I prized the discovery of a new Planet, I consider the separation of heat from light as a discovery pregnant with more important additions to science.

Believe me, my dear Sir,

very faithfully yours

JOS. BANKS.

"I shall not bring any person but those whom you have approved. Tell me whether you feel quite at ease with Count Rumford. He says that your discovery is the most important since Sir I. Newton's death."

### THE SAME *to* THE SAME

April 14, 1800

"My dear Sir,

"I saw Sir Wm. Watson yesterday and settled with him that Friday next the 18th would be a good day for our meeting. I accordingly have invited Sir Wm., Dr Gray and Count Rumford to meet you at Dinner; and I mean to add to the list Sir Ch. Blagden and Mr Cavendish if I can be so lucky as to find them disengaged. If you wish for

any private conversation with me and will be so good as to call an hour or two before Dinner, which will be at 5 o'clock, I shall certainly be at home.
                    Believe me with sincere esteem and regard,
                                                    Jos. Banks."

It has already been mentioned that Baron Von Zach and six other astronomers had met in 1800, at Lilienthal, and formed a society, under the Presidentship of Schroeter, to search for the missing planet, which, there was reason to believe, existed between Mars and Jupiter. Von Zach had just started a magazine to afford the means, rare in those days, of easy intercommunication between all the astronomers of Europe. One of these is reported to have said: "What would have been the fate of the small planets if the *Monatliche Correspondenz* had not then existed?" In January, 1801, Prof. Piazzi, of Palermo, discovered a small star which moved and whose motion he was able to follow for some weeks. He communicated his observation to Bode, at Berlin, and all the German astronomers joined in pursuit of it, but it was so small that it was not till the end of the year that Olbers was successful in picking it up again. Its orbit could now be determined, and, to the great satisfaction of the astronomers, it was found to correspond to the place assigned to it under Bode's empirical law. The next year, in March, 1802, Dr Olbers, of Bremen, discovered another little planet having the same orbit as Ceres, the name which Piazzi gave to his star. Olbers named it Pallas.

A third little planet was discovered, in 1804, by Schroeter's assistant, Harding, at Lilienthal. The orbit of this body, named Juno, was found to be decidedly eccentric, as was also that of Pallas. Three years later, Olbers added a fourth member to the group, to which he gave the name of Vesta.

Prof. Piazzi wrote, in September, 1801, to William Herschel to announce his discovery.

### PROF. PIAZZI *to* WILLIAM HERSCHEL

                                            Palermo, le 1er sept. 1801
"Monsieur,

"J'espère que vous ne m'avez pas oublié: certainement je n'oublierai jamais les bontés dont vous m'avez comblés pendant mon séjour en Angleterre.

"Le premier Janvier j'ai découvert une étoile, qui par son mouvement ressemble beaucoup à une Planète: ci-joint vous trouverez un petit mémoire sur cette découverte. Avec votre grande adresse, et vos grands moyens, je voudrais bien que vous la cherchassiez. Pour moi, faute d'instruments, je ne pourrai m'en occuper qu'à son passage au méridien pour lequel il faut de temps encore, et je ne sais pas si je serai assez heureux pour la revoir.

"Aidez moi donc à verifier cette découverte, que je ne doute pas vous intéressera autant que moi.

Je suis avec le plus sincère attachement, Monsieur,
votre très humble serviteur
Piazzi."

Herschel replied to Piazzi that he had been searching for his new star for two months, but owing to bad weather conditions had not been able to find it. He hoped it might turn out to be a planet.

There was considerable excitement in astronomical circles about these little planets. Sir Joseph Banks kept Herschel informed of everything new which reached him. In February, 1802, he wrote:

"By a letter from Zach I learn that Harding of Lilienthal, looking at Ceres with a power of 200, distinguished a little disk of the size of the 1st or 2nd satellite of Jupiter, hence he concludes the diameter of the planet to be about 2″. He saw also two lucid points near the Planet both on the west side,—which he suspects to be satellites".

He wrote again:

Soho Square. April 9, 1802
"My dear Sir,

"Dr Olbers has discovered another Ceres, on the 28th of March; it is to appearance larger than the Georgium. Messrs Schroeter and Harding have seen it since April 2nd. Mr Harding saw a star near it which he thinks will prove a satellite".

William Herschel was naturally much interested in these new discoveries. He examined both Ceres and Pallas carefully, endeavouring to ascertain their size by means of a micrometer of his own invention, which he called the lucid disc micrometer. He computed the diameter of Ceres to be only 161·6 miles and that of Pallas even less, about 147 miles.

Schroeter, on the other hand, with different instruments, found for Ceres a diameter of 1624 miles and for Pallas 1425 miles. Both are wrong, but Herschel's guess comes nearest to the modern estimate of 480 miles for Ceres and 304 miles for Pallas.

Writing to Sir Joseph Banks, Herschel said:

> "I think that my determination of the magnitude of the new planet must be much more accurate than that of Mr Harding of Lilienthal, both on account of the object with which I compared it, and of the magnifying power of my telescope".

Herschel had a passion for classification, and it seemed to him that these new bodies should be placed in a group by themselves, distinguished by certain characteristics from either planets or comets. He wrote to his friend, Sir William Watson:

> "My dear friend,    April 25, 1802
>
> "...I have now to request a favour of you, which is to help me to a new name. In order to give you what will be necessary I must enter into a sort of history. You know already that we have two newly discovered *celestial bodies*. Now by what I shall tell you of them it appears to me much more poor in language to call them planets than if we were to call a *rasor* a *knife*, a *cleaver* a *hatchet*, &c. They certainly move round the sun; so do comets. It is true they move in ellipses; so we know do some comets also. But the difference is this: they are extremely small, beyond all comparison less than planets; move in oblique orbits, so that, if we continue to call that the ecliptic in which we find them, we may perhaps, should one or two more of them be discovered still more oblique, have no ecliptic left; the whole heavens being converted into ecliptic, which would be absurd. I surmise (again) that possibly numbers of such small bodies that have not matter enough in them to hurt one another by attraction, or to disturb the planets, may possibly be running through the great vacancies, left perhaps for them, between the other planets, especially Mars and Jupiter. But should there be only two, surely we can find a name for them. The diameter of the largest of them (at present *entre nous*) is not 400 miles, perhaps much less, as I shall know in a few hours, but have not time to wait. Now as we already have Planets, Comets, Satellites, pray help me to another dignified name as soon as possible".

In the paper which he sent to the Royal Society in May, 1802, Herschel suggested the name *asteroids* for these bodies, as they differ from planets in the fact of their orbits not being confined to the zodiac, and from comets by not having any coma, and from both classes by resembling small stars so much as hardly to be distinguished from them, even by very good telescopes. He wrote,[1] "I shall take my name and call them *asteroids*".

Bode was much pleased with the discovery of Ceres, which fulfilled so satisfactorily the conditions of his law; but the appearance of a second little planet was disconcerting; it seemed superfluous. He wrote to Herschel in May 1802:

"Olbers has found a moving star, which I take to be a very distant comet. Olbers suggests most strange things; that it is a planet travelling with Ceres, in the same orbit, at the same distance round the sun. Such a thing is unheard of!"

However, the claims of Pallas to be such another planet as Ceres could not be denied, and Bode wrote again in August, 1802:

"I hold myself still convinced that Ceres is the eighth primary planet of our solar system and that Pallas is a special exceptional planet—or comet—in her neighbourhood, circulating round the sun. So there would be two planets between Mars and Jupiter, where, ever since 1772, I have expected only one; and the well-known progressive order of the distances of the planets from the sun, is by this fully proved".

In this letter Bode also makes some remarks on Herschel's paper on the two planets, Ceres and Pallas.

"I have read your interesting researches on Ceres and Pallas, not only with pleasure but with a certain astonishment; as it remains to me inexplicable how you can suggest such small real diameters for both these bodies, when you agree, as I think you do, with Dr Gauss about their real distances. How excessively small you must find the apparent diameters, of that I have no conception, nor how you were able to measure such minute diameters. How could planetary bodies send back reflected light if they...(are so small). Ceres was seen in March by several of my friends with the naked eye."

---

[1] "Observations on two newly discovered bodies (Ceres and Pallas)", *Phil. Trans.* 1802; *C.S.P.* vol. II, p. 197.

In May, 1802, Sir Joseph Banks sent to Herschel a communication he had received from Olbers. It began:

(Translated)

"With much pleasure, as you can well believe, I send you full and certain information that our *Pallas is a planet*; the own sister to Ceres, not inferior to her in dignity and importance, and perhaps on that account another remarkable discovery; as she gives rise to many speculations regarding the origin and history of our Planetary system. With this I send the elements calculated by Dr Gauss.

. . . . . .

"The orbit is therefore only as eccentric as that of Mercury. But what a great unexpected inclination! and how curious its position with regard to that of Ceres! The two cross each other like the interlocked rings of a chain.

"To my great surprise, and surely to yours too, I hear that Dr Herschel allows Ceres a diameter of only 1 second.

Leben Sie Wohl

Dr Olbers".

Olbers also wrote direct to Herschel:

Bremen. 17 June, 1802

(Translated)

"With grateful thanks I acknowledge your kindness in giving me information of your interesting observations on Ceres and Pallas. You can easily believe how curious I was to know how *you* had seen these curious bodies through your unrivalled telescope, and therefore how much pleasure your letter gave me.

"Forgive me that I do not reply in English. I can read and understand the English language, but writing it is difficult to me. I also held Ceres and Pallas to be very small (he gives examples of observations to show that by comparison with a fixed star they showed no difference)—but so small as you, honoured Sir, have found them, I have never imagined! I am curious to know how you measured the apparent diameters of these tiny bodies, for honestly I cannot understand it.

"I agree with you, honoured Sir, in your sagacious suggestion that Ceres and Pallas differ from the true planets in several respects, and the name *asteroid* seems to me to fit these bodies very well. Yet I would not lay too much stress on the difference in size, as the old planets differ from one another so much in this respect. Yet taking all the particulars together there seems to me to be a real difference between the asteroids and the true planets.

. . . . . . .

"The similarity in the period of their revolution, of their long axes, and the remarkable position of both orbits in relation to each other, have suggested to me an idea which I hardly dare to put forward as a hypothesis, and about which I should much like to have your, for me, weighty opinion. I mention it to you in confidence. How might it be, if Ceres and Pallas were just a pair of fragments, of portions of a once greater planet which at one time occupied its proper place between Mars and Jupiter, and was in size more analogous to the other planets, and perhaps millions of years ago, had, either through the impact of a comet, or from an internal explosion, burst into pieces?

"I repeat that I give this idea as nothing more than, hardly as much as, a hypothesis.

"One circumstance, about whose correctness you can doubtless instruct me, seems to confirm this view. It is this, that both Ceres and Pallas seem to vary very much in the amount of light they give, from one evening to another. Sometimes Ceres, sometimes Pallas, appeared the brightest. Other Continental astronomers have noticed this; Schroeter explains it by variations in the atmosphere of the asteroid; this seems to me unlikely; and it appears to me would be better accounted for by supposing for both these asteroids an irregular rather than a round figure; by reason of which at one time a broad surface and at another a narrow profile would be turned to us.

"From another point of view the idea deserves to be put to the test. If it is true, we may expect to find other fragments of the broken down planet; their orbits must intersect that of Ceres in the same spot.

. . . . . . .

"Forgive, Sir, this long letter, but before I close I cannot let the opportunity slip by without expressing the great respect and admiration which I have felt for you for many years. Your great, unparalleled discoveries ensure you the gratitude of this and of future generations.     I am with the greatest unbounded esteem

Your obed$^t$ serv$^t$

WILH. OLBERS".

Unfortunately, there is no record of any reply to this letter in William Herschel's Letter Book, so we do not know what he thought of Olbers' suggestion.

### PIAZZI *to* WILLIAM HERSCHEL

Palerme, le 4 juillet, 1802

"Monsieur,

"J'ai lu avec bien du plaisir l'extrait de votre savant mémoire sur Ceres et Pallas que vous avez eu la bonté de m'envoyer. Les idées nouvelles et les observations précieuses seront toujours votre partage, et je serai toujours un des plus admirateurs. Mais dites-moi, ne pourroit-on pas établir pour marque distinctive entre les planètes et les comètes, l'intersection de leurs orbites réduites à l'écliptique? Et pour la dénomination, ne pourroit-on pas appeler les petites planètes, planetoides? Car je vous avoue, le nom d'Asteroides me paraît plus propre aux petites étoiles. Mais ceci sont des riens vis-à-vis des belles choses dont doit être rempli votre mémoire.

"J'ai publié une petite histoire de la découverte de Ceres; que je prends la liberté de vous envoyer.

"Je suis avec les sentiments de la plus haute estime et sincère attachement,
Monsieur, votre très humble serviteur

PIAZZI."

Laplace also demurred to the name of asteroid, as the following letter shows:

### LAPLACE *to* WILLIAM HERSCHEL

Paris. 17 juin, 1802

"Monsieur et illustre Confrère,

"Je suis très reconnaissant de la communication de vos observations sur les deux nouveaux astres. Je suis bien curieux de connoître votre moyen pour mesurer d'aussi petits diamètres et les dépouiller de l'effet de l'irradiation; j'en ai donné un d'après Newton dans mon exposition du Système du monde; mais j'ignore jusqu'à quel point le vôtre en diffère. Quant au nom que vous donnés à ces astres, je ne vois pas encore de motif suffisant pour ne pas leur conserver le nom de planètes. La Ceres n'en diffère que par son inclination qui est un peu grande; mais il s'ensuit seulement qu'il faut étendre la largeur du zodiaque, et même y comprendre le Pallas, si, comme cela devient assez vraisemblable, son orbite est une ellipse, dont l'excentricité n'est qu'un peu plus grande que celle de l'orbite de Mercure.

"J'entends avec beaucoup de plaisir que vous vous proposez de venir bientôt à Paris; je serai bien enchanté de l'honneur de vous connoître personnellement, de vous témoigner de vive voix les sentiments d'estime que vos belles découvertes m'ont inspirés depuis longtems pour vous. J'espère qu'à cette époque vous serez associé de l'institut. La classe des sciences phisiques et mathématiques vient de vous nommer le premier de trois candidats qu'elle présente à l'institut, et il y a toute apparence que l'institut confirmera son choix comme il a fait à l'égard de Mr Banks, Maskelyne, Priestley.

&c. &c.

LAPLACE".

When Harding, at Lilienthal, discovered, in 1804, a third little planet, Herschel examined it with meticulous care and was confirmed in his opinion that these bodies constituted a separate group or species, deserving a specific name. He wrote, in the paper which he sent to the Royal Society in December, 1804[1]:

"A distinct magnifying power, of more than 5 or 6 hundred, has been applied to Ceres, Pallas and Juno; and every method we have tried has ended in proving their resemblance to small stars.

"It will appear then, that when I used the name asteroid to denote the condition of Ceres and Pallas, the definition I then gave of this term will equally express the nature of Juno, which, by its similar situation between Mars and Jupiter, as well as by the smallness of its disk, added to the considerable inclination and excentricity of its orbit, departs from the general condition of planets. The propriety therefore of using the same appellation for the lately discovered celestial body cannot be doubted.

"Had Juno presented us with a link of a chain, uniting it to those great bodies, whose rank in the solar system I have also defined, by some approximation of a motion in the zodiac, or by a magnitude not very different from a planetary one, it might have been an inducement for us to suspend our judgement with respect to a classification; but the specific difference between planets and asteroids appears now by the addition of a third individual of the latter species to be more fully

---

[1] "Experiments on the means of ascertaining the magnitudes of small celestial bodies", *Phil. Trans.* 1805; *C.S.P.* vol. II, p. 297.

established, and that circumstance, in my opinion, has added more to the ornament of our system than the discovery of another planet could have done".

Had Herschel adopted the better word *planetoids*, suggested by Piazzi, he might have saved himself from the aspersions cast upon him by some critics, but the term *asteroid* was accepted in Germany and is still used. Three years later, in 1807, Olbers discovered a fourth member of the group, which he named Vesta. Sir Joseph Banks wrote to Herschel:

"My dear Sir, Soho Square. June 3, 1807

"I trouble you with this merely to acknowledge the receipt of your paper which I shall deliver to the Secretary to-morrow and which will certainly be read to the Society before the vacation. It gives me much pleasure that more of these singular bodies should be discovered, and that the Germans should so readily and properly have adopted the distinction which you have made between them and planets.

Adieu.

Jos. Banks".

The major part of Herschel's paper on Juno is devoted to describing the methods by which he sought to distinguish the real from the spurious disc, caused by irradiation. For his experiments he required perfectly spherical globules of highly reflective material. To make these was no easy matter, and some of his descriptions of the way he proceeded may be worth quoting, as illustrating not only his ingenuity in invention, but also his marvellous patience and delicacy of touch in manipulating such small objects. He tried at first to use pins' heads, but found these too large for his purpose.

"I melted some sealing-wax thinly spread on a broad knife, and dipped the point of a fine needle, a little heated, into it, which took up a small globule. With some practice I soon acquired the art of making them perfectly round and extremely small. (They were then measured under a microscope.)

"Eight of these globules of the following dimensions, ·0466, ·0325, ·0290, ·02194, ·0210, ·0169, ·0144, ·00763, were placed upon the post in my garden, and I viewed them with the telescope.

"By this experiment it appears that with a globule so small as

·00763 of a substance not reflecting much light, the magnified angle must be between 4 and 5 minutes before we can see it round.

"As the objects made of sealing-wax, on account of their colour, did not appear to be fairly selected for these investigations, I made a set of silver ones. They were formed by running the end of silver wires, the 305th and 340th part of an inch in diameter, into the flame of the candle. It requires some practice to get them globular, as they are very apt to assume the shape of a pear; but they are so easily made that we have only to reject those which do not succeed."

He now made some from pitch, bees-wax and brimstone. With the bees-wax he found to his surprise that he always got only half a globule, when picked up on the point of a needle as with the sealing-wax.

"I took up another, that I might have a round one; but found that again I had only half a globule. It was so perfectly bisected, that art and care united could not have done it better.... The reason is not difficult to perceive; for as it melts with very little heat, it will cool the moment the needle is lifted up, and the surface which cools first will be flat.

"The roundness of the objects being a material circumstance, I melted a small quantity of the powder of brimstone, and dipping the point of a needle into it, I found that globules, perfectly spherical and extremely small, might be taken up. I had one of them that did not exceed the 1640th part of an inch in diameter."

## CHAPTER XVIII

### 1804   1817

Second paper "On the Nature of the Sun". Periodicity of sunspots—effect on climate. Letters—William Herschel to Bode, W. Watson. Brougham's article in the *Edinburgh Review*. Paper "On Concentric Rings"—adverse opinion of the Royal Society. Letter from Sir J. Banks. William Herschel's later papers "On the Construction of the Heavens". Sir J. F. W. Herschel's list of his father's most important discoveries.

HERSCHEL still continued his work on the sun, and in April, 1801, he contributed to the Royal Society a second paper, "On the Nature of the Sun", in which he gave numerous examples of his observations, extending over the five years from the date of his last paper, and confirming the conclusions he had then formed regarding the nature of sunspots and other features of the sun's disc. He also noticed that from July, 1795 to January, 1800 there was a remarkable absence of spots; that then some corrugations and luminous clouds appeared, and that by the end of the year the sun's surface "is beautifully ornamented with openings, shallows, ridges and nodules".

He now conceived the idea that there was a periodicity in the disturbances which caused the outbreak of spots and, with his usual persistence in following up a new idea, he collected all the records available of previous observations of the sun. These were lamentably few; still he found five examples of periods when spots were noticeably absent. He wrote:

"I am now much inclined to believe that openings with great shallows, ridges, nodules and corrugations, instead of small indentations, may lead us to expect a copious emission of heat, and therefore mild seasons. And that on the contrary, pores, small indentations, the absence of ridges and nodules, and of large openings and shallows, will denote a spare emission of heat, and may induce us to expect severe seasons. A constant observation of the sun with this view, and a proper information respecting the general mildness or severity of the seasons, in all parts of the world, may bring this theory to perfection or refute it if it be not well founded".

To test Herschel's conjecture properly would have required exact information regarding the recurrence of mild or severe seasons, and no meteorological records of this nature were then available; but he was not deterred by this from at least trying to get some proofs of his surmise. It occurred to him that as vegetation is affected by the amount of sunshine, records of good or bad harvests might be taken as giving indications of the variation of the seasons. If these coincided with the periods of many or few sunspots, there would at least be some ground for inferring that greater activity in the sun's atmosphere was accompanied by greater emission of heat. He applied this test to the five lean periods which he had noted in his historical survey and found that, roughly speaking, the price of wheat in England was highest when sunspots were absent. He summed up his argument in these words:

"The result of this review of the foregoing five periods is, that from the price of wheat, it seems probable that some temporary scarcity or defect of vegetation has generally taken place, when the sun has been without those appearances which we surmise to be symptoms of a copious emission of light and heat".

The occurrence of these passages in the paper read before the Royal Society exposed the writer to much criticism and some ridicule. Bode's *Jahrbuch* for 1805 contained a translation of the paper, with a foot-note objecting that the price of wheat in a single country could not supply a correct measure of the general fertility of the earth. Herschel replied in the following letter to Bode, which was printed, translated into German, in a later number of the *Jahrbuch*:

### *To* MR BODE

Slough. May 31, 1804

"My dear Sir,

"I have received the almanac you sent me for the year 1806, and am much obliged to you for the communication. I have read the remark which the translator has added to my paper. But he seems to have intirely mistaken the meaning when I referred to the price of wheat *in England* as some indication of the emission of the rays of heat. It was never intended to say that a copious emission of the rays of heat would produce a general success in the vegetation of this article. On the contrary, in climates which are full hot enough for wheat, an emission a little more copious than usual might occasion a considerable sterility. It is therefore evident that the success of the vegetation of wheat ought only to be taken in one place, and that an extensive commerce of that article with the whole world is out of the question. Nor has it been affirmed that fruitfulness is *immediately* connected with the copious emission of the sun's rays. On the contrary, I have called the sun the *ultimate* fountain of fertility; for it is well known and admitted that the solar rays pass through several modifications which give wind, rain, electrical phenomena, &c., all which affect in a different manner the fruitfulness of different vegetables. The remark of the translator therefore seems to resemble an objection that might be brought to the theory of the moon which occasions the tides. For instance, an observer shows that in the place where he lives it is high water six hours after the meridian passage of the moon, and that the tide rises 50 feet. An objector says this cannot be owing to the moon because in the place where he lives, high water is $2\frac{1}{2}$ hours after the meridian passage and amounts only to 3 feet. And yet if the movement of the waters in any one place will agree with the phenomena of the motions of the sun and moon, the argument of their being the cause of this movement will be conclusive. It is of little consequence whether you take wheat, rice, the sugar-cane or the bread-fruit for the vegetable which is to serve the purpose of the experiment, or whether it is the abundance or scarcity which happens when the sun emits its rays most copiously; if the experiment be well ascertained it will be equally conclusive. Wheat, being the only vegetable of which an authentic account was to be had in England, was chosen for the purpose, and all the imperfections to which the subject is liable were pointed out in the paper which treats of it. It should also be remembered that the whole theory of the symptomatic disposition of the sun is only proposed as an experiment to be made".

Some very adverse comments on Herschel's claim to have discovered invisible heat rays appeared in a monthly scientific magazine in 1800. Sir William Watson wrote to his friend:

"My dear friend, Bath. Sept. 9, 1800

"I have lately perused in Nicholson's last journal a paper of a Mr Leslie who controverts your experiments on heat and light and particularly denies the existence of the invisible heating rays of the sun. It is written in a petulant style and I trust will be found as void of truth as it is of candour. He says that upon trying with an instrument of his own invention which he calls a photometer he could find no heat beyond the red rays of the spectrum of a prism, when it was so placed as to receive no heat from the vicinity of other objects, and attributes the heat you found to that proceeding from the paper and board the thermometer rested upon. This position I think will be deemed groundless to whoever considers the circumstances of your experiment....If you should (think of using his Photometer) it is to be had at Cary's, Optician in the Strand....

"When shall I have the happiness of seeing you at Bath? I need not tell you how much I wish for such an event. We will then converse on these matters and also on Kant's metaphysics. I wish to know how far you have proceeded in your examination of his philosophy. I doubt, tho' we begin upon the same grounds and proceed together some way, I shall be obliged to leave him, as I suspect him to make some portion of our knowledge to proceed from sources I cannot admit.

"I often see your sister, who begins to be uneasy at not hearing from you.

"Adieu, my dear friend, give my best respects to Mrs Herschel
and believe me<br>
yours most sincerely<br>
WM. WATSON".

(Caroline was staying at Bath with her brother Alexander.)

Herschel appears to have examined Leslie's photometer, and, in a letter to Patrick Wilson, he wrote: "I have not yet seen Mr Leslie's book but hope to meet with it when I go to London".

He was evidently not much concerned with Mr Leslie's objections, but a later article in the same magazine disturbed him more. It seems to have referred to his paper on Newton's rings, which many of his friends, including Sir Joseph Banks, considered not so well supported as all his others, and about which Herschel was at this time rather touchy. Sir William Watson and Dr Patrick Wilson both took Herschel's side in the controversy which followed. But Sir William Watson wrote soothingly:

"My dear friend, Bath. April 20, 1802

"I duly received both your letters and am very sorry you should for a moment regard so insignificant a criticism as that published in the monthly review of this month. Be assured that as soon as a succeeding review appears this will be quite forgotten, and

> 'like the baseless fabrick of a vision
> leave not a wrack behind'.

"Remember the criticisms on your great discovery of the heating rays by, I forget whom. You did wisely not to answer a syllable, and he was heard of no more. There is not the slightest attempt to justify his remarks by reasoning in the whole paper.

. . . . . . .

"You sent me word by some friend (I forget who) that you was very busy about a new speculum. This makes me request that you will give me a few lines to tell me what you are doing. Remember that we can now hold only intercourse by letter, and you know what an extreme interest I take in everything belonging to you.

. . . . . . .

I remain, my dear friend,
Most sincerely yours
W. WATSON".

As we have seen, Herschel's choice of the word *asteroid* did not escape criticism. It was even hinted that he refused to honour these bodies with the names of planets lest their discoverers should rate themselves on an equality with himself. Herschel's own concluding remarks in his paper on Juno are a sufficient refutation of this accusation. His papers were also subjected to some very ill-natured criticism in the second number of the new *Edinburgh Review*, for January, 1803. The article was written by Brougham; it concluded with the following passages:

"Dr Herschel's passion for coining words and idioms, has often struck us as a weakness wholly unworthy of him. The invention of a name is but a poor achievement for him who has discovered worlds. Why, for instance, do we hear him talking of the *space-penetrating power* of his instrument—a compound epithet and metaphor which he ought to have left to the poets, who, in some further ages, shall acquire glory by celebrating his name?...

"The other papers of Dr Herschel, in the late volumes of the transactions, do not deserve such particular attention. His catalogue of 500 new nebulae, which concludes this volume, though extremely valuable to the practical astronomer, leads to no general conclusions of importance, and abounds with the defects which are peculiar to the Doctor's writings—a great prolixity and tediousness of narration —loose and often unphilosophical reflections, which give no very favourable idea of his scientific powers, however great his merit may be as an observer—above all that idle fondness for inventing names, &c., &c.

"To the speculations of the Doctor on the nature of the sun, contained in the last volume, we have many similar objections but they are all eclipsed by the grand absurdity which he has there committed; in his hasty and erroneous theory concerning the influence of the solar spots on the price of grain. Since the publication of Gulliver's voyage to Laputa, nothing so ridiculous has ever been offered to the world".

Sir William Watson mentioned this article in a letter which he wrote to Herschel about this time, acceding to Herschel's request that he would allow himself to be named as a trustee.

Bath. March 2nd, 1803
"My dearest friend,

"The warm and sincere affection I bear you will not suffer me to hesitate a moment accepting the awful trust which you design me. As our ages differ so little from each other it was highly proper to join a younger person to the number. However proper it is to prepare for all events, let us hope, my dear Sir, that we may both live happy in each other's friendship long after the time that such a trust will require. We both, I trust, see the value of life and enjoy its comforts without fearing the cessation whenever that may come; and among its chief gifts I shall ever reckon the sincere and warm affection that we have for one another. *Omnium rerum, quas ad beate vivendum sapientia comparaverit, nihil est amicitiæ possessione majus, nihil uberius, nihil jucundius.* No one so much as yourself has taught me the force of the above maxim of Epicurus as delivered by Cicero.

"The same day I received your letter I received one also from our excellent friend, Sir Joseph Banks. Among other things he tells me that there is come out an Edinburgh review of which two numbers have appeared, which appears to him to be written with a very caustic spirit, decrying our literary men, in order to raise the merits of his own countrymen. He tells me you have not escaped his lash and indeed if depreciation of merit is his aim, you must be the first to aim at. You, I am persuaded, will regard his darts with due indifference, and trust as you have hitherto done, in a calm and dignified reliance that nothing can affect and overturn truths and discoveries founded on experience and observation.

"Have we any chance of seeing you at Bath the ensuing Spring? If not, may we not hope for your and Mrs Herschel's company to spend some time with us at Dawlish in the course of next summer? Lady Watson heartily joins in this wish and we hope my dear little godson (little Trimbush) and Miss Baldwin will be of the party.

Adieu my dear friend,
and believe me yours most sincerely
WM. WATSON."

Herschel's success in demonstrating the existence of heat rays in the spectrum tempted him to venture further into a department of optics for which neither he, nor indeed anyone else at the time, was fully equipped. In 1807, and again in 1809 and 1810, he sent to the Royal Society some papers[1] on the subject of the optical appearances known as Newton's rings, which were not well received by the Committee of Publication of the Society.

Herschel devoted his first paper to pointing out the weakness of Newton's hypothesis to account for the colours; but in his next he tried by numerous very ingenious experiments to suggest another explanation, which was in fact even less convincing. As in later life he practically admitted this, and his son wished to suppress these papers altogether, it seems unnecessary to dwell more particularly on their contents. On the receipt of the second paper, Sir Joseph Banks wrote to Herschel as follows:

Soho Square. May 25, 1809
"My dear Sir,

"At the last meeting of the Committee of the R.S. for Publications, your paper on Concentric Rings was taken into consideration. It gave me pain to hear the opinions given by the members of the Committee best versed in optical enquiries; they appeared to me

---

[1] "Experiments for investigating the cause of the coloured concentric rings, discovered by Sir Isaac Newton, between two object-glasses laid one upon another", *Phil. Trans.* 1807, 1809, 1810; *C.S.P.* vol. II, pp. 368, 414, 441.

unanimous in thinking that you have somehow deceived yourself in your experiments and have in consequence deduced incorrect results. Every one of them however in stating their difference of opinion from yours, paid a first tribute to your character as an acute Philosopher in science, and while they lamented their differing from you in opinion, spoke of your intelligence and the many great advantages science has derived from your successful labours, with the highest degree of respect and attention.

"After much conversation on the subject it was resolved to defer the decision till the next Committee in the hope that in the meantime you may come to Town and discuss the matter in conversation with your friends in order that you may compare their opinions with yours on the intricate and abstruse subject of your paper.

"By what I could judge of the feelings of the Committee there was no kind of inclination in anyone to prevent the Publication of your paper in the *P. Trans.* if on mature consideration you wish it.

"...I hope, my dear Dr, that this letter will bring you to London and give me the opportunity of shaking you by the hand, which I can only do here, as I am yet a close prisoner in my house.

I beg, my dear Dr, you will believe me as always
with sincere esteem and regard,
most faithfully yours
JOS. BANKS".

Herschel did come to London, but he was not to be persuaded, even by Sir Joseph, to withdraw his paper. He wrote at great length in support of his thesis; the Committee gave way to his vehemence and the paper was printed in the *Philosophical Transactions*. But in later years he seems to have regretted his insistence. Sir John Herschel wished these papers to be omitted from any collection of his father's writings, and appears to have told Arago (who mentions it in his analysis of William Herschel's works) that it was the only occasion when his father regretted having adhered to his usual practice of communicating his researches as soon as he made them.

It must also be remembered that William Herschel, now in his seventieth year, was suffering at this time from what would now be called a nervous breakdown, due to overwork and the worry of visitors. Caroline tells in her Autobiography, from which the passages will be given in a future chapter, that in February, 1808, he had so severe an illness that his life was despaired of. From this time he never recovered the full use of his powers. One can understand that the weakness and nervous irritability consequent on such a state of health would account for the impatient temper shown in this matter, which was very unusual with him.

After this digression Herschel returned to his appointed task of "minding the heavens", to use Caroline's expressive phrase. He ceased his regular sweeps for nebulae, after presenting his last catalogue, of 500 new ones, to the Royal Society in 1802; but he continued to observe as perseveringly as ever, for other objects, till failing health at last compelled him reluctantly to desist.

There are in all seventeen papers (omitting those on the Concentric Rings) from his pen, in the *Transactions* of the Royal Society, between the years 1803 and 1821. An important one was that "On Double Stars" (1803), in which he showed that there had been changes in the relative positions of some of the pairs he had observed in former years. He concluded that these systems consisted of bodies moving round a common centre of gravity, not of a small star revolving round a greater one.

In 1811 he returned to the subject of the "Construction of the Heavens", and sent to the Royal Society an interesting paper in which he again gave reasons for his theory of the origin of stars, which he had first announced in his paper "On Nebulous Stars", in 1791.

After describing those which he called planetary nebulae, he wrote:

"When we reflect on these circumstances, we may conceive that, perhaps in progress of time, these nebulae which are already in such a state of compression, may be still farther condensed so as actually to become stars".

In 1814 he sent another paper to the Royal Society, in which he pursued the subject still further; starting with the star as the basis, he considered the formation of globular clusters. The paper begins with the following introductory remarks:

### Paper read February 24, 1814

"In my paper of observations of the nebulous part of the heavens I have endeavoured to show the probability of a very gradual conversion of the nebulous matter into the sidereal appearance. The observations contained in this paper are intended to display the sidereal part of the heavens and also to show the intimate connection between the two opposite extremes, one of which is the immensity of the widely diffused and seemingly chaotic nebulous matter; and the other, the highly complicated and most artificially constructed globular clusters of compressed stars.

"The proof of an intimate connection between these extremes will greatly support the probability of the conversion of the one into the other; and in order to make this connection gradually visible, I have arranged my observations into a series of collections, such as I suppose will best answer the end of a critical examination".

The concluding paragraph of this paper is on

*The breaking up of the milky way*

"The milky way is generally represented in astronomical maps as an irregular zone of brightness encircling the heavens, and my star gages have proved its whitish tinge to arise from accumulated stars, too faint to be distinguished by the eye.... It is, however, evident that, if it ever consisted of equally scattered stars, it does so no longer; for by looking at it on a fine night we may see its course, between the constellations Sagittarius and Perseus, affected by not less than eighteen different shades of glimmering light, resembling the telescopic appearance of large easily resolvable nebulae; but in addition to these general divisions the observations detailed in the preceding pages of this paper authorize us to anticipate the breaking up of the milky way, in all its minute parts, as the unavoidable consequence of the clustering power arising out of those preponderating attractions which have been shewn to be everywhere existing in its compass.

. . . . . .

"Now since the stars of the milky way are permanently exposed to the action of a power whereby they are irresistibly drawn into groups, we may be certain that from mere clustering stars they will be gradually compressed through successive stages of accumulation, —till they come to what may be called the ripening period of the globular, and total insulation, from which it is evident that the milky way must be finally broken up, and cease to be a stratum of scattered stars".

The last paper on this subject was written in 1817, when William Herschel was eighty-two years old; in it he gave the final estimate that he was able to make, from his observations, of the dimensions of the Milky Way. He was obliged to admit that "the utmost stretch of the space-penetrating power of the 20-feet telescope could not fathom the Profundity of the milky way".

With a rather pathetic apology for not having used the 40-foot telescope for these observations, on account of the great loss of time involved in moving the cumbersome instrument for the purpose of "sweeping", he said that he considered that though "a review of the milky way with this instrument would carry the extent of this brilliant arrangement of stars as far into space as its penetrating power can reach,—that it would then probably leave us again in the same uncertainty as the 20-feet telescope".

These words also leave the reader in the same uncertainty as to whether Herschel considered that there is a limit to the profundity of the Milky Way, and that it therefore may be classified as a nebula, though of unusual size, or whether he had abandoned this idea altogether and regarded it as a stratum of stars only. The concluding paragraph of this paper would seem to indicate that he left the question open as insoluble.

### Concluding remarks

"What has been said of the extent and condition of the milky way in several of my papers on the construction of the heavens, with the addition of the observations contained in this attempt to give a more correct idea of its profundity in space, will contain all the general knowledge we can ever have of this magnificent collection of stars. To enter upon the subject of the contents of the heavens in the two comparatively vacant spaces on each side adjoining the milky way, the situation of globular clusters, of planetary nebulae, and of far extended nebulosities, would greatly exceed the compass of this Paper; I shall therefore only add one remarkable conclusion that may be drawn from the experiments which have been made with the gaging power;...that not only our sun, but all the stars we can see with the eye, are deeply immersed in the milky way, and form a component part of it. WILLIAM HERSCHEL."

Slough near Windsor. May 10, 1817

(This paper is the first in the *Proceedings* of the Royal Society ascribed to "Sir William Herschel, Knt. Guelp., LL.D., F.R.S.")

In concluding this short survey of William Herschel's scientific work, the following list may be of interest. It was drawn up by Sir John Herschel in response to a request from Fourier for information to assist him in preparing the *Eloge on Sir William Herschel*, which he read before the French Academy in June, 1824. It is interesting as showing the estimation in which his discoveries were held by his son at the time of his decease. It will be noticed that there is no mention of the motion of the sun; at the time of writing John Herschel did not think this surmise of Sir William had been verified.

*List of the most important of Sir William Herschel's discoveries which have since been verified; drawn up by John Herschel in May, 1825*

"May 9, 1825.

"1. Discovery of a new Primary Planet and *two* of its satellites revolving in planes nearly perpendicular to the plane of its own orbit.

"2. The discovery of nearly 1000 double stars, and the *very important fact* that many of these pairs of stars form *binary systems* connected by mutual attraction;—confirmed by the observations of South and myself.

"3. The discovery of between 2000 and 3000 nebulae and clusters of Stars, and of the remarkable circumstance of the disposal of the vast number of the former in *strata* or irregularly extended beds in the heavens;—confirmed by my own observations.

"4. The complete resolution of every part of the Milky Way into stars of various sizes.

"5. The discovery of the gaseous nature of the Sun's luminous surface;—confirmed.

"6. The discovery of two satellites of Saturn. N.B. It is not probable that they will ever again be seen in a Refracting telescope of less than 2-feet aperture, or in a Reflector of less than 4 feet.

"7. The discovery of the period of rotation of the ring of Saturn;—confirmed by Laplace.

"8. Discovery of the independence of the heating and illuminating power of the solar rays;—confirmed.

"9. Discovery of the coincidence in the times of rotation on their axes of certain of the satellites of Jupiter and Saturn with (the time of revolution round) their primaries; and the extension of the analogy of our moon to all secondaries.

"10. Discovery of the true heights of the Lunar Mountains, much misconceived by former and contemporary astronomers.

"11. Discovery of a peculiar class of sidereal objects called by him Planetary Nebulae;—confirmed by M. Plana and myself at Turin last summer.

"12. Discovery of Double Stars with contrasted colours and of many extraordinary facts regarding the colours of single stars;—confirmed.

"13. Discovery of large tracts of loose nebulosity in the heavens unconnected with stars and not collected in patches of a sensible form.

"14. Discovery of extra-ordinary occasional irregularities in the distribution of the belts on the disc of Jupiter.

"15. Distinction between space-penetrating &c., &c. telescopes.

"16. Discovery of the variability of the star α Herculis."[1]

William Herschel himself believed that he had discovered four other satellites of Uranus, but this surmise has not been confirmed. John Herschel wrote, in 1871, to R. A. Proctor:

"As to the four additional satellites of Uranus, I incline to Mr Pritchard's opinion that my Father must too readily have persuaded himself that the minute points of light, which from time to time *he undoubtedly saw*, were really satellites.

"The testimony of Lord Rosse's and Mr Lassell's reflectors, which are composed of metal much more reflective than even that of the 18-inch, and very much more than that of the 4-feet reflector of my Father, I think, must be held conclusive. As regards the 18-inch I can say from my own experience of it, that at least to my eyesight (which used to be a pretty keen one) that instrument was competent only to shew the two larger ones, and that not without difficulty".

[1] Paper "On the Periodical Star α Herculis", *Phil. Trans.* 1796; *C.S.P.* vol. I, p. 559.

## CHAPTER XIX

### 1797 1802

Family life at Slough. Dr Burney's visit. Caroline's daily life—she visits Dr Maskelyne at Greenwich. John Herschel at Eton. Caroline goes to Bath. Letters—Mrs Herschel and William to Caroline. Mme Beckedorff.

A WARM friendship existed between William Herschel and Dr Burney, the father of Mme D'Arblay (better known as Fanny Burney), the authoress of the novels *Cecilia* and *Evelina*. Dr Burney was himself an authority on Music, of which he wrote a history, and this similarity of interest fostered the friendship between the two men.

Dr Burney had recently lost his wife, and his daughter encouraged him in the project of writing a history of Astronomy in verse, to distract his mind from his sorrow. William Herschel was no doubt aware of this, and it was a feeling of true friendship which enabled him to listen patiently for hours to the reading of Dr Burney's verses. Amongst Dr Burney's voluminous Diaries and Letters are several accounts of visits to Slough, which give a pleasing picture of the life of the Herschel family.

In a letter written in September, 1797, Dr Burney says:[1]

"...I went thence to Dr Herschel, with whom I had arranged a meeting by letter, but being, through a mistake, before my time, I stopped at the door to make inquiry whether my visit would be the least inconvenient to Herschel that night or next morning. The good soul was at dinner but came to the carriage himself, to press me to alight immediately and partake of his family repast; and this he did so heartily that I could not resist.

"I was introduced to the company at table; four ladies and a little boy about the age and size of Martin. I was quite shocked at intruding upon so many females. I knew not that Dr Herschel was married and expected only to have found his sister. One of these females was a very old lady, and mother, I believe, of Mrs Herschel, who

[1] Dr Burney's *Memoirs*, vol. III, p. 251.

sat at the head of the table. Another was a daughter of Dr Wilson, an eminent astronomer of Glasgow; the fourth was Miss Herschel. I apologised for coming at so uncouth an hour, by telling my story of missing Lord Chesterfield through a blunder, at which they were all so cruel as to join in rejoicing; and then in soliciting me to send away my carriage and stay and sleep there. I thought it necessary, you may be sure, to *faire la petite bouche*, but in spite of my blushes I was obliged to submit to having my trunk taken in and my carriage sent on. We soon grew acquainted, I mean the ladies and I; for Herschel I have known very many years; and before dinner was over all seemed old friends just met after a long absence. Mrs Herschel is sensible, good-humoured, unpretending, and obliging; the Scots lady sensible and harmless; the little boy entertaining, comical, and promising. Herschel, you know, and everybody knows, is one of the most pleasing and well-bred natural characters of the present age, as well as the greatest astronomer. Your health was immediately given and drunk after dinner by Dr Herschel, and after much social conversation and some hearty laughs, the ladies proposed taking a walk by themselves in order to leave Herschel and me together. We two therefore walked and talked over my subject, *tête-à-tête*, round his great telescope, till it grew damp and dusk, and then we retreated into his study to philosophise. I had a string of questions ready to ask, and astronomical difficulties to solve, which, with looking at curious books and instruments, filled up the time charmingly till tea. After which we retired again to the study, where, having now paved the way, we began to enter more fully into my poetical plan, and he pressed me to read to him what I had done. Lord, help his head! he little thought I had eight books or cantos, of from 400 to 800 lines, which to read through would require two or three days! He made me, however, unpack my trunk for my MS. from which I read him the titles of the chapters, and begged him to choose any book or the character of any astronomer he pleased. 'Oh!' cried he, 'let us have the beginning!' I read then the first 18 or 20 lines of the exordium, and then told him I rather wished to come to modern times; I was more certain of my ground in high antiquity than after the time of Copernicus.

"I began therefore my eighth chapter. He gave me the greatest encouragement, repeatedly saying that I perfectly understood what I was writing about; and he only stopped me in two places, one was at a word too strong for what I had to describe, and the other one too weak. The doctrine he allowed to be quite orthodox concerning gravitation, refraction, optics, comets, distances, revolutions, etc., etc., but he made a discovery to me which, had I known sooner, would have overset me, and prevented my reading to him any part of my work! this was that he had almost always had an aversion to poetry, which he had generally regarded as an arrangement of fine words without any adherence to truth; but he presently added that when truth and science were united to these fine words, he then liked poetry very well.

"The next morning he made me read as much, from another chapter, on Descartes, as the time would allow; for I had ordered my carriage at twelve. But I stayed on reading, talking, asking questions, and looking at books and instruments, at least another hour, before I could leave this excellent man".

Dr Burney subsequently destroyed his manuscripts, but not before he had had several more opportunities of reading to his friend. After one of these occasions Dr Burney wrote, to his daughter:

Chelsea College.[1] 10th Dec. 1798

"...Well, but Herschel has been in town, for short spirts and back again, two or three times, and I have had him here two whole days. ...I read to him the first five books without any one objection, except a little hesitation, at my saying on Bayly's authority, that if the sun were to move round the earth, according to Ptolemy, instead of the earth round the sun, as in the Copernican system, the nearest fixed star in every second must constantly run at the rate of near 100,000 miles. 'Stop a little!' cries he; 'I fancy you have greatly underrated the velocity required but I will calculate it at home'. And on his second visit, he brought me a slip of paper written by his sister, as he, I suppose, had dictated. 'Here we see that Sirius, if it revolved round the earth, would move at the rate of 1426 millions of miles per second. Hence the required velocity of Sirius in its orbit would be above 7305 times greater than that of light.' This is all I had to correct of doctrine in the first five books! And he was so humble as to protest that I knew more of the history of astronomy than he did himself and that I had surprised him by the mass of information I had gotten together.

[1] Dr Burney's *Memoirs*, vol. III, p. 265.

"In arranging another lecture, he flattered me much in a note, by saying that if I should be disengaged on a day that he mentioned, it would give him pleasure to devote it to the continuation of 'our' poetical history. This is adoption!

"He came, and his good wife accompanied him; and I read four books and a half.... And on parting, still more humble than before, or still more amiable, he thanked me for the instruction and entertainment I had given him!

"What say you to that? 'Can anything be grander?' And all without knowing a word of what I have written of himself; all his discoveries, as you may remember, being kept back for the twelfth and last book. Adod! I begin to be a little conceited!... But hold! On the first evening Herschel spent at Chelsea, when I called for my Argand lamp, Herschel, who had not seen one of those lamps, was surprised at the great effusion of light; and immediately calculated the difference between that and a single candle, and found it as Sixteen to one".

A year later Dr Burney again visited Slough, and wrote to his daughter describing his visit:

### DR BURNEY *to* MME D'ARBLAY

> Slough, Monday morning, July 22, 1799, in bed at Dr Herschel's, half past five: where I can neither sleep nor lie idle.

"My dear Fanny,

"I believe I told you that I was going to finish the perusal of my astronomical 'varses' to the great astronomer on Saturday. Here I arrived at three o'clock,—neither Dr nor Mrs Herschel at home. This was rather discouraging, but all was set right by the appearance of Miss Baldwin, a sweet timid amiable girl, Mrs Herschel's niece.... When we had conversed about ten minutes, in came two other sweet girls, the daughters of Dr Parry of Bath, on a visit here. More natural, obliging, charming girls I have seldom seen; and moreover very pretty. We soon got acquainted; I found they were musical, and in other respects very well educated. It being a quarter past four and the lord and lady of the mansion not returned, Miss Baldwin would have dinner served, according to order, and an excellent dinner it was, and our chattation no disagreeable sauce.

"After an admirable dessert, I made the Misses Parry sing and play, and sang and played with them delightfully, 'you can't think'. Mr and Mrs Herschel did not return till 7 or 8; but when they came, apologies for being out on pressing business, cordiality and kindness could not be more liberally bestowed. After tea, Dr Herschel proposed that we two should retire into a quiet room, in order to resume the perusal of my work, in which new progress has been made since last December. The evening was finished very peacefully, and we went to our bowers not much out of humour with each other, or with the rest of the world.[1]

The next day
"We all agreed at dinner to go to the Terrace,—Mr, Mrs and Miss Herschel, with their nice little boy, and the three young ladies. This plan we put into execution and arrived on the Terrace a little after 7. —When the King and Queen, arm in arm, were approaching the place where the Herschel family and I had placed ourselves, one of the Misses Parry heard the Queen say to his Majesty: 'There's Dr Burney', when they instantly came to me.

. . . . . .

"Now I must tell you that Herschel proposed to me to go with him to the King's concert at night, he having permission to go when he chooses; his five nephews (Griesbach) making a principal part of the band.—The King seldom goes into the music room after the first act. —At length he came direct to me and Herschel, and the first question his Majesty asked me was: 'How does Astronomy go on?'"

We now turn back to follow the course of Caroline's daily life, of which, after the nine years' break subsequent to her brother's marriage, she has left us full details. Up to the year 1797, when she resumes her story, it would seem that she lived at the Cottage, which she calls "The Observatory"; this, according to Mrs Papendiek, was the arrangement agreed upon when her brother married. The Cottage has been much altered in later years, so that it is difficult to guess what accommodation it afforded as a habitation at this time. Originally it had been attached to the stables and contained harness room and lofts. Part of the roof, we know, was flat, as Caroline had her telescope there, and there must have been at least two good rooms, one of which was used as a library and one as a workroom. The reason for her leaving it may have been that

---

[1] *Diary and Letters of Mme D'Arblay*, vol. III, p. 184.

she was fairly crowded out by the accumulation of books and instruments. In her Journal she merely records the fact that she shifted her quarters four times within the next five years.

At first she lived in lodgings; then she tried the experiment of taking a small cottage and having a maid to attend on her. Later on a better arrangement was made, and for nearly seven years she occupied rooms in Upton House, where Mrs Herschel had resided with her first husband, Mr Pitt. The Pitt and Baldwin properties being contiguous, Caroline could walk to and from her brother's observatory through his own grounds, but the distance is quite half a mile and must have been trying in bad weather. When one remembers that Caroline's assistance at night was chiefly required in winter, her patient courage and devotion are even more to be admired.

There is a tradition in Slough that the son of the owner of one of the houses where she lodged for a short time, in the High Street, would tell how, when a boy, he used sometimes to be roused in the night by her rap on the wall, when the weather, which had at first been cloudy, cleared up. Knowing the signal, he would get up, light a lantern, and going downstairs find her ready dressed and awaiting him. He perfectly recalled her gentle manner of saying to him: "Please will you take me to my Broder".

Every summer, when the Bath season was over, Alexander came to stay a month or two at Slough; the regular passage of the year was also punctuated in Caroline's memory by the periodical business of polishing the great mirror of the 40-foot telescope; otherwise her life ran on with very little interruption to the daily round.

In the summer of 1799, she made one of her rare visits from home in response to an invitation from Dr and Mrs Maskelyne to stay at the Royal Observatory, Greenwich. She took with her the notes of corrections which she had already entered in her brother's copy of Flamsteed's Observations, with the intention of copying these into Dr Maskelyne's volume. But Dr and Mrs Maskelyne treated her as an honoured guest, and had planned many excursions and dinner parties for her entertainment, so that, as she wrote: "the succession of amusements, &c., left me no alternative between contenting myself with 1 or 2 hours' sleep at night, or to go home without having fulfilled my purpose".

All she could do was to make a few notes in Dr Maskelyne's volume, and she records with a touch of innocent pride that "Dr Maskelyne tells me politely that he before intended to make a present of Flamsteed's

*Historia Coelestis* to the Royal Observatory hereafter, but that now he is determined to do so, since it is *enriched* with my remarks".

After her return to Slough, Dr Maskelyne sent her as a present a field-glass and a pair of binoculars. She wrote as follows to thank him:

### CAROLINE HERSCHEL *to* DR MASKELYNE

"Dear Sir,                                         Slough. Jan. 1800

"If it was not highly necessary to make you acquainted with the safe arrival of your valuable present at Slough, it might perhaps be a long while before I should think myself sufficiently collected to be able to express the greatful feelings the sight of them has occasioned me. My being pleased at having two such useful and convenient instruments has but very little connection with my present ideas, and if they were come to me from any other hands but those of the Astronomer Royal, I should use them as occasion required and think myself much obliged to the giver. But as it is I cannot help wishing I were capable of doing *something* to make myself deserving of all these kind attentions.

"I feel gratified in particular when I think of the stipulation I was making when you were taking measure of the distance of my eyes; viz. that if you in future should change in opinion and not think me worthy of the present not to bestow it upon me. Mrs Maskelyne's good-natured looks and all she said at the time come now again to my remembrance, and seeing not only the binocular (which I had but a conditional expectation of receiving) but also the night glass, makes me hope that during the time I had the honor of being in the company of such esteemed friends I have suffered no loss in their former good opinion of me; which was a circumstance I often feared might have happened; for I have too little knowledge of the rooles of Society as to trust much to my acquitting myself so as to give hope of having made any favorable impressions.

"You see, dear Sir, that you have done me more good than you was perhaps aware of; you have not only enabled me to peep at the heavens, but have put me into good humor with myself.

With my respectful compliments to Mrs and Miss M.
I remain with many thanks,
your much obliged and humble serv[t]

C. HERSCHEL".

It is one indication of the perfectly natural terms on which Caroline now lived with her sister-in-law, that in the latter's absence from home the charge of the precious little son was left to her. In her Journal for 1800, she records:

"*April* 25. My Brother (and Mrs H.) went to town for a fourthnight.

"*May* 1. Fetched my nephew from Mrs Clark and brought him to his boarding dame, Mrs Howard, at Eton.

"*May* 16. My Brother and Mrs H. returned from Town.

"*May* 26. I went to take leave of my nephew who entered at Dr Gretton's School".

Amongst other notes of Caroline's writing there is one which explains why John Herschel (who was only eight years old at this time) was so short a time at Eton. Caroline wrote:

"He was for a short time at Eton, but I cannot remember exactly how long he remained there. But this I remember; that when his Mother, in riding through Eton, saw him stripped and Boxing with a great Boy, it was thought best to take him home; for his health was but delicate in the early part of his life, having had the misfortune to be nursed by a woman who died soon after in a decline".

There is no necessity to suppose that John Herschel was actually fighting with the "great Boy". Boxing was very likely encouraged at Eton, but Mrs Herschel very naturally thought it an unsuitable exercise for her delicate child. John Herschel was next sent to a private school kept by Dr Gretton, who later became Dean of Hereford. The school was at Hitcham, very near Slough, and his father provided a private tutor in mathematics for him.

Soon after this Caroline went for some months to Bath to take charge of a house which had been taken there, but which Mrs Herschel did not intend to occupy till the winter. The Herschels had previously rented a house on Sion Hill, from which the furniture had to be removed. Caroline had the not wholly uncongenial task of sorting and arranging it all. She describes her reception in her Journal:

"1800, *July* 4. At 6 in the evening I was received by my brother Alex and his old housekeeper in a House Mrs H. had taken for the next winter in little Stanhope street. The house had been uninhabited and the furniture moved into it from the house on Sion Hill by strangers, labourers, and the things met me helter-skelter in the passage; some belonging to the drawing-room amongst currycombs and bridles and other stable utensils. My first care was to make an Inventory of the whole before I let a stranger come into the house; but by the 10th of July I hired a maid of all work to assist me to bring the house into habitable order, and by July 29th I was ready for resuming the work of recalculating sweeps, or dispatching some copying, &c., which was sent me by the Coach from Slough and from the printer in London".

By the end of the month Caroline had been able to report to her sister-in-law that the house was in good order, and received the following letter in reply:

MRS HERSCHEL *to* CAROLINE HERSCHEL

"Dear Miss Herschel, Slough. July y$^e$ 27, 1800

"I should have answered your kind letter before this but waited till our plans for the summer were quite settled. I have the pleasure to inform you that we have now fixed for going to Ramsgate for a month and shall leave Slough on Thursday for that purpose. We do not mean to hurry down but to make our journey pleasant by going short stages in a day as the weather is so very warm. John and my niece go with us. We are all in high spirits on the account and are very busy getting ready to set off. What makes us the more anxious to be there is, that all our friends seem to be going or gone already. On Friday Mrs Baldwin and Miss Ramsden set off on their journey down, and to-morrow Mrs Nash and all the Miss Adamses leave Maidenhead for the same place.

"You wish me to send you all the news of this part of the world; indeed I know of none, for since last Sunday that we were on the Terrace, I have not been from my own house. I saw all your friends at Windsor who enquired very kindly after you. Miss Planta was so good as to send us tickets for Frogmore; the one for you Dr Herschel gave to Miss Ramsden. Everything was the same as last year only that we had a most charming day instead of a wet one.

"Our good King has been so gracious as to think of us the usual time, and I make no doubt it will give you as much pleasure as it has myself.

"Your account of the Bath house I was happy to hear and hope you will have your health to injoy it. I do not expect to see it till after Xmas. The pickles you speak of I wish you to let your servant eat for they will not keep another year.

"John Herschel has promised to write to you before he goes again to school. I thank God he is quite well but as rude as ever, he desires his love to you and his uncle. Pray remember me kindly to Mr Herschel and tell him we shall hope to see him in the Autumn. I was much concerned to hear of his late indisposition. I hope he is now quite recovered. I shall send this letter in a parcel which your Brother is going to send you before he leaves Slough, and as he will write I shall make this letter but short as my time is so taken up with preparing for my journey. I shall be glad to hear from you at all times, but you must excuse me if I am not so punctual in answering your letters as I ought to be, for you know how much I hate writing, and indeed I have not time if I was ever so fond of doing it.

"Dr Herschel and my Niece and John Herschel join me in affectionate regards to yourself and Brother,

and I remain your Affect$^{ly}$

MARY HERSCHEL".

Caroline remained at Bath till November. Her brother Alexander left her alone for three weeks, when he went to Slough to assist in repolishing the mirror for the 40-foot. She made no complaint of loneliness, but only remarks characteristically in her Journal: "Some of my time during his absence I spent at his house on Margaret's Hill to clean and repair his furniture and make his habitation comfortable against his return".

About the end of October, she says: "I received notice that I should be wanted at Slough".

### WILLIAM HERSCHEL *to* CAROLINE HERSCHEL

London. Nov. 7, 1800

"Dear Sister,

"Last night my paper,[1] on which I have been so long at work, was read at the Society; I came to London to bring it, and have been so

---

[1] "Experiments on the Terrestrial Rays that occasion Heat", Part II, read Nov. 1800; *C.S.P.* vol. II, pp. 98–146.

hurried as not to be able to look out any work for you, but shall now be at liberty to do something of that kind.

"My things here are in considerable disorder, and in a short while Mrs Herschel and myself wish to come for a little time to Bath; we will let you know of it soon, that you may come here on a visit before we go, that I may point out to you the work that is most necessary to be done in our short absence. I thought it best to give you this early notice because tho' we have not fixed upon the time it will be towards the latter end of this month that we mean to come for perhaps a fortnight or three weeks according to the weather, for if that should be fine we shall return that I may have a few sweeps with you before you go back to Bath. Miss Baldwin is at Slough and stays while we are away so that you will have company, and the chaise will also be left so that you can pay visits at Windsor and shew yourself to your friends and ours.

"My last paper consisted of 80 pages so that you will have a piece of work to gather it together out of the scraps I have. Some part of it was brought together in the beginning by Miss Baldwin and Mrs Herschel, which will shew the order, but the rest remains in bits which I have gathered together and numbered.

"I had a letter from Sir Wm. Watson and have got his two volumes of the *Philosophical Transactions* which I shall bring with me. If you see him you may mention that I hope soon to see him for a short time. Remember me to our good brother Alexander, and with comp[ts] from Mrs Herschel I remain

Dear Sister, your affec[t] Brother<br>
Wm. Herschel.

"P.S. The Bacon and Cheese are very excellent. I have not had time to try Alexander's green lenses; they look beautiful."

Caroline went to Slough, but she did not return to Bath as her brother seems to have intended. William's health had suffered from the very close application which he had made, single-handed, to those experiments on Terrestrial Rays which brought him little satisfaction. Caroline says that he returned from the short visit to Bath "very ill". She never left Slough again during William's lifetime, except for a few days' visit to friends in town. Brother and sister settled down again peacefully to the occupation, which they both loved, of "sweeping" the heavens.

The next year William Herschel visited Wales for his usual summer trip with his wife and son. Caroline received the following letter from him on the return journey:

### WILLIAM HERSCHEL *to* CAROLINE HERSCHEL

Worcester. Aug$^t$ 25, 1801

"Dear Sister,

"Last night we came to this place and received your letter and Mr Penney's. We are much pleased to hear that the harvest work has gone on so well, and that all at Slough is in good order. Be so good as to give directions for Mrs Penney to brew for me the same quantity of Ale and table beer as usual, and as soon as she can; for we shall stay a few days, as long as we find the place agreeable, and then we set off directly for home; which will be three or four days' journey.

"Tell my good brother that I long to be at work again with my telescopes, and that one of the first things to be done will be to re-polish the 40 *feet* speculum. Then I shall also polish myself a beautiful 10 *feet*. The 5 feet will also be completely set right.

"I want much to look over the ecliptic again, to seek for another planet, and to find whether Piazzi's star is a real planet, and if so whether it be furnished with satellites. For this reason my 10 feet shall have all possible distinctness.

"I hope they will not want Alexander at Bath this good while, as I shall have much occasion for his help in fitting up 3 seven-feet telescopes. Let William when he goes to Windsor call on Taylor the cabinet maker, to inquire if the tube and two stands are finished.

"I am glad to hear that our brother has cleaned the time-piece. If he finds it lose or gain, by the transit of the stars, I wish he would regulate it accordingly as near as he can; and if he will be so good as to write down how many divisions he moves the adjustment we shall know how to regulate for the future. If the regular time of winding up has been changed by the cleaning, please to wind it up again by the marks I mentioned in my last, next Sunday morning. Has Alexander had time to clean the clock in the 40-feet; for that, I suppose, must be full of dust?

Remember me kindly to him and believe me
ever your faithful brother,
WM. HERSCHEL.

"Mrs Herschel and John Herschel join in Comp$^{ts}$".

One other circumstance remains to be noticed which occurred about this time and added much to Caroline's happiness. This was her meeting with Mme Beckedorff, who, when Caroline was first introduced to her at the Queen's Lodge, claimed her as an old acquaintance, reminding her of the time when they had both been pupils at Mme Kustner's school of dressmaking at Hanover. Mme Beckedorff had lately been given some position in the Queen's household, which entitled her to a separate apartment in the Royal Lodge, where she lived with her daughter and could entertain guests. Not only did Caroline find in her a steadfast and sympathetic friend, but this intimacy brought her into closer contact with the Royal family.

The atmosphere at Queen Charlotte's Court was a curious mixture of rigid etiquette with much kindly homeliness. The princesses had been most carefully educated and were naturally intelligent; they could appreciate Caroline's qualities of head and heart and evidently sought her out when she came to the Lodge. Caroline's first mention of Mme Beckedorff's name is under the date 8th February, 1802, as having dined with her. She wrote: "and the whole party leaving the dining room on the Princesses Augusta, Amalia, and the Duke of Cambridge coming to see me".

Though in general company or under critical eyes, such as Fanny Burney's, Caroline Herschel may have passed almost unnoticed as a modest and unassuming little body, one can imagine that among a small circle of appreciative friends her shrewd and rather caustic humour must have made her an amusing companion. Princess Sophia Matilda, in particular, was very friendly, and later, when William Herschel's declining health caused his sister much anxiety, she showed her sympathy by many kindly acts.

The place so often referred to in Caroline's Memoirs as the Queen's Lodge, or simply the Lodge, had been built within the precincts of the Castle by George the Third for the accommodation of the Royal family and household when residing at Windsor. It was an unpretentious building and was pulled down in the reign of his successor to make way for the more stately apartments which now occupy the site.

## CHAPTER XX

### 1802    1806

William Herschel's Diary of his visit to Paris—interview with Bonaparte, Laplace, etc. Outbreak of war. State of Germany. Dietrich's letters from Hanover. Caroline's Journal—the Spanish telescope. Dietrich visits England—his letters.

IN November, 1801, William Herschel wrote to Méchain that, when the peace made the journey to Paris easy, he hoped to come and to see again the many persons for whom he had so high an esteem.

The Peace of Amiens was not actually signed till March, 1802; it was a signal for a great crowd of English travellers to visit Paris, from which city they had been cut off for ten years or more. William Herschel wrote again to Méchain about his visit; Méchain replied, as follows:

M. MÉCHAIN *to* WILLIAM HERSCHEL

Paris, le 15 Prairial an 10 (4 juin, 1802)

"Monsieur et très célèbre confrère,

"J'ai reçu avec le plus grand plaisir la lettre que vous m'avez fait l'honneur et l'amitié de m'écrire le 22 mai dernier; je suis très sensible à cette marque affectueuse de votre bon souvenir. Je vous ai écrit deux ou trois fois depuis mon retour à Paris,... quant aux lettres, il faut s'en prendre aux difficultés qu'opposoit la guerre; mais la paix a heureusement levé tous les obstacles, la correspondance est devenue facile, et j'espère que vous me permettrez d'en profiter.

"Les observations du nouvel astre découvert par le Dr Olbers et de la planète de Piazzi, dont vous avez eu la complaisance de me donner un précis, sont infiniment curieuses et intéressantes. J'admire vos ingénieux moyens de déterminer de si petits diamètres; et je serois bien enchanté de pouvoir les comprendre, quoique je ne doute nullement de leur certitude. M. Schroeter à Lilienthal a trouvé avec ses grands télescopes le diamètre de Cerés de 2″ environ, comme vous avez appris sans doute.

"Je me félicitois déjà du plaisir de vous voir à Paris dans le cours de cet été; mais c'est avec une bien vive satisfaction que j'en recois aujourd'hui l'assurance de votre part. Hâtez donc votre voyage et notre jouissance le plus que vous pouvez; vous ne serez jamais assez tôt, ni assez longtems auprès de tous ceux qui vous désirent ici, qui ont pour vous la plus profonde estime. Vous vous y trouverez sûrement au milieu de tous vos confrères à l'Institut, car je ne doute nullement que vous ne soyez nommé à la première élection qui aura lieu le 1$^{er}$ thermidor, ou le Mardi 20 juillet prochain. Mais pourquoi ne viendriez vous pas plus tôt, pour profiter plus longtems de la belle saison et pour ne pas nous faire languir si longtems? Notre ville de Paris vous offrira encore quelques charmes, des amusemens intéressans; vous y verrez aussi les sciences, les lettres, et les arts cultivés avec succès, honorés et encouragés par le Gouvernement. Encore des vertus; la bonne amitié et l'ancienne aménité française; je saurai bien me mettre au nombre de mes compatriotes et confrères qui s'honorent de ces sentimens; croyez même que je serai très jaloux et glorieux de vous en donner des témoignages.

J'ai l'honneur d'être avec le plus sincère attachement et la plus haute estime.

Monsieur et cher confrère votre très humble et très obéiss$^t$ serv$^t$

MÉCHAIN.

"p.s. Me permettriez vous de présenter mes respectueuses hommages à Mademoiselle votre Soeur, si par hazard elle se ressouvenoit que j'ai eu l'honneur de la voir."

William Herschel seems to have been a little anxious lest, owing to the influx of visitors, his party might not find accommodation easily. He must have mentioned this in writing to Lalande, who replied:

5 juin, 1802

"Monsieur et cher ami,

"Quoique j'ai déjà répondu à votre lettre du 22 mai, par un voyageur, je reviens sur l'article où vous témoignés de l'inquiétude sur un logement; parceque je puis vous en offrir moi-même deux, pour vous quatre; un petit près de moi, et un grand à l'école militaire, mais plus éloigné. Ainsi si vous éprouviés de la difficulté dans le milieu de paris, vous auriés ces deux ressources. Ainsi je vous invite à ne pas tarder de me procurer un plaisir que je désire depuis 14 ans; vous m'avés procuré tant d'agrément, ainsi que la savante miss, en 1788 que je soupire après le moment où je pourrai vous en remercier;

mais comme je vais à Bourg en Bresse à la fin de juillet, j'aurais bien de regret si vous veniés pas auparavant.

je suis avec autant de considération que d'attachement.

Mon illustre confrère votre très humble et très obéissant Serviteur,

DELALANDE".

On the 13th July, 1802, William Herschel, with his wife and little son, and Sophia Baldwin, set out from Slough on their trip to Paris. John Herschel was now ten years old; the journey must have been full of excitement for him and hardly less for his mother and cousin, who had never before crossed the Channel. They posted from London to Dover, sleeping at Canterbury and lingering two days at Dover in hope of fine weather. William wrote to his sister from Abbeville to tell of their progress.

WILLIAM HERSCHEL *to* CAROLINE HERSCHEL

Abbeville. July 22, 1802

"Dear Sister,

"We have been travelling from the sea side by way of getting the better of the horrid sickness which you know one always has. We crossed the water last Tuesday in a day which appeared to be remarkably fine; but when we came out at sea it proved very windy, and the ship was greatly opposed to the waves. Mrs Herschel was extremely frightened and very ill; and was soon obliged to go down into the cabin with many other ladies, who all laid themselves down as you may suppose; like dead folks with sickness. We had however a very safe and short passage of little more than three hours. John Herschel was also very sick. I laid flat upon the ground upon deck and did not stir, tho' a great deal of water dashed over me. We found an excellent Inn at Calais and had a tolerable good night's rest, so as to be well enough to travel to Boulogne next day.

"This day we have come from Boulogne to Abbeville, where we now are. To-morrow, after a little rest, we shall go to sleep at Amiens, next day to Chantilly, and the day after we hope to arrive in Paris.

"Be so kind as to write to our Brother at Bath to give him this account, which he will receive as soon from you as he could if I were to write to him.

"When we arrive in Paris you will hear again from me; as we shall not stay a great while at Paris, you will write immediately after this, if you have not already written, to let us know any Slough news.

"If any inquiry is made about my being at Slough again you may say that it will be about the latter end of August. And indeed I do not think it will be much later as we really are afraid of bad weather coming on to make the return across the sea unpleasant.

"If you get information when the King is expected back from Whymouth I should be glad to hear it.

"Adieu. We all join in compliments to you, and by your means to Alexander. I hope my telescopes are all in their place.

(No signature.)"

July 23, 8 o'clock in the morning.

### WILLIAM HERSCHEL *to* CAROLINE HERSCHEL

Paris. Augt. 2, 1802

"Dear Sister,

"I received your letter of the 25th of July, and can say that it found us all in good health. We arrived at Paris on Sunday the 25th of July and have in the last week past seen a number of things well worth our notice; it would however be too long to give you an account of what we have seen, which may be done when we return to Slough. We intend to get out from Paris next Monday, and no letter from Slough can therefore reach us. Nor do I think that we can fix any place where a letter could be directed as we shall be upon the move. As soon as we have got over the water I will write again to let you know. We shall certainly stay at Calais till the wind and weather is fair, but all this is uncertain.

"Adieu in haste.

I remain, &c.

WM. HERSCHEL."

William Herschel kept a diary of his doings in Paris, which will be given here in full, so we are fortunately in possession of more information than he gave in his letter to his sister.

## WILLIAM HERSCHEL'S DIARY OF THE TRIP TO PARIS

"1802, *July* 13. Set off with Mrs H., Miss Baldwin, John Herschel. London.

"*July* 16. Dartford, Rochester, Sittingbourne, Canterbury. (59 miles.)

"*July* 17. Dover. (15 miles.)

"*July* 18. Saw a ship launched.

"*July* 19. The weather was not favourable for crossing the sea.

"*July* 20. We had a boisterous passage of three hours and arrived at Calais between 4 and 5 o'clock. Name of the Packet boat: *The Active*, Capt. Hamilton.

"*July* 21. Boulogne. Hôtel Britannique, très mauvais.

"*July* 22. Montreuil. Hôtel de france, bon.

"*July* 22. Abbeville. La tête de boef, assés bon.

"*July* 23. Chantilly. À la Poste, très bien.

"*July* 24. Paris. À l'hôtel des Indes, rue des Mailles, très mauvais.

"*July* 25. Paris. Hôtel Montorgueil, rue Montorgueil, bien bon. This day I saw Mr Méchain, Mr De Lamber, Mr Carrochet, Mr Bouvet.

"I saw the great telescope in an unfinished state. The apparatus is certainly the worst contrived that can possibly be imagined. From what I already know of Mr Carrochet's work I suppose the mirror may be a very good one, but as things are arranged I believe it impossible that it can have a fair chance for shewing objects as it ought to do. I saw through it both in the Newtonian way and by the front view, but objects appeared very ill defined. This however was by no means a proper trial; for in sunshine terrestrial objects viewed horizontally can very seldom be seen to any advantage.

"*July* 28. Dined with Mr De la Place. There were at dinner M. Chaptal, Ministre de l'intérieur, Mr Bertholet, Sir Ch[s] Blagden, Count Rumford and other persons of eminence in literature.

"We went to the Institute at 6 o'clock, where I saw Mr Hauy, the President, Mr Messier, Mr Lacroix, Secretary, Mr Fourcroy, who read a Memoir on chemistry. Mr Jeaurat the old astronomer. In the morning I saw also Mr Le Gendre.

"*July* 29. We all dined with Madame Gautier at Passi. It is a beautiful place, and there was a very elegant entertainment.

"*July* 30. I breakfasted with Mr Le Sénateur la Place. His Lady received company abed; which to those who are not used to it appears very remarkable. Mad. La Place however is a very elegant and well-informed woman.

"*July* 31. We saw the paintings and statues. I got my passport signed by the Minister. I also reviewed the observatory with Sir Ch. Blagden.

"*Augt.* 1. I dined with the Sénateur Mr Bertholet, where I also met Dr Kapp from Leipsich and M. Vauquelin the Chemist. Bertholet's assistant also is a very sensible man. The house is pleasantly situated at Arceuil, and after dinner we walked to see the aqueduct.

"*Augt.* 2. We saw the Jardin des Plantes, Museum of french monuments, Temple, the place where the Bastille has been. In the evening to one of the Gardens.

"*Tuesday, August* 3. We saw the Parade, after having breakfasted with Mad. La Place. In the evening we saw the Opera *Sémiramis*, and the Ballet of *La Dansomanie*.

"*Augt.* 4. We dined with Mon. le Sénateur Sieyes at his house one league beyond Marly. I had much conversation with M. Sieyes and found him a very excellent Philosopher.

"*Friday, Augt.* 6. I went with Genl. Komarzewski to a bookseller opposite the Thuilleries, who is a member of the Institute. He is blind but a very agreeable man and seemed well-informed.

"*Sat. Augt.* 7. We went with Genl. Komarz... to see Mr Charles the Aereonaut and lecturer. He has a very magnificent apparatus for experim$^{ts}$ in natural philosophy. We dined with the Genl. at the Gardens of the Thuilleries, and he spent the evening with us.

"*Sunday, Augt.* 8. We breakfasted with Mad. La Place, to take leave. I dined with the Minister of the Interior, Mon$^r$ Chaptal. His house is a magnificent Pallas, most elegantly finished. About 7 o'clock the Minister conducted Mr Laplace, Count Rumford and me to Malmaison, the pallas of the first Consul, in order to introduce me to him. We saw Mad$^e$ Bonaparte, who after the Minister had introduced me was pleased to accost me with great politeness. After a considerable walk with her in the Garden we met with the first Consul, who was engaged in making improvements in the garden by directing the persons employed how to conduct water for the irrigation of the plants. The Minister introduced me and Count Rumford. The first Consul addressed himself to me in politely asking some questions relating to astronomical subjects and after a considerable conversation he also addressed Count Rumford by questions of a nature which related to the character of that philosopher, well known as a man that has much contributed to the comfort of the poor at Munich, &c.

"The first Consul afterwards entered into a general conversation with the Minister of the interior, with Mr Laplace, Count Rumford and myself, on common subjects.

"After about half an hour's walk he led us towards the house, but stopping short, on meeting with some other gentlemen, he entered into conversation with them on the subject of a canal which is to be made in France. The Consul seemed to be perfectly acquainted with the subject.

"He now led us to a room where after a short time spent in conversing he seated himself in a chair, and politely desired me to sit down. As the same invitation was not given to the rest of the company, nor any of them took seats, I only bowed my thanks for the first Consul's great civility and kept standing with the rest. The first Consul then asked a few questions relating to Astronomy and the construction of the heavens to which I made such answers as seemed to give him great satisfaction. He also addressed himself to Mr Laplace on the same subject, and held a considerable argument with him in which he differed from that eminent mathematician. The difference was occasioned by an exclamation of the first Consul, who asked in a tone of exclamation or admiration (when we were speaking of the extent of the sidereal heavens): 'And who is the author of all this!' Mons. De la Place wished to shew that a chain of natural causes would account for the construction and preservation of the wonderful system. This the first Consul rather opposed. Much may be said on the subject; by joining the arguments of both we shall be led to 'Nature and nature's God'.

"Soon after the conversation took another turn, which was the breeding of horses in England. The Race horse, the Hunter, the riding horse, the draught horse, &c. In this the Minister of the interior and Count Rumford joined.

"I remarked that the 2nd and 3rd Consuls who were now also present hardly ever joined in the conversation.

"Madame Bonaparte and her daughter, who came in a short time after the 1st Consul had led us into the room, were seated on a sofa near the Consul. She is a very elegant Lady and seems hardly old enough to be the mother of her daughter, who, I hear, is married to one of the 1st Consul's brothers.

"It appeared to me pretty remarkable that neither the 2nd nor 3rd Consul sat down in the presence of Bonaparte but remained intermixed with the company; Cambaceres standing close by me and Le Brun between some General and Mr La Place.

"The conversation turned also upon the Police, which it *was remarked*, is very badly kept in London, compared to what it is in Paris. The great Licence of the English newspapers was also reprobated by some; the first Consul however did not join in expressions of that nature.

"Between the conversation Ices were handed about, which were of an excellent flavour, and the weather being extremely hot were very refreshing. Bonaparte said that the thermometer had been 28° in the shade; that is 95° of Farenheit's scale.

"After some other not particularly interesting conversation, the first Consul quitted the room.

"The minister of the interior addressed Madame Bonaparte, and after some short conversation, on the subject of the Opera and others of that nature he took his leave, and we followed his example. We then returned to the house of the Minister, where we rested a little while and where we found Mad. Chaptal and her family, and Mad. Laplace and her family who had spent the day at Mr Bertholet's. They said that the thermometer there had been observed by Mr Bertholet at 31°; that is $101\frac{3}{4}°$ of Farenheit's scale.

"As M. La Place and I went and returned in a carriage by ourselves, I led the conversation upon the subject of my last paper, of which I gave him some of the outlines. I mentioned the various possible combinations of revolving stars united in double or treble systems; when I mentioned three stars at an equal distance revolving round a center, he remarked that he had shewn in—I believe—his *Mécanique Céleste*, that six stars could turn round in a ring, about their common center of gravity.

"A few days ago I saw Mr Messier at his lodgings. He complained of having suffered much from his accident of falling into an ice-cellar. He is still very assiduous in observing, and regretted that he had not interest enough to get the windows mended in a kind of tower where his instruments are; but keeps up his spirits. He appeared to be a very sensible man in conversation. Merit is not always rewarded as it ought to be."

William Herschel's diary ends abruptly with the words last quoted. On their return journey he wrote to Caroline from Dover:

### WILLIAM HERSCHEL to CAROLINE HERSCHEL

Dover. Augt. 15, 1802

"Dear Sister,

"Last night at 10 o'clock we arrived safely in old England. It will not be necessary to say that we all were extremely sick at sea, but it is difficult to pass over such a painful thing as sea-sickness without talking of it. We are however all recovering very fast and mean to eat a hearty dinner, after which we shall set off for Ramsgate where we shall probably spend a week....

"Be so good as to let Mrs Penny know that I want her to brew me the usual quantity of Ale as soon as possible, that it may be ready for drinking by the time I get to Slough.

"Our stay being uncertain at Ramsgate I can hardly receive any letter and will therefore wait till I come home....

"We all join in comp$^{ts}$ and I remain as usual yours sincerely

WM. HERSCHEL".

It must have been at Ramsgate that John Herschel contracted the illness which Caroline mentions in her Journal. The anxiety about his son, added to the accumulation of work awaiting his return, seems to have prevented William Herschel from communicating for some months with his friends in Paris. He does not even appear to have replied at all to Lalande's civil offer to find him lodgings, nor did he meet Lalande in Paris. Lalande was not unnaturally pained, but in extenuation of Herschel's coldness it may be remarked that Lalande was rather notoriously desirous of attracting attention to himself, and Herschel may have wished to escape from his assumption of patronage, however well-intentioned. Lalande did not seriously resent Herschel's apparent discourtesy, as the conclusion of his letter shows.

### LALANDE to WILLIAM HERSCHEL

Paris, le 23 sept. 1802

"Cher et illustre confrère et ami,

"j'ai eu bien du regret d'apprendre votre arrivée à paris et de n'en avoir pas été prévenu; car je vous aurais certainement attendu. je n'ai pas su votre départ; car je serois accouru à paris pour tâcher de vous en rendre le séjour plus agréable et de vous réunir avec nos savans; mais tous se plaignent de ce que vous avés été que 8 jours à paris, et cela me parait bien extraordinaire. je vous avais offert un logement, et vous ne m'avés pas même fait de réponse.

"Enfin ce sera à Slough que je me dédommagerai de cette privation; j'y retournerai en effet dès que vous m'écrivés que votre télescope de 40 piés est en état de satisfaire ma curiosité, car vous savés que j'y ai été pour cela en 1788, et que je n'ai pu regarder dans ce beau télescope.

"M. Burckhardt vient de calculer les élémens de la planète d'olbers.

. . . . . . .

je suis avec autant de considération que d'attachement.
Monsieur et cher confrère,
votre très humble et très obt. serviteur
DELALANDE."

The fresh outbreak of war prevented Lalande from carrying into effect his intention of revisiting Slough and seeing the 40-foot telescope. Had he done so he must have been convinced of its excellence. As Herschel had not been able to announce any spectacular discoveries made with it, Lalande remained sceptical of its merits, and, in 1806, he inserted in his yearly chronicle of astronomical events a derogatory paragraph about the great telescope. Herschel's friend, Patrick Wilson, was very indignant and wrote to him:

"I don't know if as yet you have met with De La Lande's *History of Astronomy* for the year 1806.... There is a paragraph concerning you and the 40-feet Telescope, evidently calculated to impress the belief of the *total* failure of your noble Instrument and resting the proof on his correspondence with yourself. The structure of the whole paragraph appears to me very base and an outrage against all decorums which govern men who stand upon their good character. I have often wished that, for former provocations of a similar nature, you had denounced the Haters of Merit, in the face of Europe, as unworthy of your correspondence".

William Herschel took no notice of his friend's violent advocacy and maintained a dignified silence. His brother Dietrich, however, gallantly took up the defence of the 40-foot, and wrote an article which was printed in the *Neues Hannöverisches Magazin*, for August, 1804, but as he could not cite any notable discovery beyond the one of Saturn's satellites, his article had probably little effect on the opinion of continental astronomers.

General Komarzewski also wrote, after William Herschel's visit to Paris, to acknowledge the long-delayed letter from him.

Paris. March 13, 1803

"My dear Sir,

"On the 7th instant I had the greatest pleasure to receive your letter of the 27th February. You have no idea of the anxiety your long silence put in all your good friends and myself; and chiefly on account of your dear John who was known to be ill on his return to England. Mr and Mrs Greatheed took therein a very sincere share. Now we are all very glad to hear that everything goes on well in your Family.

"I had the honour to speak myself to Mr de la Place, and to Mr de la Lande. They both return their thanks for your kind rememberness. They are very well and Mr de la Place has received your letter.

"I return you my thanks, my dear friend, for your answer on several questions of curiosity. I wish too very much to ramble with you through the Highlands, but we have time enough to speak on that subject as it is not possible to perform this journey before August. Meanwhile I shall be obliged, in all appearance, to go back to Germany, and settle once for all my money business; then I will return to Paris, and from thence you shall hear from me.

"Mrs Herschel had the goodness to send me the only Receipt to make the Norfolk Punch. I should be very glad to have all others, for instance those for *Artificial Champain*, *Cordial reason wine*, Malaga, &c.

"Pray present my respectful compliments to Mrs Herschel, and Miss Baldwin, and give a pretty kiss to your charming John. I am now preparing for a long conversation with him in our pure Latine.

Believe me ever, my dear Sir,<br>
your most humble faithful friend and servant<br>
LIEUT.-GEN. KOMARZEWSKI."

It is sad to think of the good General's disappointment when the renewed outbreak of war again put a stop to all his plans for coming to

England. William Herschel received no more letters from his friend, and it was only eleven years later that he and Caroline heard with much regret that General Komarzewski had died in Paris about the year 1807.

Our story now returns for a moment to the quiet daily life at Slough, as recorded by Caroline in the extracts which she made from her Journal. After noticing the return of her brother and his family from France in the summer of 1802, and Alexander's return to Bath for the winter season, she begins the record for the next year:

"1803, *March* 25. I moved from Chalvey to Upton.

"*April* 12. Had an account of my Sister Griesbach's death. She died March 30.

"*Oct.* 18. I changed my rooms for the accommodation of Mr and Mrs Slaughter, who had taken the House and Garden at Upton, except the two rooms for my habitation.

"*Dec.* Almost the whole month I worked at Slough from breakfast till 9 in the evening".

But between the entries for April and October is one short sentence which shines out across the monotonous details like a sudden flash-light turned on to the Continent of Europe:

"*June* 25. Spent a miserable day at the Queen's Lodge; on account of the French having taken Hanover".

When, less than a year before, William Herschel had visited Paris and been so cordially received, it had seemed to many people that Bonaparte had accepted the check which his ambitious designs had received in Egypt and Syria, and having been made First Consul for life, with practically unlimited power, was applying himself to the more praiseworthy task of restoring order, of establishing a new and better Constitution in France, and of encouraging in every way the arts and sciences. Statesmen indeed were not deluded, and it soon became plain that Napoleon only regarded the treaties made as a truce, giving him time to mature his preparations for a renewed attack. But to the world in general everything seemed to point to a lasting peace, now that the agitation caused by the French Revolution had quieted down. This happy but illusory vision was reflected in the affairs of Dietrich Herschel and his family in Hanover.

Dietrich was now the only one of his generation of Herschels living in Germany. Jacob was dead; Sophia Griesbach, the eldest sister, died in April, 1803, happy in the knowledge that five of her sons were comfortably settled in England. After her death the last of her children, Charles Griesbach, joined his brothers and was enrolled in the Queen's Band at Windsor.

Dietrich was still a member of the Court Orchestra at Hanover, and as a musician and music teacher was doing well, and able to send some hundreds of pounds to his brother William to be invested in English stocks. One of his daughters married, in 1802, a well-to-do merchant named Knipping, who lived in Bremen and had business connections in America. Dietrich had consulted William about getting a situation in the service of the East India Company for his son, but later he wrote that Heinrich had accepted an engagement with a client of Knipping's in Charlestown and would soon be launched on what promised to be a propitious career. Altogether, the family prospects looked rosy enough, and no hint of coming trouble is suggested in the letter which Dietrich wrote in January, 1803, to William, to tell of Heinrich's good fortune. But in a few months all was changed; in May, 1803, war was declared between England and France, and on the 5th of June the French army entered Hanover.

In the long war which followed, Hanover suffered more than any of the German states. Bordering geographically on Prussia, it had long been coveted by the Prussian King, and in 1801 had actually been occupied by Prussian troops, at the instigation of Napoleon. Now that it was in the hands of the French, Bonaparte used it as a bait to tempt Prussia into an alliance with France. When later the Prussians occupied it, their presence was hardly more welcome to the Hanoverians than that of the French had been.

The few letters from Dietrich which reached his English relations, and from which one or two extracts will be given here, show the shattering effect of Napoleon's ambition on private life in Germany.

In September, 1803, Dietrich wrote:

"About politics and our fate here I cannot write; yet I would appeal to every Englishman to spend the last drop of blood in our defence. We are all well, which shows that anxiety does not kill. When our fortune improves I will write once more; till then farewell, dear brother".

Again, in November, 1803:

"We are living under such conditions that we know not what may befall us. Our whole fate depends on England; may Heaven grant it good luck. The conditions here are no joke".

He was very anxious about his son who had sailed for Charlestown and about whom he could get no news. He knew, through an English newspaper, that the ship had reached Charlestown after a voyage of *five months*, though the usual time was five weeks. Later he was able to write that he had good news of Heinrich.

4 July, 1804

"In unhappy Hanover there is no hope except in the conclusion of peace between England and France. What this poor country is suffering cannot be told and we can do nothing. Bewailing our fate does not help us. Farewell, dearest brother, do not forget your loving

D. HERSCHEL."

In December, 1805, Dietrich wrote to William Herschel, after the disastrous defeats of the Austrians at Ulm in October and at Austerlitz on the 2nd December. He began his letter by saying that he had delayed writing hoping to be able to send some good news, but of that, as William must know now by the newspapers, there is no more hope.

"All the efforts", he wrote, "to curb that power which aims at the subjection of all nations, have failed. Nor is this surprising if in other quarters the measures taken have been such as we see here. It is lamentable to think that so many thousand persons have had to pay with their lives and all their worldly goods, for the treachery, profiteering, and Heaven knows what more scandals.

"Dear Brother, unless an angel comes down from Heaven to lead and put life into the armies into which so many thousands of men are being drafted, the English might as well throw their money into the sea. It only goes to make some scoundrels rich, give some idlers opportunity to gorge themselves every day, and, not to omit the only benefit, allow some poor workers to earn a little. For the rest it only tempts the French the more to fall upon us. If only, when the Emperor was first attacked, there had been a *man* in authority, who was equal to the position, then Bernadotty would not have come with his army to fall on the back of the Kaiser. Is it not lamentable that not less than 50,000 Frenchmen should be wandering about in our little country, in a year when we had not enough for ourselves; and the French shut their own land and would not allow the export of a grain of corn. If everything requisitioned was not rendered up at once, then there were threats instantly of military 'execution'.

"25,000 Russians, who knows how many Swedes, are still in the country, who must cost the English much money, but how does that help us? The peasants in the country districts must feed them for nothing, and in the towns they cost the burghers much in other ways. We were lucky here in having them only for 14 days and I had only to house two men, but that cost me 15 Thalers; others had them for longer and in one house as many as ten to twenty men....

"Since the Russians have been defeated by the French I have lost all hope; and I see only too clearly that it is leadership of troops which counts, for there are no better soldiers than the Russians.

"Our only hope now is a peace. Will England make peace? Can she? The Continental Powers *must* make peace; and woe to England if she cannot come in. France is too strong.

"If England does not make peace then we here shall have the French on our necks again, and it will be worse than ever, for the land is quite exhausted."

One more extract may be of interest:

Hannover. 2 March, 1806

"The war is still going on. Peace seems still far off; indeed we are in the more danger that this place may be the battle ground, as the rest of Europe appears to be finished off. At present we are in the hands of the Prussians, after the Russians left us. The treatment we receive at the hands of the Prussians, at least in the towns, is much gentler, and for this they deserve all praise. But we abhor their policy; indeed to the uttermost, and they have got themselves into a position which will be very difficult to get out of. The hatred we have for that State makes our danger of being united to it the more terrible to contemplate. My faith remains firm in reliance on England.

"We have now a brief respite, which has been very beneficial for some arrears of payments have been made. My desire to see you, dear Brother, is so great that if the situation here does not get worse and the Weser is not closed, I will devote three or four weeks to visiting England. I long to see good old Alexander, and Caroline who is so kind to us; and to see your son would alone be worth the journey; *he* will soon be the only one of our race".

This allusion to John Herschel as the only son to carry on the name of Herschel to another generation is the only reference, in the letters from Dietrich which have been kept, to the loss of his own only son, Heinrich. We know from other sources that Heinrich Herschel had died from yellow fever at Charlestown.

Dietrich Herschel was able to pay a brief visit to England in the summer of 1806, as recorded in Caroline's Journal.

## CAROLINE'S JOURNAL

"1805, *Aug.* 14. I went to stay with Alex at Slough, while my eldest Brother went with his family from home; they had intended to have left Slough on the 12th, but were detained in consequence of a report of an invasion.

"*In Sept.* Was much hindered in my work by the packing of the Spanish Telescope, which was done at the barn and rickyard at Upton; my room being all the while filled with the Optical apparatus. The Spanish Telescope left England in October.

"*Nov.* 13. I went to Slough—the Family to Town; but in the absence of the moon my Brother was at home and much observing and work was despatched.

"*Dec.* 1. All came home and I went to my solitude again."

The King of Spain, Charles IV, was at this time an ally of France, and at war with England. Caroline was completely indifferent to contemporary events unless they affected her family; otherwise it might have occurred to her, in copying these notes from her old Day Books, to remember and remark upon the fact that the Battle of Trafalgar was being fought at the very moment when William was packing off this telescope. It seems very doubtful whether this fine telescope, on which so much labour and care had been expended, ever reached its destination.

## CAROLINE'S JOURNAL

"1806. During the winter months I suffered much from a violent cough and cold and found great difficulty in despatching the copying, &c., which dayly was sent to me when I was unable to go to my Brother.

"*April* 21. My Brother went to Bath to pay a visit to Sir Wm. Watson, and partly by way of change of air having had his share of the prevailing influence (*sic*).

"*July* 4. My Brother went to Gravesend to meet my jungest brother, who came to pay us a visit and was detained there for a passport.

"*July* 10. Alexander joined us from Bath. The same day my eldest Brother went to the Visitation of the Observatory at Greenwich and my brother Dietrich accompanied him.

"*July* 13. We all went on the Terrace and took our Tea with Mrs Bremeyer at the Castle.

"*July* 18. We went to Dr Gretton's to hear the Speaches of the young Gentlemen and afterwards dined there.

"*July* 23. Dietrich took leave of his friends at Cumberland Lodge. Alex. and I accompanied him and in Windsor I went shopping to buy presents for my Hanoverian relations.

"*July* 24. D. left us. My eldest Brother and Mrs H. accompanied him to London.

"*July* 30. Alexander left Slough for Bath."

Dietrich wrote to William on his way back to Germany:

July 28. Gravesend

"Dear and best Brother,

"My passage is all arranged. I was able to get news of my old Captain whose ship lay some three or four miles down the Thames, so I took a boat and arranged matters with him for 4 guineas. Tomorrow my things will be taken on board; I have therefore nothing more important to do than to thank you for my kind welcome in England; it was what I should have expected from the best of brothers. Very specially kind was also your dear wife to me, which I can never forget and for which I thank her again very much.

"From Wapping I went to the White Horse Cellar, where I am writing this; when I have eaten something I will go to Herr Goltermann and afterwards to F. Griesbach.

D. HERSCHEL".

DIETRICH HERSCHEL *to* CAROLINE, *from Hanover*

8th August, 1806

"Dear Sister,

"Though we were held up at Gravesend for lack of wind till midday, yet at night we were well out at sea. On the evening of the third day we anchored on still water, near Heligoland. On the fourth day we sailed up the Weser as far as Bremen; and after that all my anxiety was at an end, for up to that time there was always risk of being sent back, for two English frigates lie in the Weser, who search all ships and seize many or send them back; but we found grace in their eyes.

"In Bremen I found my daughter, Annette, preparing to go to Hanover; a carriage had been ordered, so after a good night's sleep I set out with her and by mid-day of the seventh day reached Hanover."

## CHAPTER XXI

### 1807   1813

Dietrich comes to England. Caroline's Journal. William Herschel and his family make a tour to the north of England—his letters. Trip to Scotland, 1810. John Herschel's diary. James Watt. Trip to Scotland, 1811. Visit to Brighton. Thomas Campbell.

THE years 1807 and 1808 were anxious ones for the Herschel family. William Herschel's health was beginning to give way under the constant strain of work and the boy was also far from strong. He was subject to feverish attacks, possibly symptomatic of the disease from which his stepbrother had died, and which he also had had every opportunity of catching from his wet-nurse. And in addition to these causes of anxiety the accounts they received from the brother in Hanover were more distressing than ever. The French had again overrun the country, and Dietrich found himself an utterly ruined man, obliged as a last resource to leave his wife and children behind and to come to England, where he could earn something for their support. Caroline's Journal gives a sad picture of the trouble at Slough.

### CAROLINE'S JOURNAL

"1807, *Oct.* 4. My Brother came from Brighton. The same night two parties from the Castle came to see the Comet, and during the whole month my Brother had not an evening to himself when, as he was then in the midst of repolishing the 40-ft. mirror, rest became absolutely necessary after a day spent in that most laborious work; and it has ever been my opinion that on the 14th October, his nerves received a shock of which he never got the better afterwards. For on that Day (in particular) he had hardly dismissed his troup of men, when visitors assembled, and from the time it was dark till past midnight, he was on the grassplot surrounded by between 50 to 60 persons; without having time for putting on proper cloathing or for the least nourishment passing his lips. Among whom I remember were the Duke of Sussex, Prince Galizin, Lord Darnley, Countess of Oenhausen, a number of officers, Admiral Boston and some Ladies.

"1808, *Feb.* 6. When I came to Slough to assist my Brother when polishing the 40-ft. mirror, I found my Nephew very ill with an inflammatory sore throat and fever.

"*Feb.* 9. Still very ill and my Brother obliged to go on with the polishing of the great mirror as every arrangement had been made for that purpose. Mem. I believe my Brother had reason for choosing the cold season for this laborious work, the exertion of which alone must put any man into a fever if he were ever so strong.

"*Feb.* 10. From this day my Nephew's health kept on mending.

"*Feb.* 19. My Nephew mending; but my Brother not well.

"*Feb.* 26. My Brother so ill that I was not allowed to see him, and till March his life was despaired of.

"*March* 10. I was permitted to see him only for 2 or 3 minutes, for he is not allowed to speak.

"*Mar.* 22. He went for the first time into his library, but could only remain for a few moments.

"*Apr.* 7. I went to stay at Slough, my Brother going by short stages to Bath, Mrs H., my Nephew, and Miss Baldwin going with him.

"*May* 9. My Brother returned nearly recovered, but with having catched a violent cough and cold on the journey.

"*May* 24. I went to be with my Brother till 31st.... In fine nights observing, working in the daytime and writing a paper on Comets filled up the time, though neither my Brother nor myself were well.

"*June* 7. Was the 'Montem', of course much company.

"*July*. We received very distressing accounts from our Brother at Hanover.

"*Sept.* 5. Alexander returned to Bath leaving his brother far from well. The laborious exertions required to the polishing of the 40-ft. mirror; besides the overlooking and directing the Workmen out of doors, who were during the month of August at work on the repairs of the apparatus, had again proved too much for him.

"*Nov.* 2. My Brother went to Town endeavouring to gain some information about my brother Dietrich, who (according to a message from a merchant in town) ought to have by this time been in England.

"*Nov.* 6. A letter from Harrich arrived informing us that D. was waiting there for a passport.

"*Nov.* 7. He arrived at Slough but was obliged to return for his

Trunk and to shew himself at the Ailien (*sic*) Office; and I did not see him till on the evening of the 9th.

"*Dec.* 19. Dietrich left Slough for lodgings in Pimlico. Came with F. Griesbach the day before Christmass day and returned to Town for 26th.

"Memorandum. From the hour of my Brother's arrival in England till that of his departure, which was not till nearly 4 years after, I had not a day's respite from accumulated trouble and anxiety, for he came ruined in health, spirit and fortune, and according to the old Hannoverian custom, I was the only one from whom all domestic comforts were expected.

"I hope I have acquitted myself to everybody's satisfaction; for I never neglected my eldest Brother's business, and the time I bestowed on Dietrich was taken intirely from my sleep or from what is generally allowed for meals, which were mostly taken running or sometimes forgotten intirely. But why think of it now!

"1809, *January*. Throughout the whole month I had a cough. My Nephew a sore throat and Fever! Great Flood and stormy weather! The communication between Slough and Upton was very troublesome to me.

"*Feb.* 5. I sprained my ankle in coming home in the evening from Slough by attempting to walk through the snow in pattons. My Brother was obliged to send me work to Upton, for it was not till a fourth-night after before I could walk there again; and I felt the effects of the accident for above 3 months after.

"*Oct.* 2. I was very ill and had Dr Pope to attend me.

"*Oct.* 9. Dismissed Pope and went to Dr Phips.

"*Oct.* 17. My Nephew went to Cambridge. His mother and Miss Baldwin remained in lodgings at Cambridge.

"*Nov.* 20. Phips pronounced me out of danger from becoming blind which he ought to have done much sooner or rather not to have put me unnecessarily under such dreadful apprehensions."

In one of her later letters to her niece, Caroline mentions that she was kept in a darkened room for a fortnight "to practise being blind". It may be that the doctor prescribed this as the only method of inducing her to take the much needed rest.

In the summer of the year 1809, Dr Herschel, his wife and son, with Sophia Baldwin, set out on a prolonged tour to the north of England.

They left Slough on the 12th July and Caroline then took up her abode in their house. Her brother Dietrich (John Dietrich or even John, as William sometimes called him) joined her in August, and remained at Slough till December, when he returned to London to fulfil his winter engagements.

Mrs Herschel kept a very full diary of this trip, from which one or two extracts will be given. They throw an agreeable light on the character of the writer. She was evidently very good-natured, ever ready both to be pleased and to please as a guest. It is rather amusing to notice the very secondary place which she assigns to the Doctor, especially at Cambridge. Her keen interest in the threshing machine is also worth notice; her account books show that she took careful note of the farming operations on her property.

The travellers spent some days among the Lakes; William wrote to his brother and sister from Keswick:

WILLIAM HERSCHEL *to* JOHANN DIETRICH HERSCHEL

Keswick. Aug$^t$ 3, 1809

"My dear Brother,

"You will be glad to hear that we have hitherto met with no accident to interrupt our tour. We are amidst the Lakes of Westmoreland and Cumberland, and the country is extremely beautiful; but the roads are very hilly though perfectly good for travelling.

"We are now going to take a tour up the highest mountain in this neighbourhood. Its height is 3530 feet. You that are used to the Deister will think nothing of this, but if the weather will permit us to get up our little molehill we shall think ourselves well off.

"I hope you take care of yourself; I was much uneasy when I left you in town but hope you are got strong again. The walks about London are very fatiguing. And I recommend the use of the Hackney-coach to save your legs. I should be happy to hear from you, but it will be best to write to Caroline at Slough, who will let us know how you are and what you are doing in the great town of London.

"*Penrith, Aug.* 4. Yesterday we mounted up the mountain called Skiddaw, and had very favourable weather. The country is full of mountains but this overlooks them all. It was a very fatiguing day, for although we rode upon Ponies up to the top we found it too steep to ride down and walked the greatest part of the way in descending, which is extremely troublesome.

"This place is the furthest from Slough we shall go and we are therefore now upon our return but come back a very different road.

"I send this to Slough that our good sister may see what we are about, for we have so little time for writing that all I can find is taken from travelling, which with the usual meals and the usual hours for sleeping fills up the 24 hours.

"Farewell dear brother till I see you again. My three travelling companions beg to be kindly remembered.

Yrs affectionately<br>
WM. HERSCHEL".

"Dear Sister,

"You will be so good as to cut off the preceding leaf and send it to Johann Dietrich. When I was in London he had fatigued himself so much by walking that he complained, for which reason I have advised him in my letter to use a Hackney-coach when he has much to go about.

"Please to tell Penney that M$^{rs}$ Herschel and I desire that he will once a week come to let you know how the farm goes on. Whether the Harvest is begun or what part of the Harvest they are at. Tell him that we wish to know it and that you are to write us word what he tells you about it.

"You see from my letter to John[1] that we keep advancing on our journey, and shall to-morrow begin to come nearer to Slough again.

"We are now 325 miles from my polishing room, it will therefore take some time before I can be there again. We received all your letters very safely. Please to write again as soon as you can after you receive this and direct your letter at the Post Office, Ripon, Yorkshire. Mrs Herschel, Miss Baldwin and John Herschel beg to be kindly remembered.

Adieu."

## THE SAME *to* THE SAME

Leeds. Augt. 21

"Dear Sister,

"I found your letter at the Post Office, where it had been two or three days. We have all been highly entertained by its contents, and

[1] Dietrich.

please to tell our good brother that we are delighted with his method of enjoying Slough. Most probably I shall disturb his quiet enjoyment when I come home, as then he will have to talk of politics and metaphysics and many other -ics, especially at the Levee at breakfast time in my delightful back room, which I know pleases him so well.

"Mrs Herschel wishes me to tell you to be so good as to look out for a Cook, that is to say one that would come to be engaged for six weeks or two months only. Such persons are to be had occasionally, and Mrs Herschel would then keep Sally as Housemaid in that time. Afterwards as the period for John's going to Cambridge comes on Mrs H. would then do without a Cook till her return to Slough and would probably engage a Cook in London....

"As soon as we come to Cambridge I shall write to you to mention the day of our return. From Cambridge to Slough is three of our days of travelling. For you know our motto is Slow and Sure. The chaise was not overturned at Manchester, it was only a small iron loop that broke so that the body on our side rested against the hind wheel, and we had it mended in half an hour; it occasioned only a little fright; I was walking by the side of it when it happened and helped the ladies out till the accident could be repaired.

"Mrs Herschel, Miss Baldwin and the Entomologist join in affectionate compliments to the two authors of the letter I received this morning, and I remain ever, Dear Brother and Sister,

yours most constantly<br>
WM. HERSCHEL."

In her diary Mrs Herschel mentions that, at Leeds, Dr Herschel visited his old friend, Mr Bulman, and took him a brace of "moor game" which a friend had given him. At Grantham, to which place they arrived on a Saturday, Market Day, she wrote that:

"We walked into the Market, I bought our coachman a pair of stockings. On our return from market the carriage was waiting for us to pursue our journey to the Black Bull, Witham Common. There we saw a thrashing machine, on the best construction that can be; it belonged to the master of the inn. They were thrashing beans, four horses were imployed and they could grind out more corn in one day than could be done by 6 men in a week. It is, I think, a sin that every Parish in the Kingdom has not one which would be sufficient for all the farmers in such a parish as Upton, for a few hours would thrash out a load of wheat. In the afternoon we went a stage of 11 miles of good road to Stamford.

"*Tuesday, Aug.* 29. Proceeded on our journey to Cambridge, where we arrived at one o'clock. Our Inn opposite Trinity College. After dinner we saw several colleges, the first of course was St John's, where I had the pleasure to introduce my dear son for the first time. The Rev. Mr King, a gentleman Dr Herschel was introduced to at Stantion, a small village where we only rested our horses for two hours, by a gentleman who lives in the village, on our arrival in Cambridge, was so polite as to call on us to take us to see Trinity College. The number of curiosities we saw are too numerous for me to write down. In the evening we saw Mr Hornbuckel, my son's tutor; he gave us a very pressing invitation to dine with him the next day, which we with pleasure accepted, and a most agreeable day it was to the whole party, which consisted of just a dozen of his friends, (including) Mr and Mrs Innis and son who is just entered at St John's; this young gentleman was with Dr Gretton at the later part of my son's being with Dr Gretton.

"I spent all my spare time in seeking after lodgings for myself when I return with my son; we found great difficulty in getting any as they are nearly all engaged for the students who cannot be in the different colleges.

"*Friday, Sep$^t$ y$^e$ 1$^{st}$*. We left Cambridge at 3 o'clock in the afternoon....

"*Sunday, Sep$^t$ y$^e$ 3$^d$*. We left Rickmansworth, our last stage of 14 miles to Slough. Thus ends one of the most delightful Tours that ever was taken by four people. We travelled 789 miles without any accident worth mentioning, and all of us returned home in perfect health. This journey we shall ever reflect on with great pleasure".

WILLIAM HERSCHEL *to* CAROLINE HERSCHEL

Cambridge. Aug$^t$ 29, 1809, Tuesday

"Dear Sister,

"At our arrival here I received your letter a few hours ago and am glad to read S$^r$ J. Banks's letter contained in yours; it is a very kind information, and I am glad you have let him know that I shall be at home again soon. To-morrow we stay here and intend to set off on Thursday morning for Slough; if all happens as favourable as hitherto for our journey we shall reach home by Saturday evening. Perhaps however, if we are not at home before dark, it may be Sunday morning.

"If there is not already Corn, Hay and Straw in the Stable, Cock must get it brought by the time it will be wanted. The Stable may also be set open in fine weather to air it a little.

"I hope J. D. the star gazer has got rid of his cold by this time.

"Mr Pye's letter I will answer at my return to Slough.

"Home will appear home without any additional preparation for we shall all be glad to see Slough again. I long much to go on with my work as usual. My glasses and prisms have only once been unpacked since I left home. My machines at Slough I suppose are all got mouldy and contracted rust, which will want rubbing off; but a few days will get all right again.

"*Aug*$^t$ 30. This morning we have seen some Colleges and also John's intended apartments in St John's. At present the Ladies are gone out to see what lodgings they can get for the time when they are to be here. This seems to be a very difficult task, and I do not know whether they will be successful.

"If your notable attendants can find time it will not be amiss to let them Scour the floors of my rooms and to air them a little, as that is a thing they are not often treated with.

"The Ladies are just come home but have not succeeded in getting Lodgings yet.

"We are going to dine with John's tutor at the college and the post being going out I finish."

The entries in Caroline's Journal for the next few years show that her health was seriously affected by all the extra work and the grave anxiety which Dietrich's arrival in such distressing circumstances had cast upon her. It was very evident that she must leave Upton, and in the next year, 1810, it was arranged that she should occupy a small house on the Baldwin property, between her brother's house and the Crown Inn; probably the one in which old Mrs Baldwin had lived till her death. There could be no question of her actually living in her brother's house, the accommodation was much too limited. There were only four bedrooms, two of them very small, besides the loft for the servants; one bedroom had to be kept for visitors, and Caroline could not possibly have had a sitting room to herself. The little cottage which she now took seemed admirably suited for her needs, but for some reason she moved four years later to a much less convenient abode. She wrote in her Journal:

> "1810, *May 4 or 5*. I went to Slough, my Brother going to Town with Mrs H. He returned after a short stay and I remained with him till Mrs H. came home again. Some of my last days of staying at Slough I spent in papering and painting the rooms I was to occupy in a small house of my Brother's attached to the Crown Inn, to which I removed on June 16".

One of the few letters from William Herschel to his wife refers to Caroline's occupation of this cottage.

### DR HERSCHEL *to* MRS HERSCHEL

Slough near Windsor. Tuesday, May 29, 1810

"My dear, this letter will only contain the intelligence that nothing new has happened at Slough. I go on with giving directions to have the House at the corner put in order, which I hope will soon be done; but the workmen are so very slow that I begin to grow impatient. The Glasgow mirror goes on well and is likely to be a very fine piece. This morning I expect to hear from you and according to the contents of your letter I shall take my measures.

"The coachman drives extremely well; he has carried me once to Windsor in a chaise I borrowed of Mr Baldwin for the purpose, and this morning he shall drive me again to settle the business about Mrs Grape and of many other things I want in Windsor. My sister wants to choose the pattern of some proper paper for her sitting room, and I shall take her to Windsor. If she can suit herself there I shall go on to see Osborn about a proper horse, and take my sister up again when I come back, as she intends to pay a visit to Mrs Beckedorff meanwhile.

"The post is come but, as there is no letter, I have only to say that this will come by Mrs Grape; and I have inclosed One hundred pounds for your use, if more should be required do me the favour to write immediately. I shall be glad also to know whether you have heard from our son and when he will leave Cambridge. I suppose by this time he will know for certain. The coachman says something about his fetching you home; I believe the chaise will be done by the time you want to come; if not you will be so good as to inform me how you intend to come again to your affectionate

Husband<br>Wm. Herschel.

"P.S. Do you expect another basket on Saturday next?"

On the 13th July, 1810, the Herschel family set out on a journey to Scotland, travelling in their own carriage, with their own horses and coachman; but a mishap on the first day compelled them to send back the horses with the coachman and to trust to posting arrangements for the rest of the way. William Herschel wrote to his sister:

Oxford. July 14, 1810

"Dear Sister,

"We send our horses back with the Coachman, the young horse cannot go the journey. The Coachman is regularly engaged and will do anything in the house you require, and also show the telescopes, in which you will give him the proper instructions. He is to sleep in the maid's Garret, and the Man that was engaged to sleep in the house may be discharged. You will be so good to pay him for his time one week.

"Inclosed is a letter for Mr Hausemann, who will give directions what is to be done with the horse's foot. I would not have any Farrier meddle with the horse but Mr Hausemann's directions are to be followed.

"The Coachman may deliver the letter to Mr Hausemann as soon as convenient after he comes home. He may perhaps find him at Frogmore, where he goes every morning I believe from 9 or 10 o'Clock to about 1.

"When the road requires watering the Coachman should water it as usual. If the windows are very dirty he may clean them occasionally.

"We are going on upon our journey with Post horses.

"If you can write us word how the horse came home and when Mr Hausemann has seen him what he thinks of him, a letter will be received if you write either Tuesday or Wednesday, directed to me to be left at the Post Office at Warrington, Lancashire. If you write a day or two or 3 after Wednesday it must be at Preston, Lancashire.

"The usual Compli<sup>ts</sup> from the whole family attend you, and I remain,
Dear Sister,
Yours
WM. HERSCHEL".

John Herschel accompanied his parents on this trip. Their ultimate destination was Glasgow, but they lingered by the way, stopping some days with Mr Watt at Birmingham. John Herschel described in a diary some of the wonders of the great Foundry, "famous over all Europe". One of the novelties which especially aroused his admiration was a great chandelier of gas light in the engine room. The effect of the illumination on the polished machinery he asserted to be "one of the most magnificent sights I ever remember to have seen".

The charm of Mr Watt's conversation struck the youthful undergraduate with as much pleasure as the wonders of the factory; he concluded his panegyric with the remark: "Very few are the individuals who have been of such vast utility to society as Mr Watt, assisted as he has [been] by a combination of circumstances almost unprecedented, in his partnership with Mr Boulton, a man whose liberality and public spirit will ever be remembered with gratitude".

The travellers spent a happy time with Mr Grahame and his family at Glasgow. It was then that John Herschel first met James Grahame, with whom he formed a warm and lasting friendship. He mentions that at tea-time the company were amused by the perusal of a parody of Scott's *Lay of the Last Minstrel* entitled *The Goblin Groom*.

Leaving Glasgow the travellers set out for Edinburgh, stopping the night at Carnwath, which John Herschel remembered as "a vile place, where (owing to want of horses) we were forced to sleep. Here my Father was miserably ill, and, by way of I know not what, resolutely refused to sleep in bed!! Accordingly he had 6 chairs put together and a mattress put on them and there took his repose".

In the following year, 1811, the same party revisited Scotland; leaving Slough on the 22nd July and stopping, as before, for a few days among the English Lakes. William wrote to his sister from Keswick:

"Dear Sister, Aug<sup>t</sup> 1, 1811

"Here we are, in the land of the living, safe at Keswick. Considering that this day is the *first of August* you will allow that we have lost no time in getting here. Yesterday the young folks went to the top of Helvellin. The day before they sailed upon the Lake, and so we pass our time. About *half an hour after eight* I got up, and after breakfast walked soberly to the margin of Derwentwater with Mrs Herschel. The rest of our party are again upon the Lake, but we shall meet at dinner.

"I have called at the Post Office but have not found a letter from you. By this we are to conclude that all is well at Slough. I shall however call once more at the post office this day after the post is come in and may perhaps receive one yet.

"If Mrs or Miss Grape should inquire after us please to say that we are all well and that Miss Baldwin intends to write to them.

"From this place we set off for Edinburgh which is about 130 miles from here. As soon as we get there I will let you know how long we are likely to stay there and where your next is to be directed, which however I know must be Glasgow.

"Among other things when you write let us have the latest news of the health of the King.

"I shall be glad to know the names of every person who has called at Slough either to inquire after me, or to see the telescope; and also to hear whether more papers have been sent by Bulmer for correction. If he wants to know what copies I wish to have, you will say as usual that Dr H. wishes to have 25 copies. Perhaps Dr Davy may write something about the engravings of Nebulae; of course you will mention that I shall be at home in about 6 weeks and will inspect the correction of the plates myself, which nobody can do for me.

With all our affectionate remembrances,<br>
I remain, Dear Sister,<br>
Yours very faithfully,<br>
WM. HERSCHEL".

The travellers went on to Glasgow, visiting Mr Grahame at Whitehill, and staying a few days with Mr Alexander Wilson, the brother of Prof. Patrick Wilson. Scott's *Lady of the Lake* had just been published, and had been read with delight by at least the younger members of the party; this no doubt prompted the wish to visit the scene of the poem. As John Herschel either did not keep a diary, or has destroyed it, we can only imagine the pleasure they must have derived from seeing the unspoilt beauties of the Trossachs and Lake Katrine. William Herschel merely recorded in his notebook that they went from Glasgow to Stirling, and thence by Callander to Loch Katrine and back to Callander in one day, a trip of forty miles.

William Herschel wrote to Caroline from

Newcastle. Sept. 10, 1811

"Dear Sister,

"I had not time at Glasgow to write another letter, and we have been detained there on account of the erection of the 14-feet telescope. It could not be put up till the place was properly prepared for it which was Sep$^{tr}$ 2. In the evening of that day I saw the Comet through it which you and our brother have probably observed every night, if he is at Slough. It has been very bright. If you have not seen it please to look under Ursa major in the line of the two stars called pointers.

"This day, Sept. 11, we go to Richmond in Yorkshire, and we shall now, I hope, not be detained anywhere on the road. About the middle of next week we shall reach Cambridge, and soon after Slough.

. . . . . . . .

Yours affectionately
WM. HERSCHEL".

From the names of the places in William Herschel's notebook where they stopped on their homeward journey in 1811, it would seem that he chose their route purposely to show his son the places connected with his early life. The year before they had made a detour from Newcastle in order to visit Sunderland, which had been William's headquarters when he lived in Yorkshire as a struggling musician. Thence they had gone to Leeds and doubtless again called on Mr Bulman, whom Mrs Herschel mentioned in her account of their tour in 1809. This time they took a different route, going by Boroughbridge and Ferrybridge (names which frequently occur in William's early letters to Jacob) to Doncaster. And here William makes one of his rare notes and mentions the name of Copley, as having been visited by them. This person could hardly have been the identical Mr Copley whose weekly concerts William had attended when, as a young man of twenty-two, he was conducting the Militia band, but it may have been his son.

They reached home on the 18th September, 1811.

From William's travel notebook we learn that the united family made other summer trips in the following years. In 1813, they took their way along the south coast, going by Tunbridge Wells to Hastings and thence to Eastbourne, Brighton and Bognor as far as Portsmouth. While at Brighton, William Herschel met the poet, Thomas Campbell, who has recorded, in a letter to a friend, the impression made on him by the appearance and conversation of the venerable astronomer. William Herschel was now in his seventy-fifth year.

### THOMAS CAMPBELL *to* A FRIEND

*Brighton. Sept. 15, 1813*

"I wish you had been with me the day before yesterday, when you would have joined me, I am sure, deeply in admiring a great, simple, good old man in Dr Herschel. Do not think me vain, or at least put up with my vanity, in saying that I almost flatter myself I have made him my friend. I have got an invitation, and a pressing one, to go to his house, and the lady who introduced me to him so spoke of me as if he really would be happy to see me....

"I spent all Sunday with him and his family. His son is a prodigy in science and fond of poetry, but very unassuming....

"Now for the old Astronomer himself; his simplicity, his kindness, his anecdotes, his readiness to explain and make perfectly perspicuous too his own sublime conceptions of the universe are indescribably charming. He is 76, but fresh and stout, and there he sat nearest the door at his friend's house, alternately smiling at a joke, or contentedly sitting without share or notice in the conversation. Any train of conversation he follows implicitly; anything you ask he labours with a sort of boyish earnestness to explain. I was anxious to get from him as many particulars as I could about his interview with Buonaparte. The latter, as it was reported, had astonished him by his astronomical knowledge.

"'No,' he said, 'the First Consul did surprise me by his quickness and versatility on all subjects, but in science he seemed to know little more than any well-educated gentleman, and of astronomy much less than, for instance, our own king.' 'His vain air', he said, 'was something like affecting to know more than he did know.'

"He was high and tried to be great with Herschel, I suppose, without success.

"'I remarked', said the Astronomer, 'his hypocrisy in concluding the conversation on astronomy by observing how all these glorious views gave proofs of an Almighty wisdom.'

"I asked him if he thought the system of Laplace to be quite certain, with regard to the total security of the planetary system, from the effects of gravitation losing its present balance? He said, 'No', he thought by no means that the universe was secured from the chance of sudden losses of parts. He was convinced that there had existed a planet between Mars and Jupiter, in our own system, of which all little Asteroids, or planetkins, lately discovered, are indubitably fragments; and 'Remember', said he, 'that though they have discovered only four of those parts, there will be thousands—perhaps 30,000—more not yet discovered'.

"This planet is believed to have been lost by an explosion.

"With great kindness and patience he referred me, in the course of my attempt to talk with him, to a theorem in Newton's *Principles of Natural Philosophy*, in which the time that light takes to travel from the sun is proved with a simplicity which requires but a few steps in reasoning. In talking of some inconceivably distant bodies, he introduced the mention of this plain theorem, to remind me that the progress of light could be measured in the one case as well as the other. Then, speaking of himself, he said, with a modesty of manner which quite overcame me, when taken with the greatness of the assertion:

"'I have *looked further into space than ever human being did before me*. I have observed stars of which the light, it can be proved, must take two million years to reach the earth'.

"I really and unfeignedly felt at the moment as if I had been conversing with a supernatural intelligence.

"'Nay more,' said he, 'if those distant bodies had ceased to exist millions of years ago, we should still see them, as the light did travel after the body was gone.' These were Herschel's words, and if you had heard him speak them, you would not think he was apt to tell more than truth.

"After leaving I felt elevated and overcome, and have, in writing to you, made only this memorandum of some of the most interesting moments of my life.

(Signed) T. C."[1]

[1] *Life and Letters of Th. Campbell,* by W. Beattie.

## CHAPTER XXII

## 1812   1816

Caroline's Journal. John Herschel Senior Wrangler. Letter from William Herschel to Sir Joseph Banks proposing his son as F.R.S. Last observations with the 40-foot. Alexander's accident. Letters. Alexander returns to Hanover. William Herschel receives the Guelphic Order of Knighthood.

IT is one instance of Caroline's amazing detachment from the world outside her own circle that, in writing of her brother Dietrich's return to Hanover, she makes no reference whatever to the stirring events which led to the downfall of Napoleon and the liberation of Germany. Dietrich left England in July, 1812, the very month in which Napoleon invaded Russia. Caroline wrote in her Journal:

"1812, *June* 1. Dietrich came to Slough, disengaged from all business in Town, to spend the last few weeks he had of being in England with us.

"*June* 12. I went home to look to the necessary preparations for Dietrich's precarious journey (that) he was obliged to make through Sweden.

"*July* 8. Dietrich left us. Alexander accompanied him to Town.

"*July* 14. He left Harrich and at the end of the month we received a letter dated Göttenberg, July 18, and so far we knew that he was safe; but of receiving any further account we had not the least prospect, for all communication, with Hanover in particular, was cut off.

"The latter end of September Mr Goltermann received a few lines which came through France to him, dated Sept. 4, shewing that a letter of Aug. 15 had been lost and that, at Helsingfors, D. had been robbed of his pocket book (when under examination in der Inbache). To this accident we were indebted for knowing that he was got home, as he was obliged to write for a duplicate for a Bill of exchange, as such letters were (though unsealed) allowed to pass through France".

Throughout the year, 1813, Central Germany was a battle-ground between the armies of the French and of the Allies. Prussia had abandoned the humiliating subservience to Napoleon which Dietrich

Herschel deplored in his letter to William in 1806, and had joined forces with the Russians. After the decisive battle of Leipzig, in October, 1813, Napoleon's army retired beyond the Rhine, and Hanover, as well as the rest of Germany, was at last freed from the French domination. News of their brother reached the Herschels at Slough in November; Caroline made the following entry in her Journal:

"1813, *Nov.* Mr Rehberg brought the first letter from our brother Dietrich, dated Nov. 10th; which though still written with great caution, gave us, after the lapse of 16 months, the assurance that he and his family were living".

No further allusion to Dietrich occurs in Caroline's notes till the year 1816, when in mentioning that John Herschel had received a Diploma as Member of the University of Göttingen, she added: "the packet brought very satisfactory letters from our brother at Hanover".

These letters of Dietrich's have not been preserved.

For the sake of clearness the entries from Caroline's Journal relating to her brother Dietrich have here been all collected together, but we must now go back to take up the story of the family life at Slough from the time he left England.

## CAROLINE HERSCHEL'S JOURNAL

"1813, *Jan.* 25. Congratulatory letters arrived from Cambridge on my Nephew's having obtained the Senior Wranglership. He was then contending for another Prize,[1] which a few days after he also obtained.

"*Feb.* 27. My Brother and Nephew went to town, to the Royal Society. Returned the 26th."

It must have been a proud and happy moment in the old astronomer's life when he introduced his young son to the President and Fellows of the Royal Society. John Herschel had already communicated a paper on a mathematical subject[2] which had been read and printed in the *Transactions* of the Society. William Herschel, therefore, felt justified in asking that he might be admitted as a Fellow, in spite of his youth. The

[1] The Smith's Prize.
[2] "On a remarkable application of Cotes's Theorem", read Nov. 12, 1812, *Phil. Trans.* 1813, Pt I, p. 8.

following is a copy of the letter in which he preferred this request to the President.

WILLIAM HERSCHEL *to the* RT. HON. SIR J. BANKS, BART.

"Dear Sir, Feb. 4, 1813

"You will readily forgive me if I should express a wish to see my Son's name among the list of the members of the Royal Society. I can truly say that were he not my son and I knew as much of him as I do I should readily sign the usual certificate, and in case of your approbation I am in hopes several of my friends, such as Sir Ch. Blagden, Dr Wollaston, &c., who have already personal knowledge of him, will be so good as to put their names to the required form. His name is John Frederic William Herschel, Bachelor of Arts.

"I do not know whether it is necessary for a candidate to be of age, but as he will be so in about a month's time this will not be an objection.

I have the honour, &c.

(W. H.)"

The following passages from Caroline's Journal describe the celebration of John Herschel's twenty-first birthday at Slough:

"1813, *Feb.* 27. Miss Wilson with her two nieces came to Slough.

"*March* 5. Miss S. White with her maid, Sally (one of my Nephew's nurses), came to be present at my Nephew's 21st Birthday.

"*March 7 and* 8. I joined the company who dined there on this occasion, and I must not forget that my Nephew presented me with a very hansom Necklace, which I afterwards sent to my niece Groskopf when a Bride, and I being too old for wearing such ornaments.

"*March* 11. My Nephew went again to Cambridge, to offer himself as candidate for a Fellowship, there being 3 vacant, and at the conclusion of the examination he obtained the first choice of the three.

"*May* 3. I intended to pay a long promised visit to Mrs Goltermann, but found my Brother too busy with putting the 40 ft. Mirror in the Tube. The carriage having broke down on the way between the polishing room and the Tube; therefore I postponed my journey till I was sure I should not be wanted at home.

"*May* 10. I went to London and met with a friendly reception at Mrs Golt$^n$. The 11th I went with Mrs G. and a Mrs Kramer to Kensington. I remained with Miss Wilson while they paid a charitable visit to two Ladies (attendants on the Duchess of Brunswick) who were left in a very distressed situation by the death of their Mistress. The evening we spent at Buckingham House with Mrs Beckedorff.

"*May* 12. The forenoon and early part of the afternoon were spent in shopping and visiting. The evening again at Buckingham House, where I just arrived as the Queen and Princesses Elisabeth and Mary, and the Princess Sophia Matilda of Gloucester were ready to step into their Chairs, going to Carlton House, full dressed for a Fate; and meeting me and Mrs Goltermann in the Hall, they stopt for near 10 minutes making each in their turn the kindest inquiries how I liked London, &c., &c.

"On entering Mrs Beckedorff's room I found Madame de Arblé (Miss Burney) and we spent a very pleasant evening.

"*May* 13. Paid visits at Best's family. Mrs and Miss Maskelyne, who were in Town; they came to see me next day at Mrs G.'s.

"*May* 15. I went to the Exhibition. The evening at Baron Best, where I met the Beckedorffs. On my return I found a letter from my Brother with Sir W$^m$ Watson's direction that I might give them the meeting in Town. The next morning I spent a few hours with them and next day S$^r$ W$^m$ with Lady Watson and Miss Jay called on me in Charl'$^s$ Street. Baron Best also called and brought me the place of a Comet from the *Hamburger Zeitungen*.

"*May* 18. I went home, found all well and a great deal of work prepared for me; the evening was spent with sweeping for the Comet but I could not find it; the weather was not clear.

"1814, *Jan.* 1. My Nephew, J. Herschel, brought me for a New Year's present a new publication by him.

"Mem. The winter was uncommonly severe. My Brother suffering from indisposition, and I, for my part, felt I should never be anything else but an Invalid for life, but which I very carefully kept to myself, as I wished to be useful to my Brother as long as I possibly could.

"*April* 1. My Brother went to Bath to see his Brother and S$^r$ W. Watson, his cough still very bad. And the 12th when he came home we came to know that he had been taken very ill on the road and suffered much when at Bath. And it was not till many weeks after and the warm weather coming on before he felt relieved".

William Herschel wrote to his wife from Bath, making light of his illness.

Bath. Sunday, April 3ᵈ, 1814

"My dear,

"As I suppose you will be glad to have some account of your travelling husband, I can therefore tell you that he is safely seated at a table in his brother's room writing to you.

"The Coach did not get to Bath till eleven o'Clock and I can assure you that the journey was rather too long for me, so that perhaps I may use the two day coach in coming back; but perhaps I may find myself stronger after a week's repose.

"Sir Wm. Watson, his Lady and Miss Jay are all very well, and received me on Saturday morning very cordially at their breakfast table, where according to the notice I had given I was expected. Mr Freame, knowing of my arrival, sent a note of inquiry after your and Miss Baldwin's health. Mrs Beadon came to dine with us, and in the evening we had as usual a whist party. Miss Baldwin's note to Miss Jay I delivered but could not prevail upon Lady Watson to make any charge for the trifling thing, as she called it, done for Miss Baldwin, as the whole sum of 4 or 5 things she had done together came but to so little as not to deserve mentioning. Her other letter to Miss Parry I have not delivered yet, as I wish to carry it myself, and could not go out yesterday, the weather being very rainy with violent storms.

"From what I have heard I believe that Sir W. Watson will not come to London this Spring so that we shall not have the pleasure of their company at Slough.

"My brother is now preparing to accompany me to Pulteney Street where he is invited to dinner. The account he gives me of Bath is that it is very full but that the rooms are not so much frequented as usual because they have so many routs and Assemblies. There has been no music at the Pump room this Season. To-morrow I shall enquire whether I can hear anything of Mrs Grape, for if I understand right, she must by this time be here.

"Pray take care of yourselves that I may in good time find you and Miss Baldwin as well as I left you, and when you see my sister please to say that I found our Brother well and apparently in as good health as ever. With his cordial remembrance to you and your companion as well as to my Sister, I conclude by assuring you of the constant affection of your faithful Husband,

WM. HERSCHEL."

CAROLINE HERSCHEL'S JOURNAL (*continued*)

"1815, *June* 29. Alex came to assist his brother; he was expected the 13th but had been prevented by illness.

"*Aug.* 11. Alex left Slough, my eldest Brother with him, going on for Dawlish to recruit his strength again. His declining health had a sad effect on Alexander's spirits and I was in continual fear of the consequence, for nothing but the thoughts of the yearly meeting had till then kept up his spirits, but from what is yet to follow it will be seen that our next meeting was not only the last but a very distressing one.

"*Sept.* 2. My Brother came home having a violent cough. The latter end of August my Nephew came home very unwell and the Phisicians ordered him to the sea-side, where Mrs H. and Miss B. accompanied him, and I went the same day to be with my Brother and remained with him till the 12th October. The first fourthnight of my being with him he was not able to do anything which required strength."

Caroline mentions in her Journal that during the years 1813 and 1814 she had much work to do for her brother, both by day and night. In the daytime she would have been busy copying out the paper, "On the Sidereal Part of the Heavens", which was read before the Royal Society in February, 1814, and in helping to compile the list of observations of Uranus, which were given as a paper, "On the Satellites of the Georgian Planet", read at the meeting of the Society in June, 1815.

William Herschel ceased his regular sweeps for nebulae, clusters and double stars in September, 1802, but in spite of failing health he continued to observe objects of interest for many subsequent years. The yearly repolishing of the mirror for the 40-foot telescope was continued up to June or perhaps September, 1814; though after his illness in 1808, William Herschel seems to have found great difficulty in giving the final touches himself, and the polish was very unsatisfactory. For example, in 1813, the mirror had been in the polishing room in May, yet when Herschel examined Saturn with the telescope in July, he remarked: "The mirror is so much tarnished that the image of Saturn was very imperfect".

The last observation made with the 40-foot was also on Saturn in August, 1815, and on this occasion Herschel again reported: "The mirror is extremely tarnished".

This may be considered as the final effort of the great telescope. From this time the gigantic apparatus remained in the centre of the garden at Slough, an object of curiosity to the general public and of reverent affection to all members of the Herschel family. The supporting framework was finally taken down when Sir John Herschel returned from the Cape of Good Hope, in 1839, and the tube, with the mirror inside, was sealed up and left prone on the grass-plot.

The occasion was not allowed to pass without due ceremonial. Sir John Herschel, with his wife and children, and some members of his household, assembled within the tube and sang a requiem composed by himself in honour of his father's great instrument.

When Caroline copied the entry from her old diary into the Journal which she sent to her nephew, recording the visit which William made to Bath in company with Alexander in the summer of 1815, she added a hint of the worse trouble which was soon to follow.

In February of the next year she was staying with Mme Beckedorff at the Castle, when William sent the carriage in haste to fetch her home, as "a most melancholy letter had that morning arrived (from Alexander) acquainting us of being confined to his bed, having received an injury to his knee".

Caroline wrote at once and received the following letter three days later.

ALEXANDER HERSCHEL *to* CAROLINE HERSCHEL

Bath. Feb. 6, 1816

"Dear Sister,

"I just now rec$^d$ yours and hope I shall be soon enough for the post. I will only say that Mr Edwards assured me this morning that everything goes on well, more so than could be expected. Today makes seven days and nights that my leg has laid in one position; at first he told me it would require a month or more time but today he gave me hopes that in another week I should sitt upon a proper chair in the daytime.

"I think it is well worth while to meet with an accident for I did not know I had so many friends; they have been all here, high and low; they bring me books to read and send me some things to eat and drink, but as for that I live upon bread and butter, Barly water and water grewel and Tee and that very sparingly, for I think it is better to starve myself than bring on inflammation, which I had nothing of yet and it agrees very well with me for I have not been in so good spirits for some time past; and as for pations (patience) I have a gread stock of it and have had no occasion to have resource to it for I am not in pain, at least I have had little hitherto.

"I must conclude but I shall write again soon. Remember me to our dear brother, Mrs Herschel and Miss Baldwin.

I remain affectionately yours,

ALEXR. HERSCHEL."

The next letter is not so cheerful:

Bath. Feb. 26, 1816

"Till yesterday week I got on very well in regard to my leg and so everything goes on still well in that respect, but yesterday week, that is on Sunday, in the night about two o'clock, I awoke very ill; I could not breathe and for three hours I expected every moment to be my last. Mr Edwards I sent for, but he sent the boy with a draught and towards morning I got a little better. I was sleepy but I was afraid to go to sleep for fear of such another waking. Monday was the day fixed for me to get up and I have been up every day since for 6 or 8 hours a day, but my health has been so much impaired that I have not rec$^d$ any benefit from getting up.

"I rec$^d$ your letter this afternoon and I see that you are troubled about a nurse for me. A Mr Francum, the husband of my charwoman, has been with me day and night ever since the accident, and he does very well. At night times he puts a bed on the ground in the next room to me.

"Before this severe illness I could lay in bed reading all day, but now I have lost all desire for it, but am in hopes a day or twos' pations will set everything right again. Tell Miss Baldwin that I am much obliged to her for her pretty letter, but you see what a writer I am now, and I fear it will be some time before I can venture to write to her. My best comp$^{ts}$ to her and remember me to dear Mrs Herschel and our dear Brother.

I remain affectionately yours,

ALEXR. HERSCHEL".

Caroline was so much disturbed by this letter that she proposed to go herself to Bath to nurse her brother. Alexander wrote in haste:

"My dear Sister,                                    Bath. Feb. 28

"I am exceedingly sorry that I have alarmed you so much with my last letter, quite unthinkingly and disregarding the consequence, I wrote as I felt. But I am much better and shall be quite well. Dr Heygarth has undertaken to ridd me of this nasty feverish complaint; today I am much stronger and I have eat some rabit and I may drink some wine. I sitt up for eight hours a day and sleep all night and begin to read again. Your letter has thrown me into a cold sweat and I have just time before the post goes out to write this and beg and pray hartily that you will not think of coming, for you can really do me no good but my anxiety for you will be so gread that I certainly shall get worse. I shall write often and I hope as chearfully as today and more so. The best thing you can at present do for me is to write a line or two as soon as possible to tell me that you have given up the Idea of travelling till next summer when we can take a walk in and about Bath together.

"Our dear brother I hope will also not think of leaving home till we can walk together upon Odd down.

I remain, dear Sister, affectionately yours,

ALEXR. HERSCHEL.

"Comp$^{ts}$ to dear Mrs Herschel and Miss Baldwin".

Alexander was not able to write as cheerfully as he had promised. In his next letter he reported that he could just get about the room on crutches and that the doctor continued to talk of the need of a little more "pations", when his knee would be as well as ever.

His brother and sister, however, were not hopeful that Alexander would ever be able to return to his profession. Caroline wrote to him that

"we must not hurry you before you have quite recovered (from) this last accident, and then when you are inclined we will correspond on this subject (probably his future), and you shall tell us your scheme, and we will tell you ours".

Alexander replied very despondently:

"Now, dear sister, you see I am not quite recovered and I cannot think of anything else but my terrible misfortune and at intervals read a little and that must be trifling things, such as the Arabian nights, &c. I call my accident now a terrible misfortune because I did not at first see all the bad it has and still may occasion".

He then explains that he had more pupils than the year before and had therefore expected to be able to get clear of all debts before the summer, but his accident had spoilt all these plans. There are no more letters on the subject, but the event shows that William now came to the rescue and arranged to provide for his brother for the rest of his life. Alexander had now passed his seventieth year, and as Bath had no attractions any longer for him, now that he was incapacitated from earning his living there, it was settled that he should return to Hanover, where his brother Dietrich would be delighted to welcome him and to see to his comfort.

To Caroline the parting with this brother, with very little hope of ever seeing him again, must have been a great sorrow. He had been closely associated with her life ever since she came to England, and his yearly visits to Slough, when the rest of the family were absent, had lightened her feeling of loneliness. Her whole being now was wrapped up in concern for her elder brother's failing health. She had no spirit even to feel any elation about his knighthood; she just records the fact.

## CAROLINE'S JOURNAL

"1816, *April* 5. My Brother received the Royal Hanoverian Guelphic Order.

"*May* 12. My Brother went to Town to prepare for going to a Levy at the Regent's next Tuesday, he brought me the Keys to the Library for going there to work.

"*June* 17. I went to my Brother's house and was left alone under the deepest concern for his health; he went with his family to Cambridge.

"*June* 27. He returned much the same as he went.

"*July* 24. Mr and Mrs Watt and Mrs Watt's sister were at Slough and I met them the 25th to dinner.

"*July* 31. Alexander arrived at Slough.

"*Aug$^t$* 12. At 3 in the afternoon I left my Brother's house with Alexander in company with Mr and Mrs Morsom for Wapping.

"*Aug$^t$* 13. My eldest Brother was to leave Slough for Dawlish for change of air, and for the purpose of removing him from his laborious works. The only comfort I took with me was to know that my

Nephew was to go with his Father and a promise of Miss B. to write me word how he had borne the journey, but this promiss was forgotten!

"*Sept.* 2. I saw Alex. led by Capt. Steevens on Board, of whom I had the assurance that he would see Alex. safe to Dietrich's friend, Mr Munter, in Bremen.

"A few hours after I left the place, taking with me Rec$^{ts}$ from everybody with whom I had had occasion to keep accounts. I came very ill to Mrs Goltermann where I remained a week under her care.

"*Sept.* 9. I went home.

"*Sept.* 13. My Brother returned much the same as he went.

"*Sept.* 23. We were at a Fate the Queen gave at Frogmore. I was obliged to return with my Brother soon after he had been noticed and conversed with the Queen and the Regent, being too feeble to be long in company.

"*Sept.* 26. We had letters from Hanover to acquaint us with Alexander's arrival with improved health after a pleasant journey both by sea and land."

Alexander Herschel lived contentedly at Hanover for the rest of his life and died in March, 1821, in the seventy-sixth year of his age.

## CHAPTER XXIII

### 1813   1822

Letters from William Herschel to his son on the choice of a profession. John Herschel leaves Cambridge to take up astronomy. Caroline's Journal. Sophia Baldwin's marriage and her death. Death of Queen Charlotte. William Herschel's declining health. Death of King George III. Foundation of the Astronomical Society. William Herschel first President. Caroline's narrative of Sir William Herschel's last days. Letters of condolence. Epitaph in Upton Church.

THE career of the younger Herschel does not properly come within the scope of the present Chronicle, which is intended to deal only with the earlier generation; yet some reference to his occupations at this time seems necessary to explain his relations with his father.

After taking his degree and obtaining a Fellowship at St John's College, in January, 1813, John Herschel remained at Cambridge for a year. He felt no inclination to follow in his father's steps as a practical astronomer; mathematics was still the preponderating interest with him. He was collaborating with Peacock and Babbage in the work of the Analytical Society, which had been started by the three friends in 1813, and which helped to revolutionize the study of mathematics at Cambridge. But his intention at first was to adopt a profession.

His father wished him to go into the Church; John inclined to the Law. The following letter was written by William Herschel to his son in reply to one in which John had stated his objections to taking Orders.

WILLIAM HERSCHEL *to his son*, JOHN F. W. HERSCHEL

Nov. 8, 1813

"My dear Son,

"You request my opinion upon a subject which is of so much consequence, and upon which so much may be said that it will hardly be possible to enter into it properly by letter writing; but what is more, you have already given your opinion of it, and to ask mine seems to be a mere matter of form. You say—I cannot help regarding the source of church emolument with an evil eye.—The miserable tendency of such a sentiment, the injustice and the arrogance it expresses, are beyond my conception.

"You mention the necessity of a mode of acquiring a living, and that a man should have some ostensible means of getting his bread by the labour of his head or hand, as if you already found yourself in confined circumstances, or had no expectations of assistance from your parents, which you will know is not a just idea of your situation.

"You say that Cambridge affords you the Society of persons of your own age and of your way of thinking, but know, my dear Son, that the company and conversation of old, experienced men of sound judgement, whose way of thinking will often be different from your own, would be much more instructive, and ought to be carefully frequented.

"But now let us compare the end proposed and the means of obtaining it in the two liberal professions of which it seems the choice is to be made. Your own words with regard to the church were, 'The path was wide and beaten, and I had only to pursue it'. Can anything be a higher commendation of it? Such a path must surely lead to happiness, or else it would never be so wide and so beaten. You admit that you had only to pursue it. Very true, for you are in it and have all this time been in it; nay, you are already far advanced in it, having not only carried off the highest honours of the University but being already in possession of a fellowship which insures your success in pursuit of the path.

"How is it about the path of Law? It is crooked, tortuous and precarious. It is also beaten, but how many have miserably failed to acquire an *Honest* livelihood? You are not in that path, and it is almost too late to enter it; your studies have been of a superior kind.

"You say the church requires the necessity of keeping up a perpetual system of self-deception, or something worse, for the purpose of supporting the theological tenets of any particular set of men. The most conscientious clergyman may preach a sermon full of sound morality, and no one will require of him to enter into theological subtilties. A philosopher could not scruple to recommend the forbearing disposition of Epictetus, and would as little hesitate to inculcate the Doctrine of Christ to love your neighbour as yourself and to act towards him as you would wish he should act towards you.

"Now cast a look upon the Law and Lawyers. Is it not evident that at least one half of them do act against their better knowledge and conscience? Whenever a lawsuit is decided the Lawyers on one side are always proved to be in the wrong, which must arise from ignorance, self-deception, or from a worse principle. It may be alleged, with regard to a person in the Law, that his situation is such, that in order to get his bread he must undertake the defence of a bad action, for a conscientious man may starve, where one of a different disposition will flourish, as by his eloquence he may save a traitor from the gallows. But who would envy his fame if he succeeds in a cause which all condemn?

"On the other hand a Clergyman may get his bread, and always act conscientiously and do good to every person with whom he has any connection.

"The most material circumstance is still to be considered. You say that independent of the mere drudgery of the routine of the Law it allows an unbounded scope for the exercise of the labour of the head. This I maintain to be a most egregious error. The drudgery of the Law, as you yourself call it, requires the Lawyer's whole time, or else he can get no bread. To the clerical life on the contrary you object that its natural bias lies towards habits of indolence and inactivity; which evidently admits that a clergyman is in possession of *dignified leisure*, and that he may even waste his time if he likes it. Is this an argument for one who wishes for the labours of the head? and would even take some of the pressing moments of the Law, by which he is to get his bread, to bestow them on such labours!

"A Clergyman on the contrary, without the least derangement of his ostensible means of livelihood, has time for the attainment of the more elegant branches of literature, for poetry, for music, for drawing, for natural history, for short pleasant excursions of travelling, for being acquainted with the *spirit* of the laws of his country, for history, for political economy, for mathematics, for astronomy, for metaphysics, and for being an author upon any one subject in which his most advantageous and respectable situation has qualified him to excel. He may also be a happy family man, a husband, a father, and with a paternal fortune added to his other means he will have all the real enjoyments of life within his reach.

"With my best wishes for your happiness in which all the family joins, I remain, my dear son,

Your affectionate father,

WM. HERSCHEL.

"P.S. I have never conversed with you on religious opinions, because I wished to leave you at liberty to follow your own sentiments; but when next we meet and communicate our thoughts to each other, I believe you will find that two unprejudiced persons, with natural good sense between them, cannot disagree on the subject."

John Herschel must have written a very penitent letter in reply to this one, for his father hastened to write again.

WILLIAM HERSCHEL *to* JOHN HERSCHEL

Slough near Windsor. Nov. 14, 1813

"My dear Son,

"Your letter has nearly had the same effect upon my disposition as mine seems to have had upon yours. If you are afraid of having said anything for which you would condemn yourself, I am equally afraid that I may have said what might appear to breathe a spirit of bitterness. I hope however that we both know such ideas are equally without foundation. I can as little doubt of your sincere attachment to your old philosophical father as he does of your perfect returning filial affection.

"It would be a vain attempt to enter into arguments about the subject under consideration, which can only be discussed in a few conversations, and must therefore be postponed till we can meet; for to say the truth, I can gather nothing from the contents of your letter but what amounts to, 'I do not like the Church; I prefer the Law', and for this very reason I cannot but wish to hear everything you have to say upon the subject. In hopes that you are perfectly satisfied of the affection of your father I shall only add that I am

very sincerely yours,

WM. HERSCHEL."

John Herschel spent the Christmas Vacation with his parents, and it was settled tha the should follow his own inclination; in January of 1814, he entered as a student at Lincoln's Inn. But the experiment was of short duration; the close confinement to stuffy chambers, combined with his attempts to pursue his favourite studies at the same time, so impaired his health, never very strong, that after struggling on for a year and a half he was obliged to submit to the doctor's verdict and to give up all idea of becoming a lawyer. He returned for a short while to

Cambridge, but the state of his father's health pointed out to him the path of duty. It was plain, also, that his aunt, in spite of her heroic efforts, was not fit any longer to assist her brother at night. Yet there was work to do, and William could not be persuaded to give up observing entirely. So it was on the visit to Dawlish in 1816, when he accompanied his father, that John Herschel made the final resolve to sacrifice all his other ambitions, and to settle down as his father's assistant at Slough. He wrote in very low spirits to his friend Babbage, that he was going to reside at Slough, in order, "under my Father's directions, to take up the series of his observations where he left them". He returned to Cambridge at the beginning of term only to give up his rooms, pack his things and take leave of his friends.

Caroline concluded her record of the year 1816 with the words:

"1816, *Oct.* Nothing particular happened; my Nephew remaining at home, working with his Father, and I took the opportunity of working on my MSS. Cat.[1] at those times I was left without employment".

The story of the remaining years of William Herschel's life is best told in Caroline's own words, but a brief outline will make these more intelligible. The shadow of impending loss hangs over everything that she wrote, yet the family life at Slough was often enlivened by the visits of John Herschel's friends, Peacock and Babbage. The family circle was broken in 1819, when Sophia Baldwin married, but her place, as companion to her aunt, was taken by her sister Mary, who, though she seems to have lacked Sophia's liveliness and charm, equalled her in affectionate devotion. Sophia's married life lasted little over a year; she developed consumption after the birth of her child and died a few months later. The baby was left to the care of Lady Herschel, but only survived its mother about a year.

Before these events, Caroline had had to face another separation which added much to the depression of her spirits. In November, 1818, Queen Charlotte died, and her household was consequently dispersed. Mme Beckedorff returned to Hanover with her daughter, and Caroline was thus left without the tried friend to whom she could speak in her own language and on whose sympathy she could always rely. She

---

[1] This was the "Zone Catalogue" which she completed at Hanover and for which she was awarded the Royal Society's gold medal.

returned from taking leave of the travellers a few days before Sophia's wedding; the happy bustle over the marriage preparations must have been in painful contrast to her own wretchedness.

### CAROLINE'S AUTOBIOGRAPHY

"1819, *March* 11, was Miss Baldwin's wedding day, which I spent at Slough with the Family.

"*April* 2. My Brother left Slough accompanied by Lady H. for Bath, he being so very unwell and the constant complaint of giddiness in his head so much increased that they were obliged to be 4 nights on the road both going and coming.

"The last moments before he stept in the carriage were spent in walking with me through his Library and workrooms, pointing with anxious looks to every shelf and drawer, desiring me to examine all, and to make memorandums of them as well as I could. He was hardly able to support himself and his spirits so low, that I found difficulty in commanding my voice so far as to give him the assurance he should find on his return that my time had not been misspent.

"When I was left alone I found on looking around that I had no easy task to perform, for there were packs of writings to be examined which had not been looked at for the last 40 years. But I did not lose a single day without working in the Library as long as I could read a letter without candle-light, and taking with me copying, &c., &c., which employed me for best part of the night, and thus I was enabled to give my Brother a clear account of what had been done at his return on the 1st of May. But he came home much worse than he went, and for several days hardly noticed my handyworks."

In spite of his increasing feebleness, William Herschel wrote twice to his sister from Bath. The handwriting of the first letter is somewhat shaky but very clear.

#### WILLIAM HERSCHEL *to* CAROLINE HERSCHEL

No. 36, Milsom Street, Bath

"My dear Sister,

"You know that I am not ready at writing, but as you are already used to read my scrawls you will perhaps make out a few lines, for many I cannot write, nor indeed have I much to say. I am just returned from a walk to New King Street, where I saw the House, No. 19, in the garden of which I discovered the Georgian Planet. For Bath news I must refer you to Lady Herschel, as I feel my wrist too feeble to guide the pen properly. Our journey to this place was extremely pleasant, and I am in great hopes our return to Slough will prove the same.

"I see many kind friends, but notwithstanding this I can assure you that I shall be glad to return to Slough again; for in my present state of health I do not enjoy company well enough to be fond of it. Pray write to me very soon and believe me very

affectionately yours,
WM. HERSCHEL."

The handwriting of the next letter is much more tremulous; the stay at Bath had evidently not been advantageous for his health. He wrote:

Tuesday, Apl. 20, 1819

"My dear Sister,

"I should like to give you some account of how my time is spent at Bath but this would require a longer letter than I find it easy to write. I shall leave this to be done till our return to Slough. We intend to leave this place about the beginning of next week and shall perhaps be 5 or 6 days upon the road, by which you may be able to calculate the time of our return. I shall be glad to hear how you do, and also to know whether there have been any visitors. Please to remember me to Miss Wilson, and excuse the shortness of this letter.

I remain, my dear Sister, affectionately yours,
WM. HERSCHEL".

This is the last of her brother's regular letters preserved by Caroline, but there is a note in his handwriting across which she has written:

"I keep this as a Relic! every line *now* traced by the hand of my dear Brother becomes a treasure to me.
C. HERSCHEL".

The note shows how keen the aged astronomer was to the last, when any celestial object presented itself for observation.

"Lina, there is a great comet. I want you to assist me. Come to dine and spend the day here. If you can come soon after one o'clock

**SIR WILLIAM HERSCHEL**
at the age of 81
*From an oil painting by Artaud
in the possession of Rev. Sir J. C. W. Herschel, Bt.*

we shall have time to prepare maps and telescopes. I saw its situation last night, it has a long tail."
July 4, 1819
(The date is in Caroline's hand.)

### CAROLINE'S JOURNAL (*continued*)

"1819, *May* 12. My Brother went with Lady H. to Town; he returned the 15th and from the 18th till the 26th when Lady H. came back I remained with my Brother.

"*June* 21. I went with my Brother to Town; he was to sit to Mr Artout.[1] We remained till Friday, whilst Lady H. entertained the Wilson Family at home, who were attending the Funeral of Miss Wilson at Upton.

"Mr and Mrs Watt dined one day with us at Mr Beckwith's. It was the last time the two old Friends did meet, for Mr Watt died soon after.

"*July* 8. We thought my Brother was dieing. The 9th he was persuaded to be blooded in the arm which somewhat relieved him.

"*Aug.* 10. My Brother and Lady H. took me with them to Town.

"*Aug.* 11. We went to the Bank and did what was thought necessary.

"*Aug.* 12. I went with Lady H. to see my Brother's portrait and ordered a Copy for myself.

"*Aug.* 25. Memorandum. The 13th we came home, and one day passes like the other. I have much to do and can do but little besides going daily to my Brother and oftens we are both unable to look about business; the present hot weather bears hard on enfeebled constitutions. Thermometer most days above 80 degrees.

"*Oct.* 15. I went to my Brother, his family being in Town.

"*Oct.* 29. I returned to my home."

On the 29th January, 1820, King George III died. There is no mention of this in any of Caroline's notebooks, though she does allude to the death of the Duke of Kent, which took place a few days previously. Sir William Watson wrote to John Herschel, his godson, in reply to a letter about the manner in which his father had received the news of the King's death.

[1] Artaud. This portrait, which was painted for Mr Greatheed, still hangs in the picture-gallery at Guy's Cliff. The replica, which Caroline ordered to be made for herself, is now in the possession of the Rev. Sir J. C. W. Herschel, Bt.

SIR WILLIAM WATSON *to* JOHN F. W. HERSCHEL

Feb. 15, 1820

"My dear Sir,

"Your letter was highly acceptable to me. I was beginning to be very anxious to know how my good old friend did. I am much gratified to find that he holds well, and that the loss of our late good King had not sit more heavily upon him. As a poor old man, deprived of his senses, his sight and hearing, his death must surely be deemed a relief from trouble; yet with all that consideration we all of us heard it with regret.

. . . . . .

"We were much pleased to find that there were hopes that we shall see my good friend, Lady Herschel and yourself at Bath some time in April. Now Lady W. joins with me in hoping that this time you will take up your abode with us during your stay at Bath.

. . . . . .

With our best wishes to you and your fireside

Yours very sincerely,

WM. WATSON."

Caroline's Day Book ends with an entry for October, 1821, and the melancholy words: "Here closed my daybook; for one day passed like another, except that I, from my dayly calls, returned to my solitary and cheerless home with increased anxiety for each following day".

After her retreat to Hanover she took up her pen again to record the last days of her brother's life; but before proceeding with her Autobiography some mention must be made of an event which she passes over in silence. This was the founding of the Astronomical Society.

The originator of the project of founding a Society in London entirely devoted to the prosecution of astronomy was Dr Pearson, a man of good position, who owned a private observatory at East Sheen. A committee was formed early in January, 1820, including, besides Dr Pearson, John Herschel, Babbage, Peacock and Francis Baily. At a meeting on the 29th February, the Duke of Somerset was elected President, Sir William Herschel and some others Vice-Presidents, Francis Baily and Babbage, Secretaries, and John Herschel, Foreign Secretary. All seemed going well, when the new Society received an unexpected rebuff. Sir Joseph

Banks, who had now been President of the Royal Society for forty years, saw in the formation of new societies for special objects a source of danger to the prestige and even to the existence of the Royal Society, and pressed this view so strongly on the Duke of Somerset that the latter felt obliged, out of regard to the feelings of his venerable friend, to decline the position of President and even to withdraw from membership of the Astronomical Society. The Society was placed in an awkward dilemma, as it was difficult to find a person of sufficient reputation to replace the Duke. It was decided to ask Sir William Herschel to accept the post, though it was well known that his age and infirmities would make it unlikely for him to attend the meetings. At first Sir William declined the honour and the Society remained for some months without a President. Later he consented, though on the condition that he should not be expected to take any active part in the proceedings, and in fact he was never present at any meeting of the Society, though he remained the nominal President till his death.

His election took place at the first anniversary meeting of the Astronomical Society in February, 1821. He wrote from Slough to his son:

"My dear Son, Slough. Feb. 14, 1821

"I am much obliged to you for the speech of thanks you made to the Members of the Astronomical Society, for electing me their President.

"The double star Orionis I find has four observations, in neither of which any particularity of itself has been noticed. Should the appearances you mention be deceptive, I used in similar cases to direct the telescope to the nearest star of equal brightness and colour, to see whether it was similarly affected. I also used to examine the affected star with gradually higher magnifying powers.

"I beg to be properly remembered to Mr South, and remain,

my dear Son,
your affectionate Father,
WM. HERSCHEL".

This letter, like all of William Herschel's after 1816, is in Caroline's handwriting, except the signature, which is very shaky. It shows the keen interest which he took in the observations which his son was conducting in collaboration with Sir James South, at the latter's observatory. He wrote again:

"My dear Son, Slough. March 17, 1821

"I have sent you Flamsteed's Atlas and his catalogue. In looking over your observations of Bootis I find a contradiction in the figure, to what you say in the column of remarks; where you give the position *nf*, whereas in the column of the angle it is given *np*.

"Inclosed I have sent you my observations of this double star, (see Fl. Atlas, plate 20) as they confirm what you say of the remarkable change that has taken place in the relative situation of these stars.

"With kind remembrances to Mr South, I remain, my dear Son,

Your affectionate Father,
WM. HERSCHEL".

The narrative of Sir William Herschel's last days was written by Caroline Herschel after she had left England for good, and sent to her nephew together with those "Extracts from a Day Book" from which the previous quotations have been made. It begins with the following introduction:

"Eighteen months have elapsed since I could acquire fortitude enough for noting down in my day book any of those heartrending occurrences I was witnessing during the last 9 months of the 50 years I had lived in England, and I cannot hope that ever a time will come, where I should be able to dwell on any one of those interesting but melancholy hours I spent with the dearest and best of Brothers. But if I was to leave off making memorandums of such events as either affect or are interesting to me, I should feel (like what I am, viz.) a person that has nothing more to do in this world. But to regain the thread of my narration, it is necessary to take notice of the vacancy between the present date and the ending of the year 1821; and the only way in which I can possibly fill up this vacancy must be to take a few dates with memorandums marked in my almanac and account books for the year 1822; without making any comments on what my feelings and situation must have been throughout that whole interval.

"By some letters I wrote during the first four months of 1822, to my Brother here at Han., I see that I was employed in copying from the *Phil. Trans.* the first 12 papers of my Brother's publications. The time required for this purpose, I could only obtain by making use of most of the hours which are generally allotted to rest, as during the day my time was spent in endeavours to support my dear Brother in his suffering decline. And besides the hope that we might continue yet a little longer together began to forsake me, for my own health and spirits were in that state that I was in dayly expectation of going before. Therefore each moment of separation from my dear Brother I spent in endevours to arrange my affairs so that my Nephew, J. Herschel (as the Executor of my Will), might have as little trouble as possible. And for this purpose my thoughts were continually divided between my Brother's Library, from which I was now on the point of being severed for ever, and among my unfinished work at home endevouring to bring by degrees all into its proper place.

"1822, *May* 13. Lady Herschel and my Nephew went to Town, I was left with my Brother alone, but was counting every hour before I should see them again, for I was momentarily afraid of his dieing in their absence.

"*May* 20. Lady H. came home very ill, attended by Miss Baldwin. Her complaint was a fit of the gout, from which after a forthnight's sufferings she was quite recovered.

"The Sommer proving very hot, my Brother's feeble nerves were very much affected, and there being in general much company added to the difficulty to choose the most airy rooms for his retirement.

"*July* 8. I had a dawn of hope that my Brother might regain once more a little strength; for I have a memorandum in my almanack of his walking with a firmer step than usual—above three or four times the distance from the dwelling house to his library—in his garden, for the purpose to gather and eat Rasberries with me; but I never saw the like again.

"The latter end of July I was seized by a bilious fever and I could for several days only rise for a few hours to go to my Brother about the time he was used to see me. But one day I was entirely confined to my bed, which alarmed Lady Herschel and the family on my Brother's account. Miss Baldwin called and found me in despair about my own confused affairs, which I never had had time to bring in any order. The next day she brought my Nephew to me; who promised to fulfil all my wishes which I could have expressed on paper and begged me to exert myself for his Father's sake, of whom he believed it would be the immediate death, if anything should happen to me.

"Having about this time received very distressing accounts of family misfortunes from my Brother at Hanover, I could find no rest on his account till I should have made my 500 pounds Stock over to him; but this required my presence at the Bank, and I was distracted at the thoughts of leaving Slough, till I found my Brother would be engaged for some days with his family before the departure of my Nephew who was going to accompany a Friend abroad. And besides knowing my absence would hardly be perceived as a very sensible elderly Lady (Mrs Moris) was there on a visit.

"$Aug^t$. 8. I went, and at 6 o'clock in the afternoon of the tenth I was home again. (My Nephew had left Slough.)

"I found my Brother seated by the Ladies, but so languid that I thought it necessary to take soon a seemingly unconscious leave for the night.

"$Aug^t$. 11, 12, 13, and 14. I went as usual to spend some hours of the forenoon with my Brother. The 15$^{th}$. I hastened to the spot where I was wonted to find him with the newspaper which I was to read to him. But instead I found Mrs Morsom, Miss Baldwin and Mr Bulman from Leeds, the grandson of my Brother's earliest acquaintances in this country. (Mrs Moris had left Slough for Kensington early that morning.)

"I was informed my Brother had been obliged to retire again to his room, where I flew immediately. Lady H. and the housekeeper were with him administering everything which could be thought of for supporting him. I found him much irritated at not being able to grant Mr Bulman's request of obtaining a token of remembrance for his Father; but as soon as he saw me I was sent to the Library to fetch one of his last papers and a Plate of the 40 feet telescope. But for the universe could I not have looked twice at what I had snatched from the shelf, and when he faintly asked if the breaking up of the milky way was in it, I said 'Yes!' and he looked content.

"(I cannot help remembering this circumstance, it being the last time I was sent to the Library on such an occasion. But that the anxious care for his papers and workrooms never ended but with his life was proved by his frequent wispering enquiries if they were locked and the key safe; of which I took care to assure him that they were, and the key in Lady Herschel's hands.)

"After half an hour's vain attempt of supporting himself, my Brother was obliged to consent of being put to bed, leaving no hope ever to see him rise again; and for 10 days and nights we remained in the most heart-rending situation till the 25 of August; when not one comfort was left to me but that of retiring to the chamber of death and there to ruminate without interruption on my isolated situation. And of this last solace I was robbed on the 7th September, when the dear Remains were consigned to the Grave".

John Herschel was not present at his father's death-bed. He had started a fortnight before, with his friend James Grahame, for a short tour in Holland and the Netherlands; one object of which was to visit the field of Waterloo. One letter from his mother reached him at Amsterdam in which she wrote: "Your Father, I thank God, is as well as when you left him". Freed from any pressing anxiety about his father's health, John Herschel and his friend wandered about as their fancy led them, thus missing any letters from Slough telling of his father's illness. It was therefore only on landing in England that he heard of the passing away of his loved and revered father. He posted with all haste to Slough, where he found his friend Babbage, with Mr Beckwith, doing all that friendship could suggest for the comfort and assistance of his mother. That he should not himself have been by her side at such a time must have added painfully to his own sense of bereavement. The kind and wise words in which his friend Jones referred, in a letter, to this matter should be sufficient exculpation, if any is needed, of want of filial attention on John Herschel's part. The letter may therefore fitly be given here.

### REV. M. JONES to J. F. W. HERSCHEL

Brasted. Sept. 3, 1822

"Dear Herschel,

"I got Babbage's letter this morning. I feel much for you, because I know how much you are feeling yourself, but I am not idle enough to suppose that anything I can say will either make you feel less, or see more distinctly than you do, the many grounds for consolation and pride which your father's glorious life happily presents in abundance.

"There is one thing I am a little afraid of, and that is, that you will be more hurt at having been absent than the circumstances warrant; but you must recollect, I think, that an unbroken sacrifice of your time and inclinations to your father would have had an appearance of confinement and restraint mortifying even to him on your account, while the few short absences in which you indulged yourself gave an air of freedom and choice to your usual assiduous attentions that must have very much heightened their value and grace to him,— quite enough I trust to console you for what by evil chance has occurred.

"Present my respectful condolences to Lady Herschel, remember me most kindly to Babbage and thank him for his letter, and do not let it be very long before I either see you or hear from you.

Adieu,<br>God bless you,<br>M. JONES."

From among the numerous letters of condolence received by the family two may be selected to show the attachment felt for Sir William Herschel by those most intimately acquainted with him. Planta had been Secretary of the Royal Society, and was at this time Librarian of the British Museum; Mr Bertie Greatheed has been mentioned before. In what estimation Sir William Herschel was held by contemporary men of science can be judged from the eloge pronounced upon him by Fourier, the Permanent Secretary of the French Académie des Sciences, some extracts from which will be found in Appendix I.

J. PLANTA *to* J. F. W. HERSCHEL

Brit. Museum. Sept. 6, 1822

"My dear Sir,

"It is with the greatest regret that I find myself under the necessity of denying myself the melancholy consolation of attending your late most excellent father to his last abode, a duty which, considering our long acquaintance, and the hearty kindness I always experienced at his hands, I should have been particularly desirous to have performed. Owing to a short absence from Town I received the summons too late for me to obey it. I will cherish the hope that should I be found worthy of it, I shall soon meet his beatified spirit in the everlasting habitations where our intercourse will meet with no interruption.

"In the meanwhile I shall, during the short period I have yet to live, never lose sight of the transcendent merits that distinguished his character, and have immortalized his name to the end of time.

I remain with sincere regard, my dear sir, &c.

J. PLANTA."

MR BERTIE GREATHEED *to* C. BABBAGE, ESQ.

Grey's Cliff. 5 Sept. 1822

"Sir,

"We had heard with much emotion of the death of our ever honour'd Friend, but it is a melancholy gratification to feel that his family are so well aware of our mutual attachment as to consider us worthy of a particular letter; for it is an honour to have been esteemed by such a man.

"What can we desire more for ourselves and those most dear to us, than such a life and such a death? He has enjoyed health, and competency and domestic happiness and the respect of all his fellow creatures. His days have been lengthened to see his own planet return almost to the spot from whence it shone upon his birth; and he is gone leaving an unblemished and illustrious name to all posterity.

"We beg you to assure the afflicted family of our most sincere condolence. When the first shock abates they will consider how seldom the inevitable doom is accompanied by so many sources of consolation.

I am, Sir,
your obed$^t$ Humble servant
BERTIE GREATHEED."

Sir William Herschel was buried very quietly in the little church at Upton where he had been married and where his family always worshipped. Only the immediate relatives and friends of the family attended the funeral, Dr Goodall, the Provost of Eton, being among the latter. The lengthy Latin epitaph, engraved on a marble slab, attached to the wall above the vault where lie the remains of William Herschel and his wife, was composed by Dr Goodall, the subject-matter being supplied by Sir John Herschel (see Appendix II). It contains the beautiful phrase "Coelorum perrupit claustra", which might well stand alone as sufficient epitome of William Herschel's life work. He had indeed broken through the barriers of the Heavens,

"And Lo! Creation widened to Man's view!"

## CHAPTER XXIV

### 1822    1848

Caroline retires to Hanover. Letters to Lady Herschel and to John Herschel. Her Zone Catalogue. Becomes an Honorary Member of the Astronomical Society and of the Royal Irish Academy. Her last years and death.

NO sooner had the grave closed over the earthly remains of her beloved brother, than Caroline began her preparations for leaving the country where for fifty years she had lived and worked with him. She had already resolved upon this step while she watched by her brother's death-bed, and it seems to have been accepted by the family as her wisest course. Her only surviving brother, Dietrich, was able and willing to receive her to live with him; his affection for the one whom they had lost was only less than hers, and though he would not feel as acutely as herself the daily blank in her life, she hoped to be able to rely upon his affection and sympathy. Though Caroline was sincerely attached to her sister-in-law, Lady Herschel, the two women had little in common beyond mutual affection and respect. The one tie which might still have held Caroline back was her love for her nephew and interest in his astronomical work. But John Herschel was by this time so engrossed by multifarious occupations which necessitated his living chiefly in London, that, had she remained at Slough, she would have seen little of him, while all her old friends at Windsor were either dead or, on the breaking up of Queen Charlotte's household, had left England for Hanover.

All these considerations made it very natural that Caroline Herschel should have decided to return to the home of her childhood, where she would hear her native language spoken and meet again her most valued friend, Mme Beckedorff, and her daughter. Caroline's health was at this time so enfeebled that she did not expect to live much longer; and at first she found at Hanover all the consolation that she had hoped for.

It is hard in these days of easy travel to realize how difficult of access from England Hanover was. Sir John Herschel visited his aunt three times during the twenty-six years that she spent there; on the last occasion he took with him his eldest son, Sir William Herschel's namesake; he was the only one of her nephew's children whom she ever saw, nor did she have the pleasure of meeting Sir John Herschel's wife, though the correspondence between aunt and niece was kept up with great regularity and growing affection on both sides.

Some extracts from Caroline's Journal and letters will give a sufficient idea of her life after her brother's death. She wrote:

"From the day of my dear Brother's death till Oct. 10th, I had no home! in England. I was divided between mine and Lady Herschel's house, awaiting the arrival of my brother D., and meanwhile disposing of all my property, by gift or sale".

Before this time she had made over to her brother Dietrich the £500 stock, which probably represented all the savings out of her salary, and she had now for her own support only the annual pension of £50 (which by the kindness of the King, George IV, was continued to her after the old King's death) and the legacy of £100 per annum from her brother William, which was paid to her in quarterly instalments by her nephew. This modest income seems to have been more than sufficient for her needs, accustomed as she was to living with the utmost frugality, and she pertinaciously objected to receiving the sums sent her by her nephew.

Extracts from a small book entitled "Memorandum", written by Caroline Herschel in 1823:

"1822, Oct. 3. My friends as well as myself were made easy by the arrival of my Brother Dietrich who came to fetch me.

"Oct. 7. I took leave of Princess Augusta and all my friends and connections in Windsor.

"Oct. 10. At 9 in the morning I left Slough with my Brother D.; Lady H. and my Nephew followed the next day.

"Oct. 14. The Princess Sophia Matilda sent her carriage for me, to spend the day with her at Blackheath.

"Oct. 16. I went with my Brother to Mortlake to take leave of Baron Best and Family, from whence we proceeded to Bedford place where all my Friends were assembled, among whom I had the comfort of seeing once more my Nephew's Friend, and favorit of my dear departed Brother, Mr Babbage. He had only that day arrived from the North. But I could find no opportunity for any conversation with him; but just by a pressure of the hand recommend my Nephew (in incoherent wispers) *again* to the continuance of his regards and Friendship.

"From all these sorrowing Friends and connections I was obliged to take an everlasting leave, and in the few hours we were for the last time together I was obliged to sign many papers, among which was a Rec$^d$ for a half year's Legacy I signed with great reluctance; for I knew my Nephew and Lady H. had great and many accounts to discharge, but they insisted on my taking it, according to my Brother's Will. This unexpected sum has enabled me to furnish myself with many conveniences on my arrival here of which otherwise I should perhaps have debarred myself.

"*Oct.* 17. In the morning we left our Lodging for an Inn near the Tower. Mr Beckwith joined us in Gilspur Street and settled at the Custum house what was necessary for our baggage. My Nephew came for a moment to us, and after his departure I saw no one I knew or who cared for me.

"*Oct.* 18. At 10 o'clock we went on board of a Steam packet.

"*Oct.* 20. At noon we landed after a stormy passage at Rotterdam.

"*Oct.* 21. At day break we began to proceed on our way, and

"*Oct.* 28. We arrived at the habitation of my Brother".

CAROLINE HERSCHEL *to* LADY HERSCHEL

Hannover. Oct. 30, 1822

"My dear Lady Herschel,

"We arrived here at noon on the 28th, without the least accident but not without the utmost exertion and extreme fatigue to both my brother and myself, from which it will be some time before I shall get the better, on account of the many visits of our Friends who came to convince themselfs of our safe arrival.... I found Mrs H. in personal appearance so different from what I had imagined that I could hardly believe her to be the same. She is just 60 years of age and suffers much from rewmatism, which has taken away partially the use of her hands, but is still of so cheerful a disposition and so active, by way of overcoming the disease by exercise, that I cannot wonder enough; and her reception of me was truely gratifying. The handsomest rooms, 3 or 4 times as large than what I have been used to, from which I can step in her own appartments, have been prepared for me and furnished in the most elegant stile. But I cannot say that

I feel well enough to enjoy all these good things nor be able to shew myself to those who wish to see me, at least not at present.

"Mrs Beckedorff sent to inquire after me when I had been hardly 2 hours arrived, Miss B. is confined with a severe cold. My Brother went yesterday to see them and we have postponed our meeting till Saturday when she will come to Town for the winter....

Believe me your truely and affectionate
CAR. HERSCHEL."

In her next letter Caroline gives many details about the family circle among whom she was now to spend her days. Dietrich Herschel's three daughters were all married. The eldest, Anna Knipping, had recently become a widow. She had nine children, the youngest still an infant in arms; fortunately she was comfortably provided for. Caroline wrote of her:

"She is a truly interesting little delicate creature just turned of forty, and has one daughter fit to be married, two sons preparing for the university, and the youngest weaned a month ago; she is to me a wonder when I look at her, she reads English fluently, French she was used to speak like her mother tongue from her infancy".

Mrs Knipping's affectionate devotion to her aunt compensated Caroline somewhat for the wrench of parting from those she loved in England. Of the other two nieces we do not hear often in the letters; Dorothea, the second, was married to a Dr Richter, and Caroline, the youngest, to Dr Groskopff, who was very assiduous in his attentions and helpful to Caroline. It must have been a new experience to her to be surrounded with so many small children. She wrote to Lady Herschel:

"The first time I went to them (Mrs and Miss Beckedorff) Mrs B. made all her ten grandchildren stand up before me according to their ages, and a fine healthy family it is. But all the little folks I am introduced to are disappointed at finding me to be only a *little* old woman; which I suppose must be owing to having been told the *Great Aunt* Caroline from England was coming".

When Caroline left England she took with her eight volumes of Sir William Herschel's "Book of Sweeps", and the "Catalogue of 2500 Nebulae", with the intention of preparing for the use of her nephew a fresh catalogue of nebulae arranged in zones for the convenience of an observer, beginning from the North Pole. During the first year of her stay at Hanover her health was too much shattered for her to undertake any regular work, but in August, 1823, she wrote to her nephew that she hoped to begin the catalogue "as my next winter's amusement".

In September of the next year John Herschel paid her a visit on his return from a tour in Italy, and she was then able to show him what progress she had made in the work. The catalogue was finished and sent to John Herschel, together with all the papers she had taken away, by the Easter messenger the following April. Her anxiety about these precious papers was greatly relieved when she received the following letter from her nephew:

JOHN HERSCHEL *to* CAROLINE HERSCHEL

April 18, 1825

"Dear Aunt,

"I received this afternoon your most valuable packet containing your labours of the last year which I shall prize, and more than prize —shall use myself and make useful to others.

"A week ago I had the 20-feet directed on the nebulae in Virgo and determined afresh the Right ascensions and Polar distances of 36 of them. These curious objects (having now nearly finished the double stars) I shall now take into my special charge—nobody else can see them".

Of Caroline's Zone Catalogue[1] Dr Dreyer writes in his Introduction to the *Collected Scientific Papers of Sir William Herschel*:

"This work has remained unpublished and can only have been seen by very few people. It was very properly rewarded by the Astronomical Society by the bestowal of their gold medal in 1828....

"The whole work is carried out with the same scrupulous care evinced in her work on the current sweeps. Of course the resulting positions are not quite as accurate as those afterwards derived by Auwers, since Caroline Herschel did not pay any attention to the difference of precession of nebula and comparison star, and also neglected the influence of nutation and aberration; but still, on the

---

[1] The full title was: "A Catalogue of the Nebulae which have been observed by William Herschel in a Series of Sweeps; brought into Zones of N.P. Distance and order of R.A., for the year 1800, by applying to the determining stars the variations given in Wollaston's or Bode's Catalogues".

whole, the results may be said to do justice to the original observations.... It certainly is to be regretted that Caroline Herschel's catalogue was not at once given to the public, since the publication of the separate results of her brother's observations would have cleared up many difficulties of identification, and prevented the continuance of many an error.... (Sir John) never published it; partly, no doubt, because he shared the universal opinion at the time, that very few of his father's nebulae could be seen, or at least usefully observed, with any but the largest telescopes; but chiefly because he always intended to bring out a General Catalogue of all known Nebulae and Clusters, a task which the vast amount of valuable work he carried out did not allow him to complete till 1864".

John Herschel did not lightly abandon the idea of publishing the Zone Catalogue. He wrote to his aunt in the autumn of the same year:

"I have already found your Catalogue of Nebulae in Zones very useful in my 20-feet sweeps and I mean to get it in order for publication by degrees, but it will take a long time, as it will require a great deal of calculation to render it available as a work of reference".

In January, 1825, Caroline wrote to Sir John that she had been lent a copy of the *Moniteur*, containing the full report of Fourier's eloge of Sir William Herschel. As she could not read French, Miss Beckedorff had kindly translated it for her.

"We could not", she wrote, "withhold a tear of gratitude to the Author for having so feelingly adhered to the truth in the details of your dear Father's discoveries, &c. But I could point out 3 instances where too great a stress is laid on the assistance of others, which withdraws the attention too much from the difficulties your Father had to surmount....

"Of Alex and me can only be said that we were but tools and did as well as we could; but your Father was obliged to turn us first into those tools with which we could work for him; but if too much is said in one place, let it pass; I have perhaps deserved it in another by perseverance and exertions beyond female strength! well done!"

When Caroline joined her brother's family at Hanover, they occupied a flat, and an additional room seems to have been taken for her, which was sufficiently comfortable. But unhappily she did not find Dietrich's society so congenial as she had hoped. Dietrich himself was suffering much from some complaint which rendered him nervous and irritable. He died less than three years after her arrival in Hanover, and she then removed to a separate lodging, where she could be more independent. Though she made an exception for the niece, Anna Knipping, and her children, to whom she was sincerely attached, she found the rest of the family unsympathetic and pined ever more and more for the affection of those she had left behind in England. To Lady Herschel she wrote contrasting her kindness with the treatment she received from her brother Dietrich's widow:

"The following hint is only to you as a dear sister, for as such I now know you. All I am possessed of is looked upon as their own when I am gone. The disposal of my brother's picture[1] is even denied me; it hangs in Mrs H.'s drawing-room, where a set of old women play cards under it on her club day".

In another letter to Lady Herschel she wrote:

"Out of my family connections I can boast to possess the esteem and love of all who are great and good in Hanover, but to a lonely old woman, who is seldom able to go into or to receive company, this does not compensate for the want of sympathizing relations. (She signs herself to Lady Herschel):

Your most affectionate sister,

C. HERSCHEL".

Caroline had still the society of her devoted friend Mme Beckedorff and of her daughter; and the high esteem of all the eminent men of science who visited Hanover and were eager to make her acquaintance. The members of the Royal Family who had known her at Windsor paid her many kind attentions. The regular concerts at the Theatre were a source of enjoyment which she thoroughly appreciated. For fifteen years she subscribed for a place and her well-known little figure was greeted with respect and affection whenever she occupied it. In a letter to John Herschel, dated 5th May, 1825, she wrote:

"Last Saturday, between the acts of the concert, the Duke of Cambridge asked me many questions about you. He enquired if you were much engaged with astronomy? I said you were a deep mathematician, which embraced all, &c., then he asked if you studied

[1] See p. 355.

chemistry? answer, very much! you had built yourself a laboratorium at Slough, had a house in town for three years, was secretary of the Royal Society, would probably in the vacation be at Slough, &c., &c., and in return he told me that he heard from everybody you were a very learned philosopher; and if I tell you that the Duke of Cambridge is the favorite of all who know him, I think I have made you acquainted with one another".

When George IV died and Hanover was separated from the British Crown, the Duke of Cumberland becoming King of Hanover, Caroline lost a kind friend, for the Duke of Cambridge then left Hanover to reside in England. But the new King and especially the Crown Prince and Princess were very friendly and gracious.

In her eightieth year Caroline had her portrait painted to comply with a request from her nephew; it was executed by a German artist, Prof. Tieleman. About this picture she wrote[1]:

"Whatever you may think about my looking so young I cannot help; for two of the days I was sitting to him I received agreeable news from England. One day Lady H.'s likeness was thrown in my lap (Mr Tieleman taking it out of the Box), and four days after the account of your approaching happiness arrived. No wonder I became a dozen years younger all at once!"

The news which had so transported her with joy was that of her nephew's intended marriage. She wrote on the day fixed for the wedding:

"I can find no words which would better express my happiness than those which (after having read the letters) escaped in exclamation from my lips according to Simon; see St Luke, Cap. 2, verse 29: 'Lord, now lettest thou thy servant depart in peace'.

"...I have now some hopes of passing the few remainder of my days in as much comfort as the separation from the land where I spent the greatest portion of my life, and from all those which are most dear to me can admit. For, from the description Miss B. has given me of the dear young Lady of your choice I am confident my dear Nephew's future happiness is now established.... It is now between 11 and 12 and perhaps you are at this very moment receiving the Blessing of

[1] This portrait of Caroline Herschel is now in the possession of Rev. Sir John C. W. Herschel, Bt.

Dr Jennings, in which I most fervently join by saying: God bless you Both!

Your happy and affectionate Aunt<br>CAR. HERSCHEL".

After his mother's death in 1832, Sir John Herschel[1] was able to carry out his project of removing his telescope to the Cape of Good Hope, there to complete his father's work by the survey of the stars and nebulae in the southern hemisphere. Before starting he went to Hanover to take leave of his aunt, and wrote as follows to his wife describing how he found her:

Hanover. June 19, 1832

"...I found my aunt wonderfully well and very nicely and comfortably lodged; and we have since been on the full trot. She runs about the town with me and skips up her two flights of stairs as light and fresh at least as *some folks* I could name who are not a fourth part of her age.... In the morning till eleven or twelve she is dull and weary, but as the day advances she gains life, and is quite 'fresh and funny' at ten or eleven p.m. and sings old rhymes, nay even dances to the great delight of all who see her".

Sir John Herschel left England for the Cape in November, 1833. Caroline was kept fully informed, by long letters from him and his wife, of all their new experiences, and of the progress of his astronomical work. She was even a little jealous of discoveries and inventions which might seem to dim the lustre of her brother's fame. But the accounts which her nephew was able to give her of his confirmation of his father's observations filled her with delight. When she first heard of his intention to take his telescope to the Cape her enthusiasm found expression in her exclamation: "Ja! if I was but thirty or forty years younger, and could go too! In Gottes Namen!"

Before he started she wrote to him:

"Dear Nephew,

"As soon as your instrument is erected I wish you would see if there is not something remarkable in the lower part of the Scorpion to be found, for I remember your Father returned several nights and years to the same spot, but could not satisfy himself about the uncommon appearance of that part of the heavens".

[1] John F. W. Herschel was made a Knight of the Guelphic Order in 1831.

Soon after his arrival at the Cape Sir John Herschel wrote in reply:

"I have not been unmindful of your hint about Scorpio. I am now *rummaging* the recesses of that constellation and find it full of beautiful globular clusters".

Caroline was not satisfied:

"It is not *Clusters of Stars*", she wrote, "I want you to discover in the body of the Scorpion (or thereabouts) for that does not answer my expectation, remembering having once heard your father, after a long awful silence, exclaim, 'Hier ist wahrhaftig ein Loch im Himmel!'"

During her nephew's absence at the Cape, Caroline had a gratifying proof that her services to science were not forgotten. In March, 1835, she received a letter from the Secretary of the Royal Astronomical Society, informing her that the Society had conferred the distinction of Honorary Membership on herself and Mrs Somerville. In her reply she said that she could "only regret that at the feeble age of 85 I have no hope of making myself deserving of the honour of seeing my name joined with that of the much distinguished Mrs Somerville".

Three years later she received the Diploma of Membership of the Royal Irish Academy, sent her by the president, Sir William Rowan Hamilton, Astronomer Royal at Dublin, and her nephew's warm friend. Though with characteristic modesty she deprecated the bestowal of this honour "on one who has lived eighteen years without finding as much as a single comet", she added, in writing to Sir John, "more I cannot say on this subject, only just so much; it has given me pleasure".

By the time Sir John Herschel had been two years at the Cape of Good Hope he was able to write to his aunt:

"I have now very nearly gone over the whole southern heavens, and over much of it often; so that after another season of reviewing, verifying and making up accounts (reducing and bringing in order the observations) we shall be looking homewards".

The expedition to the Cape had aroused a good deal of interest both in England and abroad; Sir John Herschel received an ovation from his scientific friends on his return, and was created a baronet at the Queen's Coronation. The Duke of Cambridge had been the first to announce to Caroline the happy news of the safe landing of her nephew and his family, which occurred on the 11th of May, 1838. As soon as his public engagements permitted, Sir John crossed over to Hanover, taking with him his eldest son, William James Herschel, a child of five years.

They could only stay for five days, and Caroline experienced more pain than pleasure from the agitation of the meeting and pain of parting.

Little of importance remains to be told about the closing years of Caroline Herschel's life. She lost her most valued friend, Mme Beckedorff, some years before her own death, but was consoled by the attentions of the latter's daughter, and of her own niece, Anna Knipping. She had no lack of friends, indeed she sometimes complained in her letters of the fatigue of conversing with so many visitors. When the exertion of guiding a pen became too much for her, Miss Beckedorff acted as her secretary, and from her letters we are able to give an account of these last years.

In 1846, she received from the King of Prussia the large gold medal for Science, sent her through Baron Alexander von Humboldt: "In recognition of the valuable services rendered to Astronomy by you as the fellow-worker of your immortal brother, Sir William Herschel".

Lady Herschel wrote to congratulate her as follows:

Collingwood. Oct. 7th, 1846

"My dearest Aunt,

"We congratulate you with all our hearts on the receipt of the gold medal from the King of Prussia. I had heard of his gracious intentions towards you and therefore longed to know of your having received this mark of his Majesty's regard and esteem. You are more likely, my dear Aunt, to grow weary of the world than the world is to grow weary of you, and this testimony to your value may prove to you how little chance you have of being forgotten. We have drunk the King of Prussia's health in a bumper! Now in return for all that has been done in your day, here is a specimen of what the next generation are doing.

"*A new planet is discovered beyond Uranus!* I might almost say in consequence of the discovery of Uranus; for its perturbations were not sufficiently accounted for by the gravitation of the hitherto known planets, and when all the calculations were gone into by a Frenchman named Leverrier and a young Cambridge man, Mr Adams, they asserted that another planet must exist.... This is one of the greatest triumphs which have ever been achieved by Theoretical

# CAROLINE LUCRETIA HERSCHEL
at the age of 97

*From an engraving made at Hanover*

Science, for without looking at the heavens at all, two men in their studies direct stargazers to point their telescopes at a particular portion of the sky and there they will find the new Planet.—I wished Herschel himself to announce to you this glorious discovery but he is just in the swing of *theorizing* about Halley's comet, which we saw at the Cape, and he deputed me to be the bearer of the good news.

"All join with me in the fondest love to you, our beloved Aunt, believe me always your very affectionate niece,

M. B. HERSCHEL".

It was a great satisfaction to Sir John Herschel that Caroline was able to see in her lifetime the completed volumes of his "Cape Observations", which brought to a conclusion Sir William Herschel's vast undertaking, the "Survey of the Nebulous Heavens". To the letter which accompanied this gift, Sir John added this interesting postscript: "Louisa sends you all our news and the autographs of Struve and Adams, who with M. Le Verrier are now at Collingwood".

It was at this visit, purposely arranged by Sir John Herschel to allay the painful dispute between French and English astronomers for the honour of the discovery of Neptune, that Le Verrier and Adams met for the first time and shook hands.

Two likenesses were taken of Caroline Herschel in her old age; of one of these (reproduced here) Miss Beckedorff wrote:

"I am sorry to say the drawing which I saw did not do justice to her intelligent countenance; the features are too strong, not feminine enough and the expression too fierce; but I hear the picture which I did not see is more like her".

A month later Miss Beckedorff wrote, sending what was probably a print of the second drawing[1]:

"I am commissioned by dear Miss Herschel to send to you and for her dear nephew with her best love, the accompanying print, which I fear will at first sight not satisfy you. The artist has, I believe, imitated the style of the old German school of Albert Durer, resembling more a 'woodcut' than a print, nor does it do justice to her fine old

[1] This engraving is now in the possession of Rev. Sir John C. W. Herschel, Bt, at Slough.

countenance. Yet it is extremely like in feature, expression, and deportment, her eyes have taken the languid expression more from fatigue occasioned by her *sitting* for the picture whilst she is used generally to recline on her sofa, and I see them very frequently sparkle with all their former animation".

This drawing was made when Caroline had just kept her ninety-seventh birthday; she did not live to celebrate another. On this occasion she received a large velvet armchair from the Crown Prince and Princess of Hanover, who came themselves to present their gift. Miss Beckedorff states that "they stayed nearly two hours, Miss Herschel conversing with them without relaxation, and even singing to them a composition of Sir William's, 'Suppose we sing a Catch'".

On the 10th of January, 1848, Miss Beckedorff wrote to Sir John Herschel:

"Your excellent aunt, my kind revered friend, breathed her last at eleven o'clock last night, the 9th January.—She suffered little and went to sleep at last with scarcely a struggle. Up to the last moment she has had the most indeniable proofs of the affection and veneration of her own family and a number of friends, English and German. Mr Wilkinson, the English clergyman, has been unremitting in his visits, and so kind and judicious was his manner, that she received them to the last with unfeigned satisfaction".

The funeral service over Caroline Herschel's mortal remains was conducted by the pastor of that same Garrison Church in which, nearly a century before, she had been christened and afterwards confirmed. She was buried near her father, in the Gartenkirchhof, on the Marienstrasse, at Hanover. The grave is well preserved; it is covered by a heavy slab on which is engraved a long epitaph,[1] composed by Caroline Herschel herself.

[1] Given in Appendix III.

# APPENDIX I

## FOURIER'S ELOGE OF WILLIAM HERSCHEL

At the anniversary meeting of the "Institut Royal de France" on the 7th June, 1824, Baron Fourier, the Permanent Secretary of the Institute, pronounced the usual tribute to the memory of a departed Member, William Herschel. The great length of the oration indicates the high estimation in which the subject of it was held by the Members of the Institute. A few extracts have been made, translated from the French, chosen particularly to show what aspects of his career and which points in his contributions to science attracted the greatest interest at the time.

Beginning with an account of Herschel's early life, of his career as a musician and of his struggles to supply himself with an instrument for the pursuit of astronomy, Fourier next described the discovery of Uranus and the sensation which this caused among astronomers throughout Europe.

"Great interest was excited in England and throughout Europe about the musician at Bath: his discoveries in astronomy, the perfection of his instruments, all made by himself, the singular circumstances of his life, and the noble use which he made of his moments of leisure. All these details came to the ear of the King, George III, who called Herschel to him, granted all his requests and desired him to take up his residence in the near neighbourhood of Windsor.

"This retreat at Slough became one of the most remarkable spots in the civilized world. Herschel resided there with his family and was visited by many illustrious travellers. It was there that he completed his long and memorable career.

"Herschel was assisted by one of his brothers, who was skilled both in theoretical and practical mechanics. Their sister, Miss Caroline, very soon acquired advanced knowledge in astronomy and mathematics; and prompted by her constant affection and a lively desire to contribute to the glory of her brother, and no doubt by a disposition of mind natural in this remarkable family, she made astonishing progress in these studies. She revised and published the observations; she herself discovered several comets. She shared all the vigils and assisted in all the literary compositions of her brother; certainly no astronomer has ever had an assistant more intelligent, more faithful or more attentive."

Referring to the object which Herschel had in view in constructing his large telescopes, M. Fourier said:

"He recognized that by exercising the eye by degrees it becomes much more susceptible to faint illumination, and in consequence he was able to make use of magnifying powers much beyond the limits of other observers. He also distinguished two different qualities in telescopes which had hitherto not been discerned: that which consists in increasing the apparent dimensions of objects, and that of penetrating further into the profundity of space, in order to reveal objects hitherto entirely imperceptible. Numerous examples leave no doubt of the truth and usefulness of this distinction. Herschel, then, undertook to carry to the furthest possible limit the range of his instruments, and studying less such considerations as would facilitate their use than those which would increase their optical power, he constructed a telescope of an extraordinary size. It is the largest instrument of the sort that has ever been made.

. . . . . .

"Herschel's observations are too numerous and of too varied a character to be described on this occasion. They have been repeated and have mostly received complete verification. Further, it should be noted that the instruments he used, which have so many remarkable advantages, are subject to difficulties which limit their use. His very large telescopes must not be considered as instruments of precision and measurement but rather of discovery; in this respect they afford all that human invention has yet supplied in the utmost perfection".

Glancing rapidly at other aspects of Herschel's work, M. Fourier laid particular stress upon Herschel's discovery of the invisible rays in the solar spectrum, and their heat-producing power.

"These singular discoveries", he said, "excited attention in all the Academies of science and roused much opposition. The inventor was exposed to contradictions which passed the limits permitted to literary criticism. But this great physicist, having published all the explanations necessary, held his peace; his experiments were repeated in England, in Germany, and in France, under the eyes of the most skilful observers in Europe, and the truth of his conclusions was generally recognized."

M. Fourier then passed on to describe in some detail Herschel's astronomical work: the catalogues of nebulae, the structure of the Milky Way, and his discovery of the time of rotation of the ring of Saturn. In connection with the latter he pointed out that Laplace had arrived at the same result as a consequence of the general principle of gravitation.

"The history of science", he said, "offers nothing more worthy of the attention of philosophers than this admirable accord between theoretic deduction and the application of human invention."

He then alluded briefly to Herschel's theories on cosmogony and to Laplace's agreement with them. In his peroration M. Fourier alluded to William Herschel's philosophical convictions.

"The most highly endowed mind fails to grasp the immensities of the Universe and can only find rest in the thought of an Order even more sublime. We are reminded of the feeling which Sir William Herschel has often expressed and which the contemplation of the marvels of the heavens incessantly recalled to him. In every one of the grand phenomena which he has observed he recognized the imprint of a Wisdom, eternal and creative, which animates and conserves and has given immutable laws for the whole of Nature.

"The influence of great men is prolonged into the future and the full fruition of their labours cannot be fully appreciated at their death. The picture of the heavens traced by William Herschel will be compared, by future generations, with more recent observations; some portions of the spectacle of the heavens may have changed, but in those distant ages the memory of Herschel will still endure. His name, confided to a grateful science, will for ever be safe from oblivion."

# APPENDIX II

## EPITAPH ON SIR WILLIAM HERSCHEL
(on a tablet in Upton Church, Slough, Bucks)

(Translated)

William Herschel, Knight of the Guelphic Order.
Born at Hanover he chose England for his country.
Amongst the most distinguished astronomers of his age
      He was deservedly reckoned.
For should his lesser discoveries be passed over
He was the first to discover a planet outside the
      orbit of Saturn.
Aided by new contrivances which he had himself both
Invented and constructed he broke through the
      barriers of the heavens
And piercing and searching out the remoter depths of space
He laid open to the eyes and intelligence of astronomers
      The vast gyrations of double stars.
      To the skill
With which he separated the rays of the sun by prismatic
      analysis into heat and light
      And to the industry
With which he investigated the nature and the position of nebulae and of the luminous apparitions beyond the limit of our system, ever with innate modesty tempering his bolder conjectures, his contemporaries bear willing witness.

Many things which he taught may yet be acknowledged by posterity to be true, should astronomy be indebted for support to men of genius in future ages.

A useful blameless and amiable life distinguished not less for the successful issue of his labours than for virtue and true goodness was closed by a death lamented alike by his kindred and by all good men in the fulness of years on the 25th day of August, the year of our Salvation 1822, and the 84th of his own age.

## APPENDIX II

The epitaph in its original form contained no mention of William Herschel's discovery of binary stars; but after Struve had conclusively proved the rotation of $\eta$ Coronae, John Herschel considered his father's claim to the discovery of rotation in these bodies sufficiently confirmed and important to be mentioned. He wrote to Caroline Herschel at Hanover:

Slough. April 17, 1832

"My dear Aunt,

"Struve writes me word that his observations entirely corroborate the rapid rotation of $\eta$ Coronae which I first pointed out. This was a favourite star of my Father's for trials of his telescope, but he little suspected that it would turn out the finest specimen among all the double stars of a revolving binary system; the period in which the two stars revolve about each other being only 42 years! This branch of his Astronomical discoveries is now so completely established that I have thought it right to alter the inscription on his monument".

## APPENDIX III

### EPITAPH OF MISS CAROLINE HERSCHEL
(written by herself in German; translated)

Here rests the earthly veil of
Carolina Herschel
Born at Hanover, March 16, 1750
Died January 9, 1848

The gaze of her who has passed to glory was, while below, turned to the starry Heaven; her own discoveries of Comets and her share in the immortal labours of her Brother, William Herschel, bear witness of this to later ages. The Royal Irish Academy of Dublin and the Royal Astronomical Society in London numbered her among their members. At the age of 97 years 10 months she fell asleep in happy peace, and in full possession of her faculties; following to a better life her father, Isaac Herschel, who lived to the age of 60 years 2 months 17 days and lies buried near this spot since the 25th March, 1767.

## APPENDIX IV

Papers communicated to the Royal Society by William Herschel, which are referred to in the foregoing chapters.

| | |
|---|---:|
| Observations relating to the Mountains of the Moon (1780) | page 76 |
| Account of a Comet (discovery of Uranus) (1781) | 81 |
| Catalogue of Double Stars (1782) | 94 |
| On the Parallax of the Fixed Stars (1782) | 94, 97 |
| Account of the Discovery of Two Satellites of the Georgian Planet (1787) | 162 |
| Account of the Discovery of a sixth and seventh Satellite of Saturn (1790) | 164 |
| On the proper Motion of the Sun and Solar System (1783) | 183 |
| On the Construction of the Heavens (1784) | 190 |
| Catalogue of Double Stars (1784) | 193 |
| On the Construction of the Heavens (1785) | 193–205 |
| Catalogue of One Thousand new Nebulae and Clusters (1786) | 222 |
| Catalogue of a Second Thousand of new Nebulae and Clusters (1789) | 224 |
| On Nebulous Stars, properly so called (1791) | 226 |
| Catalogue of 500 New Nebulae, nebulous stars, Planetary Nebulae and Clusters of Stars (1802) | 229 |
| On the Nature of the Sun and Fixed Stars (1795) | 261 |
| Investigation of the powers of the prismatic colours to heat and illuminate objects (1800) | 262 |
| Experiments on the refrangibility of the invisible Rays of the Sun (1800) | 264 |
| Additional Observations on the variable Emission of the Light and Heat of the Sun (1801) | 265 |
| Observations of the two lately discovered Celestial Bodies (Ceres and Pallas) (1802) | 271 |
| Experiments for investigating the Cause of coloured Concentric Rings (1807) | 281, 284 |
| Account of the changes that have happened during the last twenty-five years in the relative situation of Double Stars (1803) | 286 |
| Astronomical Observations relating to the Construction of the Heavens (1811) | 286 |
| Astronomical Observations relating to the Sidereal part of the Heavens and its connection with the Nebulous part (1814) | 286 |
| Astronomical Observations to investigate the Local Arrangement of the Celestial Bodies in Space and the Milky Way (1817) | 287 |

# INDEX

Algol, periodicity of, discovered by Goodricke, 184
Arago, F., Biographical Notice of W. Herschel, the house at Slough, 145
Asteroids, Ceres and Pallas discovered, 268; measuring diameters, 269; term suggested by Herschel, 271; disapproved by Laplace and Piazzi, 274; defended by Herschel, 275; Juno, 275; Vesta, 276
Astronomical Society, 356
Aubert, A., introduced by Hornsby, 75; Letters, 103, 120, 155, 179, 180, 187

Babbage, C., Secretary of Astronomical Society, 356; at Slough, 361
Baldwin, Mrs, mother of Lady Herschel, 172
— Thomas, brother, 176
— Mary, niece, character of W. Herschel, 179
— Sophia, niece, 294; her marriage and death, 352
Banks, Sir Joseph, President of Royal Society, 183; Letters, 95, 108, 112, 121, 260, 266, 267, 269, 284
Bath, in W. Herschel's time, 42
Beckedorff, Mme, first mention, 303; with Mme D'Arblay, 340; at Hanover, 367
— Miss, Letters, 375
Blagden, Sir C., Letter to Caroline, 154
Bode, Letters, name Uranus, 124, 207; Ceres and Pallas, 271
Brougham, article in the *Edinburgh Review*, 282
Bulman, Mr and Mrs, 29; at Bath, 40; return to Leeds, 68, 72
— grandson visits W. Herschel at Slough, 360
Burney, Dr, describes Herschel's telescope at Walmer Castle, 143; visits Slough, 291, 295
— Fanny, extracts from *Diary*, 169; description of Caroline Herschel, 170; W. Herschel's marriage, 177

Cambridge, Duke of, meeting with Caroline at Hanover, 370
Campbell, T., meeting with Herschel, 335
Cavendish, Henry, seeing stars round, 102; orbit of Uranus, 107; friendship with W. Herschel, 184
Ceres, *see* Asteroids
Colebrook, Mrs, takes Caroline to London, 43, 53, 55
Cook, Sir Brian, patron of W. Herschel, 35, 39
Copley, Mr, concerts at Doncaster, 14

De Chair, Dr, engages W. Herschel as organist, 38; W. Herschel visits at Nancy, 48, at Risington, 55
— Mrs, remark about Caroline's manners, 33
De Luc, M., 243
Dettingen, battle of, 6

*Edinburgh Review*, criticism of Herschel's paper on Heat, 282
Edwards, Rev. J., composition for mirrors, 92

Ferguson, J., 60, 61
Fourier, eloge of W. Herschel, 369, Appendix I
Front-view, 63

Gerstner, Letter, 94
Globules, making small, 276
Goltermann, Mrs, Caroline visits in London, 339
Goodricke, observed periodicity of Algol, 184; Letter, 195
Greatheed, B., Letters, 241, 363
Griesbach, George, musician in the Queen's band, 115, 117
— Sophia, née Herschel, date of birth, 3; marriage, 6, 7; her death, 315

Harding, assistant to Schroeter at Lilienthal, 216; discovered Juno, 275
Harvests, good or bad, indications of the sun's heat, 279
Herschel, Abraham, grandfather of William Herschel, 1, 2
— Alexander, brother, date of birth, 3; education, 32; joined William at Bath, 45; praise of his solos on the violoncello, 67; assisted his brother in making

## INDEX

Herschel, Alexander (*cont.*)
   telescopes, 142; letter from William on the death of his wife, 178; accident at Bath, 343; retires to Hanover, 346; his death, 347
   Caroline, date of birth, 3; early recollections, 8, 10; her confirmation, 33; life after her father's death, 45–47; accompanies William to England, life at Bath, 49–52; visit to London, 53–55; dancing lessons, 56; sings in public, 56; her capabilities, 57; move to Datchet, 133; accident, 137; her telescopes, 150; discovers her first comet, 153; life at Slough, 170–173; salary of £50 from the King, 172; her comets, 245–252; her telescopes, 253; eighth comet, 254; Catalogue of omitted stars, 255; life after her brother's marriage, 295; visits Dr Maskelyne at Greenwich, 296; moves from Upton to Slough, 330; visit to London, 340; account of her brother's last days, 358–361; returns to Hanover, 365; Letter to Lady Herschel, 366; Catalogue of Nebulae in Zones, 368; Hon. Memb. of R.A.S., 373; receives Diploma of Membership of Royal Irish Academy, 373; gold medal from the King of Prussia, 374; her death, 376; epitaph, Appendix III
— Dietrich, date of birth, 3; his character, 32; runs away from Hanover, 69; his son, 316, 319; Letters describing the troubles from the French invasion, 317; takes refuge in England, 323; returns to Hanover, 337; comes to England to fetch Caroline, 365; his daughters, 367; illness and death, 370
— Hans, of Pirna, 1
— Isaac, his autobiography, 2, 3; love of astronomy, 5; present at the battle of Dettingen, 6, of Hastenbeck, 9; his discharge, 11; his exertions for his family, 33; his death, 34; Letter to William, 34
— Jacob, date of birth, 3; Caroline's recollections, 4; joins the regimental band, 7; discharged, 8; escapes to England with William, 9; returns to Hanover, 11; member of Court Orchestra, 11; Caroline's estimate of his character, 12; his musical talent, 12
— John, date of birth, 233; at Eton and Dr Gretton's school, 298; his delicate health, 322; goes to Cambridge, 324; accompanies his parents to Glasgow, 332; Senior Wrangler, 338; F.R.S., 339; choice of a profession, Letters from his father, 348; studies law, 351; settles at Slough to continue his father's work, 352; absent abroad at the time of his father's death, 361; visits Caroline at Hanover, 368; his marriage, 371; goes to the Cape, 372; created a baronet on his return, 373; visits Caroline with his son, 373
— Lady, *see* Pitt
— Sophia, *see* Griesbach
— William, date of birth, 3; his education, 3–5; his French teacher, 7; joins the regimental band, 7; first visit to England, 7; Campaign of 1755, 9, 10; escapes to England, 10; life in London, 13; Yorkshire, Lord Darlington's militia band, 13; Miller's story, 14; Letters to Jacob, 16–28; settles at Leeds, 29, 30; receives his official discharge, 31; visits Hanover, 31; competition for post of organist at Halifax, Miller's story, 37; chosen organist for the Octagon Chapel at Bath, 38; life at Bath, 39, 40, 43; fetches Caroline from Hanover, 48; turns to astronomy, 50; Letter to Dr Hutton, 59; making telescopes, 65–70; goes to Hanover in search of Dietrich, 69–70; meeting with Dr Watson, 73; prize question in *The Ladies' Diary*, 74; joins Bath Philosophical Society, 74; first meeting with Drs Maskelyne and Hornsby, 75; paper sent to Royal Society, *On the Mountains in the Moon*, 76; discovery of Uranus, 78; paper *Account of a Comet* sent to Royal Society, 81; first *Catalogue of Double Stars*, 94; *On the Parallax of Fixed Stars*, 94, 97; observations of Uranus, 107; goes to London, 113; Letters, 114–117; shows his telescope to the King, 118; name for the new planet, 124; moves to Datchet, 133; ague, 139; moves to Clay Hall, 143, to Slough, 144; grant of £2000 from the King, 144; goes to Göttingen, 147; Letters to Caroline, 147, 148; making the 40-foot telescope, 156–160; King's second grant of £2000, 157; discovery of two satellites of Uranus, 161, 162, of two new satellites of Saturn, 164, 166; bust taken by Lochee, 172; marriage, 173, 175; portrait by Abbott, 181; paper *On the Motion of the Sun*, 186; method of

# INDEX

Herschel, William (*cont.*)
"gaging", 190; first paper *On the Construction of the Heavens*, 191; second paper *On Construction etc.*, 193; his theories considered irreligious, 197; mathematical attainments, 198–199; Letters to Bode, 203, to Schroeter, 210, 212, 214, to Lalande, 201, 205, 239; method of taking "sweeps", 222; discovery of nebulous matter, 226, 228; planetary nebulae, 229; nebular hypothesis, 225–232; tour to Glasgow, 233–237; observations on Venus, Jupiter, Saturn, 259; paper *On the Nature of the Sun*, 259, 262; discovery of heat-rays, 263; term "radiant heat", 202; diameters of Ceres and Pallas, 269; term "asteroids" suggested, 271; paper *On Concentric Rings*, 284; List of W. Herschel's important discoveries, 289; trip to Wales, 302, to Paris, 306, to the Lakes and Cambridge, 325, 328, to Scotland, 331; meets Campbell, 335; his declining health, 342; President of the Astronomical Society, 357; his death, 360; epitaph in Upton Church, Appendix II; funeral, 363
Hornsby, Rev. Thomas, first meeting with W. H., 75; Letters, 79, 80, 93, 106
Houses inhabited by W. Herschel, at Bath, 49, 68, 72, 78, at Datchet, 125, 133, 138, Clay Hall, 143, Slough, 145, 146
Hume meets W. Herschel at Edinburgh, 18

Jones, Rev. M., Letter to J. F. W. Herschel on his father's death, 361
Juno, *see* Harding

King George III summons Herschel to London, 113, 114, 118; grants salary, 125; £2000 for the great telescope, 144; another £2000, 157; grants salary to Caroline Herschel, 172; his death, 355
Klosterzeven, Capitulation of, 10
Knipping, Mrs Anna, 367
Komarzewski, Count, godfather to John Herschel, 233; accompanies Herschel to Glasgow, 233; Letter, 314

Lalande, Letters from, 126, 201, 217, 218, 238, 239, 251, 305, 312
Lambert, Motion of the Solar System, 186
Lanchester and Bell system, 3
Laplace, Marquis de, *Mécanique Céleste*, 196; Letters, 274; nebular hypothesis, 230

Lexell, orbit of Uranus, 93, 129
Lichtenberg, Letters, 130, 204, 208
Linley, T., at Bath, 39, 42
— O., taught the violin and mathematics by W. Herschel, 42; opinion of Herschel's music, 23
"Loch im Himmel", Caroline's letter to J. F. W. Herschel about, 373
Lothian, Marchioness of, at Bath, 41, 67

Magellan, J. de, Letters, 129, 138
Maskelyne, Rev. N., first meeting with W. Herschel, 75; Letters, 79, 80, 81, 109, 111, 246, 249
Mayer, Christian, Letter, 127
Mechain, Letters, 219, 240, 304
Messier, Letter, 86
Michell, Rev. J., Letter, 91; his telescope, 92
Milbanke, Sir Ralph and Lady, 14, 21, 22
Milky Way, W. Herschel's theories regarding the, 192, 193–195, 223; breaking up of, 287
Miller, Dr, stories told by, 14, 37
Minden, battle of, 11
Mirrors, casting one at Bath, 88–90; polishing, 141; composition for, 88; Pictet's description of polishing, 158; for the 40-foot telescope, 163; polishing machine, 163
Moon, paper on the mountains in, 76; volcanoes in, 218

Nebulae, classified, 193; Catalogues of, 221, 224; nebulous matter, 228; planetary nebulae, 229
Nebular hypothesis, 225; Laplace's approbation, 230; Sir John Herschel's comment, 231
Neptune, discovery of, 374

Observations, how recorded, 160
Octagon Chapel, 39
Olbers, discovers Pallas, 271; Letter, 272
Organs built by Snetzler, 36, 40

Pallas, *see* Asteroids
Papendiek, Mrs, house at Slough, 145; her friend, Mrs Pitt, 173; W. Herschel's marriage, 174, 176; his music, 179
Papers, by W. Herschel, read to the Royal Society, Appendix IV
Parry, family, 294, 295
Piazzi, discovers Ceres, 268; Letters, 268, 274

# INDEX

Pictet, *Journal de Genève*, description of Herschel's workshop, 158
Pitt, Mrs Mary (née Baldwin), 174; her character, 176
— Paul Adee, 238

Saturn, satellites of, discovered, 164, 166
Schroeter, Letters, 131, 202, 209, 213, 215; observatory destroyed by the French, 216
Snetzler, 37, 40
*Sweeps*, method of taking, 222

Telescope, the 20-foot, 167, 222
— the 40-foot, King's first grant, 144, second grant, 157; anecdote of the King and the Archbishop, 157; dazzling effect of its magnifying power described, 159; Oliver Wendel Holmes' description, 168; drawbacks to its use, 167
Telescopes, description of, 63; front-view, 64, 160; space-penetrating power, 64
— the Spanish, 177, 261, 319

Uranus, discovery of, 78; planetary nature established, 94; orbit calculated by Lexell, 93, by Cavendish, 107; name Georgium Sidus, 122, Uranus, 124; Bode's sign for Uranus, 124, Lalande's, 201; previously observed, 129; two satellites discovered, 161

Vince, Prof., 198
Volcanoes in the moon, 217, 218
Von Zach, 268

Watson, Sir W., senior, prescription for ague, 140
— Sir W., junior, Letters, 76, 85, 93, 96, 99, 100, 105, 122, 156, 198, 280, 281, 283
Wilson, Prof. Alexander, paper on the motion of the sun, 186
— Patrick, Letter to W. Herschel, 313
Wright, T., Motion of the Solar System, 188

Zone Catalogue of Nebulae, made by Caroline Herschel, 368, 369